MARCEL PROUST
Selected Letters
1904–1909

MARCEL PROUST

Selected Letters

VOLUME 2
1904–1909

Edited by Philip Kolb
Translated with an Introduction by
Terence Kilmartin

New York
OXFORD UNIVERSITY PRESS
1989

Oxford University Press

Oxford New York Toronto
Delhi Bombay Calcutta Madras Karachi
Petaling Jaya Singapore Hong Kong Tokyo
Nairobi Dar es Salaam Cape Town
Melbourne Auckland
and associated companies in
Berlin Ibadan

First published in 1989
Correspondence de Marcel Proust, Volumes IV–IX,
edited by Philip Kolb, copyright © Librairie Plon 1978, 1979, 1980, 1981, 1982
English translation and Introduction
copyright © Terence Kilmartin 1989

Published in 1989 in the United States by
Oxford University Press, Inc.
200 Madison Avenue, New York, New York 10016

ISBN 0-19-505961-1

Set in Linotron Janson by Rowland Phototypesetting Ltd
Bury St Edmunds, Suffolk
Printed and bound in Great Britain by
Hartnolls Ltd, Bodmin, Cornwall

CONTENTS

ACKNOWLEDGEMENTS

The preparation of this volume was made possible by three grants, for which the publishers, editor and translator are most grateful: from the Translations Program of the National Endowment for the Humanities, an independent agency of the United States Government; from the French Ministry of Culture; and from the Wheatland Foundation, New York.

Special thanks are owing to the University of Illinois Research Board for continuing support of this project, and to the staff of the University of Illinois Library, for the use of its vast resources and the expert assistance of its personnel.

INTRODUCTION

Writing to his friend Antoine Bibesco towards the end of 1902, Proust lamented the fact that the work he was doing was not real work, only translation and documentation, which was enough to arouse his thirst for creation without slaking it. 'A thousand characters for novels,' he wrote, 'a thousand ideas urge me to give them body, like the shades in the *Odyssey* who plead with Ulysses to give them a little blood to drink to bring them back to life.'

Over a year later he is still immersed in Ruskin. In the first few weeks of 1904 we find him putting the final touches to the proofs of *La Bible d'Amiens*, appealing to his English adviser Marie Nordlinger for help in clearing up 'a few last-minute grammatical or archaeological misgivings' (letter 13), then writing to all and sundry to drum up reviews for the book when it appears in March. In the meantime he has embarked on a translation of *Sesame and Lilies*, which will occupy him for much of the next year. But his 'thirst for creation' has not subsided – witness letter 68, in which he tells Miss Nordlinger of a request he has had from an Italian publisher to translate *St Mark's Rest*, and adds: 'I think I'll refuse, for otherwise I shall die without ever having written anything *of my own*.'

These letters to his English helpmeet enable us to follow very closely the course of his Ruskinian labours. It is clear that he put an immense amount of thought and effort into them, and, in spite of the appalling disorder in which he worked, was an extremely meticulous translator and annotator. The Ruskin translations may have been undertaken at his mother's suggestion as a stop-gap after the abandonment of his early novel *Jean Santeuil* (itself a very considerable labour amounting to some three or four hundred thousand words), but it was no mere hackwork. His preface to *La Bible d'Amiens* ran to over a hundred pages, and the extensive notes refer the reader to countless other volumes in Ruskin's vast *oeuvre*, which Proust seems to have got to know almost by heart. Moreover, his long apprenticeship to the English sage was immensely fertile in a variety of ways. Not only was Ruskin's thought a crucial influence on his own aesthetic, but in order to understand and interpret it he undertook a vast and heterogeneous programme of reading, deepening and widening his knowledge of the Bible, both the Old and the New Testaments, of medieval history, art and architecture, of the natural sciences, of English poets, novelists and philosophers.

By 1905, the Ruskin period was over. The essay 'On Reading', which he used as a preface to his translation of *Sesame and Lilies*, was by way of being a valediction: he must choose between reading and writing, between the works of others and his own. As he wrote in March 1904 to Maurice Barrès (letter 28), 'I still have two Ruskins to do, and after that I shall try to translate my own poor soul, if it doesn't die in the meantime.' This preoccupation with mortality was an inevitable product of his deteriorating health, which now becomes a constant refrain. The very first letter in this volume, to Madame de Noailles, begins with the words: 'Too ill yesterday to be able to write even a line.' His long letters to his mother from the yacht *Hélène* in the summer of 1904 are blow-by-blow accounts of his condition from hour to hour, and in September he writes to a specialist, Dr Georges Linossier (letter 65), giving a minutely detailed account of his symptoms down to his daily output of urine and the quality of his stool – a letter which, it appears, he finally decided not to send. The following spring he begins letter 102 to Marie Nordlinger: 'The words "I have been so ill, I am still so ill" have been pronounced by me so often – signifying an almost habitual state, painful but not excluding from time to time the possibility of epistolary relations – that I'm very much afraid they must reach your only too accustomed (I certainly don't mean incredulous) ears somewhat faded and lacking in exculpatory and absolving power.'

There is indeed something monotonous in his perpetual lamentations, and his friends seem to have been extraordinarily tolerant of his noctambulist hours and his constant excuses for not being able to meet them. While the genuineness of his ill-health need not be doubted, he used it as a pretext for warding off importunate visitors (notably Robert de Montesquiou) and avoiding social engagements. For he was beginning, in his thirty-fourth year, to withdraw from the salon life he had so assiduously cultivated in his twenties, and to be sufficiently detached from social ambition to tease his friend Lucien Daudet, with only the slightest tinge of envy, about all the grand dinner-parties he is invited to (letter 41). The musical *matinée* he himself arranged in March 1905, the guest-list for which reads like a roll-call of the faubourg Saint-Germain, may be regarded as a sort of farewell to society. (It should be noted, however, that at the same moment he was applying for membership of the exclusive Cercle de l'Union, only to be blackballed and to vow henceforth 'to give up any idea of ever joining a club' (letter 114) – a vow he failed to keep, for in the spring of 1908 he allowed his friend the Duc de Guiche to put him up for the Polo Club, as a result of which, he ruefully remarks in letter 281, he gets nothing but offers of ponies for sale.) Meanwhile his circle of young aristocratic friends was gradually dispersing, some, like Antoine Bibesco and Bertrand de Fénelon, to diplomatic posts abroad, others, such as Guiche, Gabriel de La Rochefoucauld and Louis d'Albufera, to get married.

Something of his ambivalent attitude, not only towards social life but also towards friendship, can be read between the lines of these letters. In his relationships with his male friends, wherever his emotions were engaged, Proust was at once intensely giving and intensely demanding, and his acute sensitivity to slights, betrayals, coolings-off on their part sometimes explodes into furious denunciation, sometimes expresses itself in mournful reproach. For a combination of the two reactions, see letter 29, in which he upbraids Antoine Bibesco for some unspecified betrayal and tells him why 'in the true sense of the word I am no longer your friend', but goes on to confess the enormous pleasure he still gets from seeing him, a pleasure he explains thus, in a sentence of characteristic subtlety and charm: 'This is because your person, your physical person itself, retains the unconscious memory of the marvellous qualities which you once had and which, materialized, bidden by the wicked fairy of your self-destructive character to be now merely looks, gestures, inflections of the voice, preserve nonetheless for someone who has known them the charm of a particularly touching image, that which objects leave of themselves, make of themselves, an imprint full of meaning and beauty.' Bibesco was sufficiently untouched by this letter (more likely he was amused and flattered) to include it in his own selection of letters from Proust, though not a subsequent one (letter 87) which gives an even harsher sketch of his character. His sin on the first of these occasions may well have been some playful insinuation to do with homosexuality; his brother Emmanuel and Louis d'Albufera are elsewhere rebuked for similar transgressions (letters 280 and 281). This is a subject on which Proust seems to have been almost pathologically touchy. It is difficult to know what to make of his weird letter (303) to Georges de Lauris in the autumn of 1908 after the latter had broken a femur in a motor-car accident, in which, after hymning the beauty of his friend's limbs, he complains of the 'cruelty and incomprehension' of which he has so often been the victim.

As time went on and he gradually withdrew into his self-imposed, asthma-ridden claustration, Proust was to become more and more disillusioned with friendship, which he saw as having no spiritual significance, as the negation of the irremediable solitude to which every human being is condemned, but which the artist knows how to put to account. (Sexual love and its torments are another matter, but whatever was going on below the surface, they figure scarcely at all in the letters of this period.) Proust continues to write to his friends, but sees them less and less. The outstanding exception is the composer Reynaldo Hahn. After their passionate liaison in the 1890s had come to an end, they remained devoted companions. For Proust, Reynaldo was second only to his mother as intimate and confidant, constant at his bedside, especially after Mama's death, in spite of his musical commitments and his friend's unsocial hours. Proust's letters to Reynaldo are among the most enjoyable to read, full of wit and fantasy, private jokes and scurrilous gossip. Once one has recovered from the initial embarrassment induced by their private language or baby-talk (incidentally not easy to render in English) one is soon captivated by the humour and charm. Although Proust, with his characteristic modesty, deferred to Reynaldo on musical matters, from time to time he teases him with affectionate malice – see his description of Hahn's conducting style in letter 207.

Proust's letters to women are less fraught with emotional undercurrents than those to his closest male friends – apart from the special case of Reynaldo. For all the dithyrambic hyperbole of his outpourings to Anna de Noailles, and whatever the admixture of romantic snobbery in his attitude towards the 'Eastern princess' whose poetry was 'quite as fine as Hugo's or Vigny's' (as he describes her in *Remembrance of Things Past*), his letters to her are affectionate and good-humoured and show a genuine fellow-feeling: she had been his ally at the time of the Dreyfus case, and would more than repay his advocacy of her poetry by her loyal and passionate support for his own work. His adulation now seems to us over the top, but she was by no means a negligible poet and Proust was genuinely impressed by her lyrical gift and feeling for nature. In any case there is always an element of ambivalence and self-parody in his flattery, and it is typically disingenuous of him to pass on to her, in letter 25, the even more comically extravagant claims which Robert de Montesquiou makes for her work. Moreover it must be emphasized that Proust always erred on the side of generosity when commenting on the writings of others – not only famous *maîtres à penser* such as Barrès and Anatole France, or distinguished historians like Albert Sorel and Georges Goyau, but friends, whether bourgeois or aristocratic, such as Robert Dreyfus, Fernand Gregh, Maurice Duplay, Gabriel de La Rochefoucauld or Georges de Lauris, and even unknowns. He reads their books from cover to cover as soon as he receives them and takes infinite pains to offer them criticism and advice.

His most constant female correspondent throughout his adult life (she was twenty years his senior but outlived him by four years) was Madame Emile Straus, whose brilliant salon, composed of writers, artists, journalists and politicians as well as the more enlightened members of the *haut monde*, he had frequented in the 1890s. By the time this volume opens, her health, like Proust's, has deteriorated, and she seems to have lapsed into a sort of vague neurasthenia. Proust's letters to her are full of concern for her health, as from one valetudinarian to another (letter 94 is a particularly moving example), and he tried to cheer her up with titbits of news and gossip, often mildly improper – see letter 123 about the unintentional *double-entendres* of the Comtesse de Saint-Paul – and about the books he is reading. It is to her that he confides many of his thoughts about literature and politics – she too had been an ardent Dreyfusard, and he recalls those heady years in July 1906 when the last act of the drama is played out in the courtyard of the Ecole Militaire with Dreyfus's rehabilitation (letter 163).

There is a poignant contrast in his relationship with two younger women, Marie Nordlinger and Louisa de Mornand, as revealed in these letters – the one a serious, well-brought-up, dedicated artist, the other a pretty, frivolous, mercenary actress, the mistress of his friend Albufera. The letters to his 'dear, dear Mary' are affectionate and comradely and full of gratitude for her help with his Ruskin translations. They exchange presents, she is a frequent visitor to his room during the Ruskin period, he commissions from her a medallion to commemorate his father; but there is no hint on his side of any deeper feeling. Not so on hers: it is clear that she read more into his amiable gallantries (see for instance letter 17, which ends with the words: 'I am all on fire for *Sesame* – and for you') than he intended, and must have betrayed signs of frustration and unhappiness, for in August 1904 he writes to her in reply to a letter in which he had detected 'a sort of vague allusion to some deep sadness which you don't express', and advises her 'to feast on life . . . and not weep sadly beside an urn containing nothing but regrets' (letter 62). A fortnight later, however, he chides her for boasting of 'overflowing with health, life, strength, the power to be happy', which he nevertheless interprets as 'an offer of a moral blood transfusion'. Finally he confesses: 'I understood the melancholy of your solitude; but I was very much the last person who could dispel it', and ends up by advising her to accept a job in America (letter 68).

By contrast, the letters to Louisa are intimate, amorous, suggestive, sometimes even salacious. What, for instance, are we to make of the dedicatory note he inscribed in her copy of *La Bible d'Amiens* (letter 21), or the erotic innuendoes in the verses of letter 32? *Pace* George Painter, who in his classic biography postulates actual 'carnal relations', it seems much more likely that Proust's flirtations with Louisa were a surrogate for his unavowable passion for her lover Albufera. There is an obvious parallel with the Baron de Charlus in *Remembrance of Things Past* in his role as go-between for Swann and Odette, and an even closer one with the narrator of the novel as gooseberry to Saint Loup and Rachel. Although at the time, 1904–5, the novel was still in the future, no doubt the 'thousand characters' mentioned in his 1902 letter to Antoine Bibesco were constantly revolving in his mind. Albufera represents only one aspect of Saint-Loup (with his slightly asinine philistinism he is much closer to the Duc de Guermantes) but Louisa has all the characteristics of 'Rachel when from the Lord'. She seems to have been through the hands of several of Proust's friends before Albufera, to whom she is blatantly unfaithful while extracting considerable sums of money from him. Proust's fascination with this grasping cocotte is an aspect of his novelist's curiosity about the sleazier side of life.

Proust's relationship with 'the fatal Count', Robert de Montesquiou, has often been a source of embarrassment to his admirers. But when we remember the use he was to make of it in the creation of Charlus we are inclined to forgive him. Besides, their epistolary exchanges make entertaining reading. Many of Montesquiou's letters have survived, and a number have been included here, so that we can enjoy their contrasting styles and the comedy of their continuous battle of wits. Proust's letters are a mixture of unctuous flattery and feline insolence, the Count's lofty, patronizing and imperious. Montesquiou was a good deal more formidable (as indeed is Charlus) than his reputation as an aristocratic dilettante and poetaster would suggest. One feels that he has the measure of his 'dear Marcel' and sees through his stratagems for dodging the lectures and readings he expects him to attend. How is it, he testily demands, that 'you recover when it's a question of the La Rochefoucauld orangeade' (letter 119), provoking a wail of injured innocence from a mortified Proust (letter 120). The series of letters exchanged between them in the spring of 1905, when Proust is finally trapped into arranging a reception for the Count to give a reading from his new book, is a particularly rich example of their recurrent serio-comic duologues.

On 26 September 1905, after a short illness, Proust's mother died. During the two years since her husband's death she had been closer than ever to her elder son. His létters to her during this period – even when he is giving her (at her insistence) running commentaries on the state of his health – show no signs of the selfishness and petulance that had sometimes marred them in the past. Now, with her death, he writes to Montesquiou in reply to a message of condolence, 'my life has lost its only purpose, its only sweetness, its only love, its only consolation' (letter 152). To Madame de Noailles he writes, in terms reminiscent of those that will be applied in his novel to the narrator's dead grand-mother: 'She has died at fifty-six, looking no more than thirty since her illness made her so much thinner, and especially since death restored to her the youthfulness of the days before her sorrows . . . She takes away my life with her,' he adds, 'as Papa had taken away hers' (letter 151). And for a long time thereafter he indeed shows little sign of life: 1906 seems to have been a year of mourning, silence, physical and moral prostration. Writing to Marie Nordlinger in America, he talks of 'my ravaged life, my shattered heart' and goes on to ask: 'Are you working? I no longer am. I've closed forever the era of translations which Mama encouraged. And as for translations of myself, I no longer have the heart' (letter 170). And in letter 172 to his mother's old friend Madame Catusse, who is organizing the move to his new flat in the boulevard Haussmann while he himself takes refuge in a Versailles hotel, he writes: 'As for the desks to be kept . . . mine will be my uncle's (but what use will a desk be to me if I never get down to work?).'

What finally aroused him from his torpor was a newspaper report of a macabre crime committed by a family acquaintance, Henri van Blarenberghe, with whom Proust had recently been corresponding: he had killed his mother, and then shot and stabbed himself, and was interrogated by the police as he lay dying on his bed, his left eye lying on the pillow. On reading the story Proust proceeded to write a remarkable essay for the *Figaro*, 'Sentiments filiaux d'un parricide', reflecting on the crime in terms of Greek tragedy and Dostoevskian psychopathology, and shocking the *Figaro* sub-editor, who omitted the last paragraph because it seemed to show 'insufficient disapprobation for the unfortu-nate parricide's deed' – see the amusing exchanges with the *Figaro*'s editor Gaston Calmette in letters 185–187. This article is significant, partly for its circular form ('the word parricide', he tells Calmette, 'having opened the article, closed it: the article acquired a sort of unity thereby') which prefigures the circular unity of *Remembrance of Things Past* but also, and principally, as a means of exorcizing his guilt for his own mother's death. Something of what he has been going through for the past eighteen months is revealed a month later when his friend

Georges de Lauris's mother dies: 'When you have become accustomed to the terrible experience of being forever thrown back on to the past, then you will feel her gently returning to life, coming back to take her place again, her whole place, beside you' (letter 192). Letters of condolence to his friends for the loss of their parents and grandparents, expressed in terms of the most extreme anguish, are a constant feature of his correspondence, and he even wrote an obituary article about Robert de Flers's grandmother. But the most striking instance of the intensity of his filial piety is the strange and violent letter to a critic who had written a favourable review of a book entitled *Amours* by Paul Léautaud which was apparently somewhat cynical about family feelings: Proust talks of its 'moral baseness' and denounces the author as 'the most despicable creature, the most devoid of intelligence, of style, of grammar, of sensibility, of originality' (letter 179).

Having at last emerged from his grief-stricken, guilt-ridden silence, Proust went on to write several more articles in 1907 and began to reappear in society. In April he attended a musical *soirée* at the Princesse de Polignac's, where he heard Hahn's orchestral suite *Le Bal de Béatrice d'Este*, and his letter to Reynaldo recalling the occasion is full of lively and malicious observation of a kind that presages *The Guermantes Way*. He notes with astonishment that all the people he used to know have aged – the dowagers, 'rudimentary and ferocious old divinities . . . immutable in the barbaric hideousness of their Lombard effigies' (letter 207). In July he organized a grand dinner at the Ritz as a thank-offering to Calmette for publishing his articles, with an elaborate musical entertainment afterwards. Then, at the beginning of August, he set off on his first adult visit to Cabourg which, as Balbec, would play such a significant role in his novel. His letters from there, amusing and even ebullient, show to what extent his morale and even his health had improved. He writes to the great medievalist Emile Mâle for advice on churches and other historical monuments to visit (letter 242), and, sustained with caffeine, hurtles round Normandy 'like a flying cannon-ball' in a motor car hired from Madame Straus's son, his old schoolfriend Jacques Bizet, and driven by Alfred Agostinelli (of whom much will be heard in subsequent years). He describes these excursions in long letters to Madame Straus and Georges de Lauris, and later in an article for the *Figaro*, 'Impressions de route en automobile', passages from which (in particular the evocation of the shifting perspective of the steeples of Caen which was to inspire the narrator's essay on the steeples of Martinville seen from Doctor Percepied's carriage) will eventually be adapted for the novel. To Reynaldo he reports entertainingly on the raffish social life of Cabourg, and describes *inter alia* a visit to the painter Vuillard (letter 249) in terms which remind one of his great fictional creation, Elstir, and in a letter to Robert de Billy (252) about this 'astonishing' summer he adds 'watching polo and losing at baccarat every evening' to the list of his activities. However, in replying to a remarkably percipient letter from Montesquiou (236), describing him as one of 'those valetudinarians whose economy of physical expenditure allows them prodigalities of mind and heart', he wonders whether the active life he has been leading at Cabourg is not 'using up and wasting what remains of my strength', and whether, in order to be able to write, he may not need the physical reserves which the Count had rightly observed that he built up in bed (letter 250).

At all events, back in Paris in the autumn of 1907, Proust seems to have reverted to his old routine. But one senses a new vitality, almost a new optimism. He has completely recovered at last from the trauma of his mother's death, and has put Ruskin well behind him; one feels he is at last groping his way towards the book he knows he has it in him to write. One of his first letters in the New Year of 1908 suggests that he is preparing to write a novel (see note 5 to letter 261 to Auguste Marguillier). In February he writes to Madame Straus to thank her for some 'little almanacs' she has sent him as a present, and tells her that he is anxious to settle down to 'a long piece of work' (letter 264), and in a letter to Louis d'Albufera a few months later he lists the projects he has in hand: 'a study on the nobility, a Parisian novel, an essay on Sainte-Beuve and Flaubert, an essay on women, an essay on pederasty (not easy to publish), a study on stained-glass windows, a study on tombstones, a study on the novel' (letter 281). It may be noted that all these subjects eventually found their way into *Remembrance of Things Past*, but the list suggests that Proust was still hesitating over the form he should give to the work he had embarked on. 'Should I make it a novel or a philosophical study – am I a novelist?' he noted in one of the 'little almanacs' given him by Madame Straus. This working notebook, superbly deciphered and elucidated by Philip Kolb and published under the title *Le Carnet de 1908* (though it also contains notes from 1909 and later), complements and elucidates the cryptic and often contradictory references to his work which we find in the letters. A list in the notebook of 'pages written' corresponds to the novel begun in January 1908 and mentioned in the letter to Albufera quoted above. The name Sainte-Beuve begins to appear more and more frequently in the notebook, until in December he writes simultaneously to Lauris and Madame de Noailles telling them that he intends to write something about Sainte-Beuve and asking for their advice as to whether it should be in the form of a traditional essay or a quasi-narrative (letter 315). We can see from his letters during 1909 which course he eventually adopted, and how the critical project gradually gave way to the fictional until the book, which he still referred to as *Sainte-Beuve* or *Contre Sainte-Beuve*, transformed itself into a first draft of *Remembrance of Things Past*.

Although the letters of 1908–9 are especially significant for the light they throw on the gestation of the novel, they are by no means devoid of other interest. Early in 1908 he published in the *Figaro* a series of extraordinarily accomplished and very amusing parodies of famous French writers, the common theme of which was a sensational diamond swindle, *l'affaire Lemoine*, which had tickled his imagination. Writing to Robert Dreyfus (letters 270–271) he describes them as 'literary criticism in action' and explains, à propos of the Renan pastiche: 'I adjusted my inner metronome to his rhythm and could have written ten volumes like that' – a facility he was to make marvellous use of in creating such characters as Norpois, Bloch and Legrandin. If there are signs of a revival of social life, one senses it is with the object of sniffing out raw material and local colour for his book. He attends several balls in the hope of meeting a mysterious young beauty called Mademoiselle de Goyon, and is disappointed when he eventually does (though he remembered her Christian name, Oriane, when it came to naming the Duchesse de Guermantes). He asks Albufera to lend him a family photograph album and to provide him with genealogical information for 'what I'm working on', and the name Guermantes crops up for the first time in a letter to Lauris in which he wonders whether the title is extinct and available to an author, and asks: 'Do you know any other pretty names of châteaux or of people?' (letter 334). In a charming note to his cook Céline he compliments her on her *boeuf mode* and adds: 'Would that I might bring off as well as you what I am going to do tonight, that my style might be as brilliant, as clear, as firm as your *gelée*, that my ideas might be as succulent as your carrots and as nourishing and fresh as your meat' (letter 339) – words which remind us of the passage in *Within a Budding Grove* where Françoise's artistry as a cook is compared to Michelangelo and her *boeuf en gelée* is greatly admired by the Marquis de Norpois.

In the summer of both these years Proust revisited Cabourg, whence his letters reveal that he has taken up with a group of new young friends, both male and female – perhaps the Balbec 'little band' is beginning to take shape. He seems to have spent most of his evenings in the Casino – see the amusing letter (300) to the playwright Henry Bernstein acknowledging the repayment of a gambling debt. Proust's gambling was not, alas, confined to the baccarat table. After his mother's death he began to take an active interest in his financial affairs, but instead of relying on the advice of his friend and man of business, Lionel Hauser, he could not resist speculative investments which invariably ended in disaster. 'Did you see that in my *Figaro* pastiches I mention my discomfiture over De Beers?' he asks Albufera in letter 274. Reynaldo is deputed to consult the Belgian banker Baron Lambert about possible investments, and writing to thank him Proust adds: 'At the present moment my subtle mind, lulled by the waves, is sailing between the mines of Australia and the railways of Tanganyika and will alight on some goldmine which I hope will really deserve its name' (letter 307). His correspondence with Lionel Hauser offers a hilarious commentary on his financial incompetence and incurable punter's optimism.

Altogether, then, these letters give us a fascinating insight into a mind of great depth and complexity, and a strange, elusive, multifaceted, often exasperating but ultimately very engaging personality. Even without the benefit of hindsight, they would be worth reading for what they tell us about a period and a society, for their many incidental felicities, for their observations on literature, art and music. Knowing as we do that their author was to go on to write the greatest novel of the twentieth century, we can read them in the light of the masterpiece to come, which was already, at the end of 1909, well under way. Proust was being over-optimistic when he wrote to Madame Straus in August of that year to say that he had 'begun – and finished – a whole long book' (letter 343) and at the same time approached a publisher, Alfred Vallette, offering him the manuscript of a novel, 'and an extremely indecent one in places', for serialization in his review the *Mercure de France* and publication in book form in the New Year (letter 341). Much, in fact, still remained to be done, but a letter in early December to Georges de Lauris, to whom he had sent the first two hundred pages of the book (evidently a first draft of *Combray*), shows that he is now completely sure of himself: 'All I ask is that you shouldn't mention the subject or the title or indeed anything that might be indicative of what I'm up to (not that anyone would be interested). But more than that, I don't want to be hurried, or pestered, or ferreted out, or anticipated, or copied, or discussed, or criticized, or knocked. There will be time enough when my thoughts have run their course to allow free rein to the stupidity of others' (letter 356). The final letter in this volume, to one of his new young Cabourg friends, Max Daireaux, talks of alterations to his flat 'which are essential for my peace and quiet' – an allusion, no doubt, to the cork-lined room in which for the next few years he would continue to pursue his quest for Lost Time.

TERENCE KILMARTIN

TRANSLATOR'S NOTE

The letters in this volume have been selected by the editor from volumes IV to IX of *Correspondance de Marcel Proust*, texte établi, présenté et annoté par Philip Kolb (Paris, Librairie Plon, 1970–   ).

Persons mentioned are identified in a footnote attached to the letter in question, for the most part where the name occurs for the first time. In the index, page references for such footnotes are indicated by an italic *n*; in the entries for Proust's correspondents, page numbers in bold indicate the first page of a letter from or to Proust.

Proust rarely dated his letters; the dates given here in square brackets have been established by Professor Kolb by dint of meticulous research (the portion of the date that remains dubious being followed by a question-mark).

References in the notes to Proust's novel are to the three-volume English translation, *Remembrance of Things Past* (London, Chatto & Windus; New York, Random House; 1981) and indicated by the initials *RTP*.

The notes and index have been compiled by my wife Joanna Kilmartin, to whom I owe a huge debt of gratitude.

TK

1 *To Madame de Noailles*[1]

Madame,

Too ill yesterday to be able to write even a line, I was unable to include with the little vase[2] my wishes for the coming year, in which, I hope, your joy, your strength, your fame and your genius will not cease to flourish. As I am still unwell today I beg M. de Noailles to forgive me if I don't write to him separately and to be good enough to accept through you my most affectionate good wishes. I crave the same indulgence from Anne-Jules,[3] who is often in my thoughts.

Please accept, Madame, my respectful regards.
Marcel Proust

I really must write to your husband, for I insist on wishing him a happy New Year.

1. Comtesse Mathieu de Noailles, *née* Princesse Anna-Elisabeth Bassaraba de Brancovan (1876–1933). Anna de Noailles, prolific poet and novelist, was assiduously cultivated and panegyrized by Proust (cf. *RTP*, II, 104–5: 'a young eastern princess said to write poetry quite as fine as Victor Hugo's or Alfred de Vigny's') and she in turn became his devoted friend and admirer. Both had been ardent Dreyfusards (see vol. I).

2. Vase by Emile Gallé (1846–1904), of the distinguished Lorraine family of decorative artists in the Art Nouveau style, notably in glass (cf. *RTP*, I, 860), from whom Proust would commission presents.

3. Mme de Noailles's son, aged three.

2 *To Joseph Primoli*[1]

Dear Sir,

I called at the rue de Berri and asked to see you. I was told that it was not possible. Let me simply tell you that I weep bitterly with you, because I loved the Princess[2] with an infinite respect – and because it distresses me so much to think of you being so unhappy, you who are so kind that with all one's heart one wishes you nothing but happiness, you whose heart is sore and wounded, when one so wants you to be spared all cruel blows. I have long foreseen the pain that this loss was bound to cause you. I suffered physically at the thought that you were going to suffer yet again, that all the sorrows of your life would coalesce to weigh upon you more heavily still when this final misfortune struck. You have at least the consolation (a consolation that was denied me since by my ill health I cast a shadow over my father's last years[3]) of knowing that you have been the joy – the grave joy, the happiness – and also the charming gaiety, the perpetual diversion – of the Princess's life. No one made her feel as you did the charm of affection and the value of gaiety. It was you who always brought to her lips the exquisite smile which we shall never forget and for which all those who cared for the Princess's happiness, all those for whom the delightful, smiling image they have of her will always be precious, will forever remain grateful to you.[4]

Accept, dear Sir, my grief-stricken respects.

Marcel Proust

1. Comte Joseph ('Gégé') Primoli (1851–1927), nephew of the Princesse Mathilde. He is unflatteringly described in letter 123.

2. Princesse Mathilde (1820–1904), daughter of Napoleon's brother Jérôme Bonaparte (King Jérôme of Westphalia). News of her death on Saturday 2 January filled the *Figaro* front page. Her Wednesday receptions, attended by most of the leading writers of the middle and late nineteenth century, had been the subject of the first of Proust's pseudonymous articles on Paris salons for *Le Figaro* (see vol. I).

3. Dr Adrien Proust, eminent physician, pioneer in the new science of hygiene and member of the Légion d'Honneur since 1870, had died in November 1903, aged sixty-nine.

4. Proust was to recall this 'smiling image' in the passage in *RTP* (I, 583–6) where Swann and Odette introduce the narrator to her in the Bois de Boulogne.

3   *To Antoine Bibesco*[1]

[First week of January 1904]

My dear Antoine,

May I tell Lucien Daudet[2] that you accused me of having a 'Parisian accent', the story of your telegrams to Bertrand[3] and your projected trip to Petersburg, or is it silence tomb[4] for him?

I'm sad not to be able to come and say good-bye to you, but my back pain has reached such proportions that I can't budge. If you see your cousins[5] would you be good enough to tell them so as to explain my absence. Forgive my scruples about the tombs but I should be sorry to offend you. Good luck for your trip to Egypt (although I'm certain you won't go).[6] Admire

> The sphinxes brooding in the wilderness,
> Seeming to sleep in an eternal dream,[7]

Cheops

> Who sleeps in his immutable splendour.[8]

Indeed the whole of Egypt,

> Her forehead wreathed with science and with pride,[9]

Egypt, 'the parent of Astronomy, Geometry and Chivalry, the tutress of Moses and the Hostess of Christ'.[10]

Fond regards,
Marcel

1. Prince Antoine Bibesco (1878–1951), Romanian diplomat and Proust's close friend. He and his brother Emmanuel (see letter 49) would accompany Proust on long excursions to look at churches and monuments (see vol. I).

2. Lucien Daudet (1878–1946), Alphonse Daudet's handsome younger son with whom Proust had had a passionate friendship; when, in 1897, the malicious journalist Jean Lorrain hinted at a sexual relationship between them, Proust challenged him to a duel (see vol. I).

3. Vicomte Bertrand de Salignac-Fénelon (1878–1914), a young diplomat about to be transferred from Constantinople to St Petersburg. He and the Bibescos formed a secret circle to which Proust was admitted and for a few years he and Proust had an intense friendship, travelling to Holland together in 1902 (see vol. I). Mentioned by name as the narrator's 'dearest friend' in *RTP* (II, 799), he was also a model for the dashing, aristocratic Saint-Loup and, like him, was to be killed in the First World War.

4. Among Proust's intimate friends, *silence tombeau*, often abbreviated to *tombeau* (tomb), meant 'not a word to anyone'.

5. Mme de Noailles and her sister, the Princesse Alexandre de Caraman-Chimay (*née* Princesse Hélène Bassaraba de Brancovan), to whom Proust was also deeply attached.

6. He left for Egypt on 9 January, later joining Fénelon in Constantinople. In this letter Proust doubtless had in mind the abortive plans he had made to visit Egypt and Constantinople with Bibesco the previous year. Cf. vol. I.

7. 'Les grands Sphinx allongés au fond des solitudes / Qui semblent s'endormir dans un rêve sans fin.' Baudelaire, 'Les Chats', in *Les Fleurs du Mal*.

8. 'Dans sa gloire, Chéops inaltérable dort.' Proust paraphrases the last line of Sully Prudhomme's 'Cri perdu' (the cry is that of a builder of the pyramids).

9. 'L'Egypte au front bandé d'orgeuil et de science.' Anna de Noailles, 'L'Ombre des jours', in *Les Voyages*.

10. John Ruskin (see letter 11) in *The Bible of Amiens*: chapter III, 27. Proust quotes him in English.

4 *To Antoine Bibesco*

[First week of January 1904]

My dear Antoine,

It is completely untrue that I violated that tomb.[1] Your assertion is of the kind that one terms lies. I refrain from doing so for many reasons (first of all because you don't say it seriously, but simply in order to appear perspicacious – or whatever). However, I've spoken to no one about it; perhaps I shall because of the tone of your letter but perhaps not. Besides, I never violate tombs. These I would have had the right to since they weren't tombs (and anyhow I didn't violate them) but merely some humorous remarks of mine about the naïvety and extravagance with which for some time you've enjoyed drawing your own caricature. But the compunction I have always had up to now in this regard, which has prevented me from ever talking about you to anyone *cum grano salis*, certainly deprives me of an element of humour that is important in conversation. But I take no credit for this, as I haven't yet reached the point of enjoying seeing you in a comical light and when I'm particularly struck by the partial blindness with which you seem to be stricken (and which seems only to increase your arrogance in judging others) the only person whose attention I try to draw to your absurdity is yourself. In fact this wasn't even the object of my last letter, as those tombs would involve reflections of a more absolute significance and a more general character, something that cannot be discussed by correspondence.

Fondest regards to your brother with whom you've made me fall out.

Good-bye, dear Antoine, may your journey do you good: *that would give me great pleasure*, though I don't think you will go. Thank you for the information about iodide, but you know it's like antipyrine for migraine or poultices for stomach-aches: as old as the hills.[2] My asthma isn't bad enough, or *normal* enough, for me to force myself to take this boring treatment which makes one's eyes water, causes sneezing fits, gives one a temperature, etc.[3] I warn you that on your return from Egypt, like you on my return from Evian (but on your return from Egypt I shall perhaps not be here as Mama wants me to go away for a bit without her), I shall tell you as you told me then a few things once and for all which you'll have to accept or we shan't see each other again. But from now on all that's over: I forgive you all your stupidities of the other night and send you all my devoted regards.

Marcel

1. See letter 3, n. 4.
2. The use of potassium iodide is discussed in *L'Hygiène des asthmatiques* (1896) by Dr Edouard Brissaud. This treatise had a preface by Proust's father, Dr Adrien Proust.
3. Proust's asthma was to become progressively worse.

5   *From Madame de Noailles*

[Paris, 8 January 1904]

My dear Marcel,

I must tell you again of my pleasure and my pride in those dear and touching ferns on that glass vase.[1] They make a pattern on my heart, the pattern of their sweet little outstretched grassy bodies – and of my friendship for you.

I am getting better. I shall be happy to come and see you at home; I long to hear that you are no longer unwell.

Affectionately yours,
Anna de Noailles

1. See letter 1, n. 2. Proust had asked Gallé, who was a botanist, to engrave his present with intertwined ferns in tribute to an image used by Anna de Noailles in her short story 'L'Exhortation'.

6   *To Madame de Noailles*

[Friday evening, 8 January 1904]

Madame,

How thrilled I always am when I catch sight of the disciplined tumult of your handwriting, those magnificent volutes as of an infinite and pulsating sea from the bosom of which your thought emerges sparkling like Aphrodite, as divine and as beautiful. But when, through some excess of kindness or refinement of graciousness as in your letter of this morning, it literally tortures me by arousing in me a gratitude I feel I shall never be capable of expressing, then this joy is painful and mingles

The foam of pleasure with the tears of pain.[1]

When Papa, who was as active as I am lazy and went out every morning, brought up the mail, he would say to me, knowing my joy: 'A letter from Madame de Noailles', and Mama[2] would scold him, saying: 'Don't spoil his pleasure by telling him in advance.' And I assure you it was a very touching comedy (not for you, perhaps, but for me, especially now), the air of supreme indifference Mama would assume when she brought up a letter from you, as if to say: there's nothing but trivial papers, so that my pleasure would be unalloyed.[3]

No, I shall come to *you*, but have you been told that I'm unwell, rheumatism in my back, which hurts only when I make certain movements, but then excruciatingly, and gives me a fever! I don't think I shall be able to go out for two or three days at the earliest, and my first outings will be fairly cautious and perhaps not a proper visit like going to see you. As I've never had rheumatism I don't want to start now.

I believe Constantin[4] will be back before long and since he has behaved not at all nicely towards me I'm going to write him some unkind things to make sure that I don't see him beforehand, either here or at your house, because when speaking to somebody, if one gets angry, one sometimes says more than one intended. It's true that I do intend quite a bit. One shouldn't get angry with people over matters to do with reviews. And moreover I don't feel in the least angry with him. But his behaviour towards me on three successive occasions within a month has been so egregious that, being in the habit when people are nice to me of dissolving into tears of gratitude and affection, I feel that when someone is too much the opposite it's essential to say so too, if only so as not to devalue the gratitude I've shown to the nice and the kind (no need to tell you that you are not among the nice ones to whom I show gratitude! – you are the sublime goddess who will be no less blessed on the day when she withdraws her favours from me). So don't be cross with me if you hear that I've written Constantin a somewhat harsh letter. I assure you that, liking and admiring him as I do, and precisely for that reason, I was *obliged* to do so in order to get the matter over and done with, and moreover it was necessitated by the fact that I wrote him an almost insanely amiable one three weeks ago, two days before I learned of his outrageous behaviour!

As for Antoine,[5] I have more serious problems with him, since they are to do with sentiment. But those are not to be settled or corrected through letters.

Silence alone is greatness.[6]

La Renaissance latine (I speak of it as an editorial administration not as a sanctuary forever illuminated by the perpetual adoration of *La Nouvelle Espérance* and 'L'Exhortation'[7]) is not worth the effort of *silence*.

Your respectful friend, your adoring admirer,
Marcel Proust

1. 'L'Ecume du plaisir aux larmes des tourments.' Baudelaire, 'Femmes damnées', in *Les Fleurs du Mal*.

2. For Proust's closeness to his mother, see letter 55 and subsequent letters to her.

3. Proust was to evoke this comedy in *RTP* (III, 578).

4. Prince Constantin Bassaraba de Brancovan (1875–1967), Mme de Noailles's brother. As editor of the literary magazine *La Renaissance latine* he had promised Proust a regular column and then later reneged.

5. Antoine Bibesco.

6. 'Seul le silence est grand; tout le reste est faiblesse.' Alfred de Vigny, 'La Mort du loup'.

7. Prince Constantin had published his sister's first novel *La Nouvelle Espérance* in serial form and her short story 'L'Exhortation', causing Robert de Montesquiou (see letter 12) to dub *La Renaissance latine* 'the Brancovan chamber-pot'.

7 *To Madame de Noailles*

[Saturday evening, 9 January 1904]

Madame,

I see that I forgot to tell you in my letter what a marvel worthy of the vegetable kingdom were the two phrases on the fern[1] in your letter.

How delightful for me to have provoked what will enable a scholarly edition of 'L'Exhortation' (of the kind I'm preparing for *The Bible of Amiens*[2]) to include a note at the word 'fern'.*

*Cf. by the same author, etc.

As I find it exceedingly tiring to write (not because of my rheumatism – that has nothing at all to do with it and moreover I no longer feel it at all this evening, I think it's over), would you be so kind as to tell the Princesse de Chimay[3] that she has overwhelmed me with embarrassment by thanking me for the cast, which wasn't a real present and is worth five centimes![4] It's the plaster cast of one of the Months from Notre-Dame de Paris which she had admired and thus a small token of affectionate remembrance from her great admirer for the clear, sweet words that have fallen from her incomparable lips.

Your

Marcel Proust

1. See letter 5.
2. Proust was correcting the proofs of his translation of Ruskin's book (see letter 11).
3. See letter 3, n. 5.
4. One of a series of bas reliefs representing the labours of the months, in the porch of the St Anne doorway on the west front of the cathedral.

8 *To Madame de Noailles*

Wednesday, on returning home
[13 January 1904]

Madame,

When one has just taken one's baccalaureate one realizes that one has forgotten to say three-quarters of the things one knew. No sooner had the door of the avenue Henri-Martin[1] closed on the staircase whose lights I didn't know how to switch off than I realized that the improvised defence of my conduct which I had offered (in response to your vaguely reproachful air) was wildly incomplete. And one after another, forgotten faults on the part of the Adversary,[2] and my own long-suffering merits, came surging back. On reaching home I returned, like a drunkard to the drink that inflames him, to the chalice of *Constantin's Letters*. In the most recent, a sentence which I had not savoured with sufficient bitterness represents incidentally and insidiously the decision to deprive me of the critic's appointment *as having been taken in my interest*. I quote it word for word (the letter is dated Tuesday, yesterday): 'You have doubted me, but I don't hold it against you. I should be happy for you to contribute as frequently as possible to the Review which I edit. I feel however that for many reasons *it would be better for you not to take on the responsibility of a regular column which would give you a great deal of fatigue, hard work and boredom.*' It's a perfect example of La Fontaine's

Your scruples betray excessive delicacy.[3]

But how. is it that Constantin didn't have the same scruples last September when hc offered me that 'regular column' and only thinks of them in January. For to contribute by submitting articles and having them accepted is the right of any individual who writes a piece and brings it to him without even knowing him, a right from which his friends are not a priori excluded, after all. (Although this letter is dated Tuesday and I only received it on Tuesday, I think it dates from Monday since our quarrel dates from Monday and when I saw him he said to me: You'll find a letter from me when you get home.)

Madame, forgive this latest flood of 'impotent rage'. The absurdity of the paper on which (for lack of any other to hand) I'm writing to you adds an even more litigious air to my complaint: 'If it please the court, etc.' Thank you again for having received me at an hour which was so inconvenient to you, beneath the flowers of those antique veils, you who have

The body of Iphigenia and the heart of Virgil.[4]

I found you in such good physical condition and I was so pleased to see what stamina you had, that after a pultaceous angina (no I don't know what it's called) you were so well. I hope that the change of routine, getting up, etc. did not upset your sleep, your well-being, your appetite, the mechanism of your habits as Saint-Simon calls it. I felt that you thought me very boorish on two or three occasions (when I asked if I could telephone, and whether the staircase could be lit, and when I said I didn't know how to turn the lighting off), and it pained me.

Good-bye, Madame, I admire you and love you respectfully.

Marcel

1. 31 avenue Henri-Martin was the Paris address of the Comte and Comtesse Mathieu de Noailles.

2. Prince Constantin. Proust had visited Mme de Noailles in order to enlist her sympathy in his quarrel with her brother. Cf. letter 6, n. 4.

3. 'Vos scrupules font voir trop de délicatesse.' La Fontaine, *Les Animaux malades de la peste*.

4. 'Le corps d'Iphigénie et le coeur de Virgile.' Anna de Noailles, 'La Terre', in *Le Coeur innombrable*.

9   *From Madame de Noailles*

Thursday morning [14 January 1904]

My dear Marcel,

I am distraught, appalled, you are prodigiously right, and Constantin's thoughtlessness is so overwhelming that I seem to feel it covering me a little, giving me a sort of mask.

I am very sad, because I feel what joy it would have been for me, on the dismal morning of the fifteenth of every month,[1] to read what you alone can write, that marvellous mixture of irony and sweetness, like two opposing rivers gliding close to one another.

I must have seemed doleful and torpid last night. I was a little tired, but so happy, dear Marcel, to see you again. I felt that beneath my tiredness there stole a great contentment, which I shall treasure. I am very fond of you, you know.

Anna de Noailles

1. Publication day of *La Renaissance latine*.

10   *To Madame de Noailles*

[Friday evening, 15 January 1904]

Madame,

The day before yesterday I wrote you a long letter to thank you (I had intended to go and see you, I had telephoned you, you had gone out – and now once again I'm not very well). It was so kind of you to write to me, I can't tell you how much pleasure it gave me. Read the admirable pages from *Walden*[1] in the perfidious *Inconstance latine*, the ungrateful *Jactance latine*, the outrageous *Inconvenance latine*,[2] which in spite of everything is still the best review in Paris. It's as though one were reading them inside oneself so much do they arise from the depths of our intimate experience. The end of a friendship is a sad thing and obliges one to replace old friends by new. I'm beginning to grow very fond of Gabriel de La Rochefoucauld.[3] Truly. I'm infinitely sad to see how few people are nice deep down, and how profitlessly one opens one's heart. The article on Comte[4] in *L'Indécence latine* seemed to me very interesting. The three-quarters-of-an-hour-long prayers to Clotilde de Vaux are a prodigious revelation.[5] Lucien[6] would call it a punishment from heaven. But really it seems that all those who have been too superhuman, who have committed the crime of Prometheus or Nebuchadnezzar, must end up eating grass like Nietzsche, or becoming besotted by a crackpot religion like Comte. Nevertheless my admiration for *La Méconnaissance latine* doesn't extend to the poems. They struck me as stupid, though I only glanced at them.

Good-bye, Madame, silence tomb about Horatio.[7]

Your respectful friend,
Marcel Proust

1. Extracts from the French translation of Thoreau's *Walden* by Winaretta Singer (the Princesse Edmond de Polignac (1865–1943), American sewing-machine heiress, writing under her maiden name).

2. Parody titles for *La Renaissance latine* reflecting Proust's irritation at being refused a regular column: *Inconstance* = fickleness; *Jactance* = conceitedness; *Inconvenance* = impropriety.

3. Comte Gabriel de La Rochefoucauld (1875–1942), novelist. An elegant young aristocrat who was to become a lifelong friend and a model for Saint-Loup in *RTP*.

4. Auguste Comte (1798–1857), French philosopher, founder of Positivism.

5. Clotilde de Vaux (1815–46) was Comte's mistress for the last two years of her life. His positivist philosophy developed into a sort of religion of humanity with himself as its high priest and Clotilde its patron saint.

6. Lucien Daudet, who was a very pious Catholic.

7. Proust wished his pseudonym Horatio, under which he was writing articles on Parisian salons for *Le Figaro*, to be kept secret.

11 *To Emile Picot*[1]

45 rue de Courcelles
Monday [18 January 1904]

Dear Sir,

On correcting the final proofs of a translation of Ruskin[2] that I have made I notice that I've left three letters blank, which I hope can be filled in without necessitating a new make-up. But I didn't know where to find the information I needed, and since I know of no library that is as well stocked and as hospitable as your admirable memory (not, incidentally, that that is what I most admire in you) – since, too, you are always so kindly disposed towards me, I am taking the liberty of telling you what is troubling me. Ruskin speaks of the town of Eisenach. And he says 'its name, Iron-ach, recalling the Thuringian armorial bearings of old'.[3] *Iron* means *fer*. But *ach*? Perhaps you know. The authentic etymology of Eisenach is, I confess, a matter of some indifference to me, and even if Ruskin was mistaken I cannot, now that everything is set, correct him in a footnote. But if however the etymology of Eisenach could provide me with the meaning of *ach*, I would have all I want to know. I believe that an old word *ach* used to mean burg or town. But I don't see how an iron town could recall armorial bearings, even Thuringian ones, since it would be impossible to distinguish on those bearings whether a town was iron or not. If by chance this is something that you happen to know, you would be doing me a great service by telling me. Not, I repeat, by giving me the *true* etymology, which in itself has nothing to do with the case. But what Ruskin meant, true or false.

Forgive my importunity, dear Sir, and please accept on behalf of Madame Picot and yourself my most sincere respects.

Marcel Proust

1. Auguste-Emile Picot (1844–1918), professor at the Ecole des Langues Orientales and member of the Institut de France.

2. *The Bible of Amiens* by John Ruskin (1819–1900). Proust had completed the translation by the end of 1902 when extracts appeared in *La Renaissance latine* (see vol. I). A false start with one publisher caused a year's delay (see letter 63) and *La Bible d'Amiens* was finally published by Mercure de France in February 1904.

3. Ruskin in fact wrote 'Thuringian armouries' but Proust mistranslated him, taking 'armouries' to mean *armoiries* (armorial bearings).

12 *To Robert de Montesquiou*[1]

[Tuesday 19 January 1904]

Dear Sir,

You know that your precious existence is essential to French literature, to all the minds you enrich through the written and spoken word, and also to the hearts which cherish you and which, like mine, are wounded by your wounds. What joy it is therefore to see you emerge so magnificently from that combat[2] in which you excelled yourself, because you are in all things excellent, but from which poets should be spared because their battles are of a different nature which demand that they hoard their blood and conserve their strength. But in your case some help from your ancestors must descend in the hour of combat bringing you a magical increase of valour and audacity, if I understand the report I read, as eloquent as an oration by Bossuet (which happily in this case has nothing funerary about it) for some Prince de Condé – 'Three times we saw this valorous prince, etc. . . .' How magnificent it all is, that courage, the vivid picture of that joust which accords uncannily with the portrait you gave of yourself in *Les Perles rouges*.[3] But now we no longer want you to go on putting yourself at risk and living the flamboyant life of a Byron. We now want 'order and genius' and the seemingly lacklustre existence of some solitary of Port-Royal.[4] Tell Yturri[5] that I thought of him, knowing that he must be wretched and tormented, and accept, dear Sir, my affectionate regards.

Marcel Proust

1. Comte Robert de Montesquiou-Fezensac (1855–1921), prolific poet, connoisseur, dandy and wit. Inordinately proud of these roles and of his lineage, he cultivated an extravagant personality and indulged what Proust called his 'steadfast cult of Enmity'. Huysmans portrayed him as Des Esseintes in *A Rebours* and Proust caricatured his speech and mannerisms in the great comic character of *RTP*, the Baron de Charlus. Throughout a lifelong correspondence in which Montesquiou addresses him as 'Cher Marcel', Proust, as if to keep this importunate friend at arm's length, replies formally 'Cher Monsieur'.

2. He had just been tempted into his second duel, having ridiculed a *nouveau-riche* patroness of the arts, Ernesta Stern (pen-name 'Maria Star'), and been challenged by her son Jean, a skilled fencer. Despite Proust's remarks, he was reported in *Le Figaro* as having tamely lost all three bouts.

3. A volume of some of Montesquiou's best sonnets, one of which eulogizes his ancestors.

4. An abbey in the Chevreusé valley near Paris, headquarters of Jansenism.

5. Gabriel de Yturri (1864–1905), Montesquiou's devoted amanuensis, an exotic-looking Latin American who matched him in panache and extravagant tastes.

13   *To Marie Nordlinger*[1]

Sunday [24 January 1904]

Dear friend,

I told Reynaldo[2] (but I don't much rely on him to pass on the message!) that I had mislaid almost all the recommendations you were kind enough to give me for my Ruskin. And now that I'm sending back my proofs, I need your help (it's only because you're so kind to me that I dare ask you) to remove a few last-minute grammatical or archaeological misgivings. Would you be free at about half past nine tomorrow, Monday, evening (that's to say this evening when you get this note), and would you mind coming round to me? Otherwise I could come round to you, also tomorrow, but as I can't go out early I wouldn't be there much before eleven and fear that might incommode you. – If Monday is inconvenient what about Wednesday? – And if it's all too much of a nuisance, what about not at all?

Your respectful friend,
Marcel Proust

1. Marie Nordlinger (1876–1967), a young English artist and jeweller who was greatly attached to Proust and, being bilingual, gave him considerable help with his Ruskin translations. She had met Proust with her cousin Reynaldo Hahn when visiting Venice, chaperoned by her aunt, in 1900.

2. Reynaldo Hahn (1874–1947). Born in Caracas, Venezuela, he entered the Paris Conservatoire as a child prodigy aged eleven, and became a successful composer and conductor. He and Proust first met in 1894 and their long and intimate friendship lasted until Proust's death. See vol. I.

14　*To Madame de Noailles*

Sunday evening [24? January 1904]

Madame,

I need for my Ruskin,[1] for a quotation, the *Revue des Deux Mondes* in which you said that one should hurry up and love nature while one was still a child because afterwards, once love had come, one no longer saw anything at all. I was in no hurry to look out the issue. Then, on the point of sending back my proofs, I simply could not put my hand on it. So if you would be good enough to give me the date of that number if you happen to know it (you remember, it's the one with the blue rhubarb, the tender veins, the watering-can, etc.[2]) that would be most helpful and would enable me to dash round to the *Revue des Deux Mondes* and still send my proofs back in time.

I am as devoted to you as ever, Madame.
Your respectful admirer,
Marcel Proust

Searching through old numbers of the *Revue des Deux Mondes* (much too old for those verses of yours, but I hadn't seen the date) I found some lines of Madame de Régnier's infinitely better than I'd thought and really charming.

> Violet butterfly flecked with pale gold
> Thought-fostering flower, so dear to pensive hearts!*
> I love your double wing and threefold petal[3]

*The exclamation mark is hers, and not intended to indicate an enthusiasm which I don't feel. What a distance there is between her *pensive heart* and your *innumerable heart*![4]

1. Proust was correcting the proofs of his preface to *La Bible d'Amiens* and wished to quote from Anna de Noailles's poem 'Déchirement', published in the *Revue des Deux Mondes* in June 1903.

2. Various images from 'Déchirement' and another poem published in the same issue.

3. 'Papillon violet que veloute un or pâle / Pensée en coeur, ô fleur si chère aux coeurs pensifs! / J'aime ton aile double et ton triple pétale.' (Note that in French *pensée* means both 'thought' and 'pansy'.) Mme de Régnier, wife of the novelist and poet Henri de Régnier (see letter 25, n. 6), was Proust's old friend Marie de Heredia (see vol. I).

4. A reference to Anna de Noailles's collection of poems entitled *Le Coeur innombrable*.

15   To Marie Nordlinger

[First week of February 1904]

Dear friend,

Do you know that I am still, and more and more, unwell and thus prevented both from working and from having a young girl visit me? But I miss you very much and being unable to tell you so in person I send this brief message of impatient friendship. You left Whistler (*Gentle Art*[1]) behind here. I shall return it to you with the beautiful pictures[2] of sunbeams and shadows, the lovely twilight scenes which you sent me and which have given me great pleasure and done me much good. I particularly appreciated the views of Senlis, the waves, the flowers, the clouds. But that means all of them.

Dear friend, the *Kings' Treasuries* are precious. But they appeal to one in vain when one lacks too many other treasures, such as good health – to which one would readily add a few lilies plucked from the *Queens' Garden* [sic].[3]

I look forward to seeing you soon.

Your

Marcel Proust

I've corrected the mistakes you pointed out to me and fifty or rather three hundred others.

1. *The Gentle Art of Making Enemies* by James McNeil Whistler (1834–1903) which Proust had given her. He greatly admired the American painter – a reproduction of Whistler's portrait of Thomas Carlyle hung in his bedroom – but could never spell his name, varying between Wisthler and Wistler.

2. Her own watercolours.

3. *Of Kings' Treasuries*, about the hidden treasure in books, is Part I of *Sesame and Lilies*, the Ruskin lectures in three parts which Proust had begun translating with Marie Nordlinger's help. He was to translate only Part I and Part II, *Of Queens' Gardens*, which was about the education of women. Part I, *Des Trésors des rois*, first appeared in *Les Arts de la vie* in March–May 1905; Part II, *Des Jardins des reines*, was included with the completed translation, *Sésame et les lys*, published by Mercure de France in June 1906.

Proust's long preface to *Sésame*, which supplanted Ruskin's own, was first printed in *La Renaissance latine* in June 1905. Called 'Sur la lecture' (On Reading) it is the important essay in which Proust makes his first attack on the critic Sainte-Beuve. Cf. letter 315, n. 1.

16   *To Marie Nordlinger*

[First week of February 1904]

Thank you for the non-balsamic balsam. All the more balm to me since it exudes no balm. Seeds, a witty [*spirituel*] present (in the humoristic sense of the word *spirituel* which it's losing because of your cursed English infiltration) for someone who loves flowers but fears perfumes. Seeds, flowers of the imagination, as the dwarf trees at Bing's[1] are trees for the imagination – in the one case the future of flowers, in the other the past of trees, in a small present. In the Whistler itself[2] I don't want to write anything. The copy is a rare one and it would be criminal to deface it in any way. Besides, it really is the book of too great a man to tolerate a name like mine. Or I'd be like the cockney who signs his name in Amiens cathedral (see *The Bible IV*).[3] Ruskin and Whistler thoroughly despised one another, because their *systems* were opposed. But the truth is *one* and they both perceived it. In the very court-case against Ruskin[4] Whistler said: 'You say I painted this picture in a few hours. But the fact is that I painted it with the knowledge gained through a lifetime.' And at that very moment Ruskin was writing to Rossetti: 'I prefer the things you do quickly, immediately, your rough sketches, to what you work over. What you work over you spend, say, six months on. But what you draw at one go is the outcome of many years of dreams, of love, and of experience.'[5] In this respect the two stars strike the same point with a ray that is perhaps hostile but identical. There's an astronomical coincidence.

Thanks to Reynaldo (everything I've ever done is always 'thanks to Reynaldo') I met Whistler one evening. And he told me that Ruskin knew absolutely nothing about pictures. It's possible. But in any case, even when he was rambling on about other people's works, his misjudgments described and recreated marvellous pictures which should be loved for themselves. And to exhibit and acclimatize them in France, in the shape of scrupulous and impassioned copies, will be our joint endeavour.

Affectionately yours,
Marcel Proust

Could you tell Coco[6] at once that his walking-stick has been found in my house.

I am *very* ill.

1. Marie Nordlinger's employer, the German ceramicist Siegfried ('Samuel') Bing (1838–1905), pioneer of Art Nouveau, friend of Whistler and dealer in the Japanese prints and curios which were the height of fashion. The Japanese dwarf trees (Bonsai trees) reappear in Albertine's dithyrambic monologue in *The Captive* (*RTP*, III, 126).

2. *The Gentle Art of Making Enemies* (see letter 15, n. 1). Proust seems to have given Marie Nordlinger to understand that the copy in question, richly bound in gold-stamped red leather, had been given him by Montesquiou, Whistler's devoted admirer. Rather than inscribe it to her on the fly-leaf, Proust wrote her a long poem – he described it as 'a bit of nonsense' (*chose absurde*) – which she could slip inside the cover.

3. In *The Bible of Amiens* Ruskin says of the builder of the cathedral: 'He signed his name nowhere, that I can hear of. You may perhaps find some recent initials cut by English ... visitors ..., but Robert the builder – or at least the Master of the building, cut *his* on no stone of it.'

4. Famous libel suit of November 1878 following Ruskin's attack on Whistler's painting *Nocturne: the Falling Rocket*; he had called it 'a pot of paint flung in the public's face'.

5. Proust appears to quote from memory a letter tentatively dated autumn 1854 from a collection published in London in 1899 (George Allen, ed. W. M. Rossetti).

6. Coco was the nickname of the painter Frédéric de Madrazo (1878?–1938), whose
father, the Catalan painter Raymond de Madrazo, had married Reynaldo's sister,
Maria Hahn, *en secondes noces.*

17   *To Marie Nordlinger*

[Saturday evening, 6? February 1904]

Dear friend,

I decided at half past five to go round to Durand Ruel,[1] I arrived five
minutes before closing time, and I saw absolutely nothing. I didn't have
time to warn you in advance and I won't be going back there since I'm
retiring to bed with a terrible flu – temperature, sore throat, etc. A
charming blond gentleman[2] from Bing's (he has a cousin living in) was
extremely kind to me at Durand Ruel's. I've worked like a black on
Sesame and redone the whole of the beginning and the entire first
notebook *nec varietur.* There is only one thing I'm still dissatisfied with
and that is *careless writing.*[3] I've replaced *entretien fortuit* with 'dans le
négligé de la causerie' – it's worse but at least it isn't pretentious, which
would be the worst kind of mistranslation. I've written comments on
several passages in this first notebook, comments intended to serve
either for a preface or as notes. As soon as I'm in a fit state to receive you,
I shall write, for I'm all on fire for *Sesame* – and for you.

Your

Marcel Proust

1. An important Paris art gallery. The exhibition was mainly of Japanese works.
2. A man called René Haas who worked for Siegfried Bing.
3. In English in the original.

18   *To Reynaldo Hahn*[1]

[Around 27 February 1904]

Since little Ruskin cannot thank you himself for the ravishing tears you
caused the Muses to shed on the anniversary of his death,[2] he has asked
me to express to you his gratitude and his admiration for your fraternal

genius. And it is I who thank you, O my little Reynaldo, O my greatest affection in life; you know that this little book was desdicated[3] to you, as long as my little Papa was still alive. But he so wanted to see it appear that, now, I've decided to take it back from you to offer it to him.[4]

Come, my little Master, don't make funs of the pony[5] in his new Anglomaniac 'exercises' which are pretty perilous for him.[6] And above all very boring.

Marcelch.

1. This dedicatory letter to his most intimate friend (see letter 13, n. 2) was inscribed by Proust in Hahn's personal copy of *La Bible d'Amiens*, which had just been published.

2. An allusion to *Les Muses pleurant la mort de Ruskin*, composed by Reynaldo Hahn in 1902 and dedicated to Marcel Proust.

3. 'Desdié' in the French, for *dédié*. One of the features of the private idiom Proust often used in his letters to Hahn was the insertion of 's' or 'h', or sometimes 'ch', in various words. Cf. 'make funs' ('mosquez' for *moquez*) further on in this letter, and the signature 'Marcelch'. Cf. letter 64.

4. The book was dedicated to the memory of his father (see letter 2, n. 3). Part I of his translation of Ruskin's *Sesame and Lilies* (see letter 15, n. 3) was to be dedicated 'To M. Reynaldo Hahn, composer of *The Muses Mourning the Death of Ruskin*'.

5. A familiar term of endearment between them.

6. 'Exercises' in translation, 'perilous' for someone who doesn't know much English.

19   *To Henry Gauthier-Villars*[1]

[Sunday, 28 February? 1904]

Dear Sir,

My notes on *La Bible d'Amiens* have prompted you to reflections, sometimes friendly, sometimes grudging, always wittily erudite, for which I thank you.[2] So you must have read Ruskin in the original to have been able to point out so quickly (the book was taken round to the rue de Courcelles[3] only yesterday) an error in my translation regarding the respective positions of Jupiter and Minerva![4]

You laugh when M. Huysmans calls the *Beau Dieu*[5] a 'fop'; you laugh when I call M. Helleu 'a great painter'.[6] Must I congratulate you on this 'oecumenical' mirth, as M. d'Humières[7] calls it . . .

Will you please tell Madame Gauthier-Villars, etc.[8]

1. Critic and novelist under the pen-name 'Willy' (1859–1931), the first husband of the famous Colette (1873–1954).

2. Presumably in a letter Proust had just received.

3. Willy and Colette lived at 177 *bis*, rue de Courcelles.

4. Proust quotes a passage from *The Queen of the Air* in which, à propos of English public schoolboys' Latin versification, Ruskin remarks: 'Minerva was only a convenient word for the last line of an hexameter, and Jupiter for the last but one.' In his translation Proust reverses the names of god and goddess.

5. A famous statue on the west front of Amiens Cathedral. In his preface to *La Bible d'Amiens* Proust refers unfavourably to Huysmans's epithet (*bellâtre*) for the statue.

6. In his preface, Proust praises Paul Helleu (1859–1927) in a reference to his cathedral interiors.

7. Robert, Vicomte d'Humières (1868–1915), poet and translator of Kipling and Barrie, whose help in the translation of *The Bible of Amiens* Proust acknowledged in a foreword.

8. The end of the letter is missing.

20   *To Lucien Daudet*

[Monday, 29 February 1904]

My dearest Lucien,

You are an angel. I'm sure Albu[1] must be delighted and with his warm-heartedness will be thoroughly grateful to you and like you all the more for it.[2] I cannot promise you as much for my part, as I don't feel capable of liking you more than I do. However your niceness imparts to my friendship something even more tenderly fond (which irritates you) or at any rate more touched and more affectionate.

One thing worries me after re-reading your note as I was about to send it to Albu, and that is the phrase 'I had two seats'. If you had them, was it not with the object of going yourself? I cannot however believe that you gave up going in order to enable Albu to go. That would be going beyond all conceivable bounds and would be quite contrary to my wishes. I said 'procure some seats', meaning in addition to your own. But since your kindness, infinite though it is, is counteracted by the most solid common sense, I refuse to be worried by such an absurd hypothesis. I haven't seen Albu since, but I imagine he must be delighted to have what he wanted, as I am to have given him pleasure and to have appreciated your kindness. The day before yesterday I signed your copy of Ruskin[3] at the same time as those of Madame Daudet[4] and of Léon.[5] I didn't put the word admiration in yours, for fear that, since you don't write, it might appear m.g.[6] in the eyes of imbeciles. But now I think of it, quite apart from all the reasons there are for admiring you, there's a concrete motive which is your beautiful painting, your *Golden Loire*, your *Paradise pas-de-quatre of Tintoretto Degas*, your *Mysterious Garden*, your *Mother* – so many masterpieces. I shall add it privately if you will allow me.

Your

Marcel

Mama returned from the country yesterday, still very poorly.

I saw that you had dined out at the Noailles's and that everyone there except the nobles (and you) was a nationalist. I suppose it must have been Mme G.L.[7] who announced it in the *Figaro*.

1. Albu was the nickname of Louis-Joseph Suchet, Marquis (later Duc) d'Albufera (1877–1963), a Napoleonic title. Proust became not only his close friend but also the go-between with his mistress Louisa de Mornand. Albufera was yet another of the models for Saint-Loup in *RTP*, though in some of his more absurd upper-class mannerisms he resembles the Duc de Guermantes.

2. Proust having failed, Daudet had provided the tickets Albufera wanted for Feydeau's *La Main passe*.

3. *La Bible d'Amiens*, Proust's Ruskin translation published the previous day by Mercure de France.

4. Mme Alphonse Daudet, *née* Julia Allard (1844–1940), poet and essayist, widow of the novelist and mother of Léon and Lucien.

5. Léon Daudet (1868–1942), Alphonse Daudet's eldest son. Novelist and right-wing journalist, he was to be co-founder with Charles Maurras of the monarchist paper *L'Action française*. Proust dedicated the Preface to *La Bible d'Amiens* to him, and later *Le Côté de Guermantes*.

6. Initials probably standing for *mauvais genre* ('bad form'), but in this context signifying 'homosexual'.

7. Mme Gaston Legrand, *née* Clotilde de Fournès (nicknamed 'Cloton'), reigning socialite of the racing world and probably Proust's model for the absurdly snobbish Mme Blanche Leroi in *RTP*.

21   *To Louisa de Mornand*[1]

[End of February–March 1904]

Dedication not to be left lying around
To Louisa Mornand
Ringed by the blaze of her adorer's eyes[2]

Mornand is certainly not the present participle of the verb *morner*, for this archaic verb had a meaning which I don't remember exactly but which was extremely improper.[3] And God knows! . . . Alas! – For those who have had no success with you – that is to say everyone – other women cease to be attractive. Whence this couplet:

> He who Louisa cannot win
> Must be content with Onan's sin.[4]

I love you and admire you with all my heart.

Marcel

1. Louisa de Mornand (real name Louise Montaud: 1884–1963), minor actress, cocotte, mistress of Louis d'Albufera (cf. previous letter, n. 1). Proust made her the principal model for Saint-Loup's mistress Rachel (the name of her maid) in *RTP*. This letter was in the form of a dedication inserted in a copy of *La Bible d'Amiens*.

2. 'Cypris, sur la blancheur d'une écume qui fond, / Reposait mollement, nue et surnaturelle, / Ceinte du flamboiement des yeux fixés sur elle.' Victor Hugo, 'Le Satyre', in *La Légende des siècles*. (In Greek mythology Cypris is an appellation for Aphrodite.)

3. Archaic French verb meaning to blunt a lance or to render it harmless by fitting a ring to its tip.

4. 'A qui ne peut avoir Louisa de Mornand / Il ne peut plus rester que le péché d'Onan.'

22   *To Madame Alphonse Daudet*

[Early March 1904]

Madame,

I should like to be able to do a translation every day in order to receive more letters like yours.[1] And if only I had written the phrase about the charm 'which glides from one language to the other, in the space between the words', or the phrase 'dream visit' to describe someone who arrives very late and departs very early (did Madame de Sévigné ever invent anything so delightful), or the one about Ruskin 'who went as far as Tours' and who with 'his visionary eye on Gaul' displays an original vision which makes the whole of *The Bible of Amiens* new for me, as though hitherto unknown – if for one minute I had been able to write such delightful phrases I should have no need of flattering testimonials, I should have something that would be more precious to me than any of them, even such illustrious ones as yours, namely the satisfaction of my literary conscience, which nothing can replace, and which alas I have not!

I am no longer capable of posting a letter addressed to a Daudet without its being a hymn of gratitude, and if I had to declare everything I owe to you, to Leon and to Lucien I should have to write to you every day. Have no fear that this threat will be carried out, but allow me once more, Madame, together with my deepest admiration, to express to you my respectful gratitude.

Marcel Proust

1. Mme Daudet's letter, from which Proust goes on to quote, had acknowledged a complimentary copy of *La Bible d'Amiens* inscribed to her by Proust.

23 *To Adolphe van Bever*[1]

[Sunday evening, 6 March 1904]

Dear Sir,

I cannot tell you how annoyed I am by the way the book[2] has been distributed. Some friends of mine, or rather mere acquaintances without the slightest connection with the Press or even Literature, have received the 'Please Insert'![3] Which makes me look utterly ridiculous. On the other hand by this evening, Sunday, several people (there may well be a large number but I can speak only of those I see) had not yet received the book. I still need (whenever you can, there's no great urgency) another ten copies. I can do the distribution myself to save you further trouble; at any rate I've done so for about thirty copies. But I am possibly soon to be obliged to appeal once more to your kindness for help with the distribution. I should be enormously grateful if *as soon as you receive this letter* you could send a 'Please Insert' which I should like to send round to a newspaper *this very evening*. If at the same time you could send me the exact surname, Christian name and address of M. de Miomandre[4] that would be marvellous. But do you think he was the man who was interviewed by the *Weekly Review*, and not M. de Maiomande? Do you know the address of M. Louis de Robert?[5]

I should like to think that you are well, and should be happy to know that you have recovered.

Your devoted
Marcel Proust

I happen to have a copy ready and inscribed for M. Mourey,[6] so I shall send it to him. I was waiting to get his private address, but I shall simply address it to him at the rue de la Chaussée d'Antin.

1. Adolphe van Bever (1871–1925) was on the editorial staff of the review produced by Proust's publishers, Mercure de France.

2. *La Bible d'Amiens*.

3. *Prière d'insérer*: publication slip enclosed with review copies sent to newspapers and periodicals.

4. Pen-name of a critic (Francis Durand) who had contributed to a symposium on the contemporary novel for the *Weekly Critical Review* and mentioned Proust's name among a small group of young writers whom he admired.

5. Louis de Robert (1871–1937), writer, Dreyfusard, who was evidently on Proust's long list for complimentary copies of *La Bible d'Amiens*.

6. Gabriel Mourey (1865–1945), writer and art historian, publisher of the periodical *Les Arts de la vie* whose offices were at 6 rue de la Chaussée d'Antin.

24 *To Georges Goyau*[1]

Monday [March 1904]

Dear Sir,

Permit me to thank you most warmly for the beautiful letter with which you have honoured me. Your compliments are particularly precious to me:

> For only three or four
> Among whom you are numbered
> Have power to bestow
> Praise that lasts forever.[2]

Forever would only apply if your praise was addressed to Ruskin. Everlasting praise for an object that is not calculated to last, if indeed it exists, would only give rise to the ambiguity of the 'dubious fables' of which Baudelaire speaks.[3] Unfortunately I feel that you only half like Ruskin, and I blame myself for this. I began by translating only a few extracts from *The Bible of Amiens* which were chosen to give the loftiest idea of his work, those most worthy of his reputation and most characteristic of his genius. But a publisher, hearing of this, commissioned me to translate the whole of the *Bible*. Later this publisher went bankrupt[4] (it isn't the Mercure!). But I finished the work, for which I had acquired a taste in the meantime, for another publisher and I did not have the heart to sacrifice a single one of those great nebulae which I had tried to bring to relative daylight. And yet I would have been rewarded for the sacrifice. Every boring passage, every obscure page omitted would have been transformed at once into a pure and breathable air which would have circulated between the chosen pages and the splendid passages, setting them in their place and in their atmosphere – like pedestals heightening the noble and lofty pages – like magic mirrors which would have repeated and multiplied *ad infinitum* the beauties of the passages retained.

But I did not do this, and I am well aware that the book is hardly calculated to initiate people to Ruskin and to win him admirers. However I am busy translating another which is much more appealing, a simple lecture – a long lecture – on Reading (*Sesame – Of Kings' Treasuries*), with no longueurs, no fallings-off, no obscurities, none of that mishmash of superficial archaeology and fanciful history, and which I think you will enjoy. But it's difficult to do. It is above all of translations that one can say, as Verlaine says of the humble life, that, full of 'tedious and simple labour, it is a work of choice that needs a great deal of love'.[5]

Thank you again, dear Sir, I am most sensitive to your compliments and those of Madame Goyau and I send you my most grateful respects.

Marcel Proust

1. Georges Goyau (1869–1939), historian and critic. His reviews of *La Bible d'Amiens* appeared anonymously in the *Revue des Deux Mondes* (15 September) and *Le Gaulois* (18 December).

2. 'Car trois ou quatre seulement / Parmi lesquels on vous range / Savent donner une louange / Qui dure éternellement.' Proust paraphrases from an 'Ode' by Malherbe, 'to the queen, mother of the king, on the happy results of her regency'.

3. 'Fables incertaines', from *Spleen et Idéal* XLI, in *Les Fleurs du Mal*.

4. The publisher was Paul Ollendorff, subsequently editor of the daily newspaper *Gil Blas*, who had had Proust's translation for a year before releasing it so that a contract could be signed with Mercure de France (cf. letter 11, n. 2).

5. 'La vie humble aux travaux ennuyeux et faciles / Est une oeuvre de choix qui veut beaucoup d'amour.' Paul Verlaine, *Sagesse*.

25   *To Madame de Noailles*

[Saturday evening, 12 March 1904]

Madame,

I no longer write to you to tell you the things that I constantly think about you since I'm well aware that they must leave you cold. But tributes to which you are a little less indifferent I hasten to bring to your attention.

I dined this evening with Montesquiou, who with an almost pathological exaltation in his high-pitched voice, underlined with gestures that had to be seen to be believed, said: '"L'Exhortation"[1] is not only sublime, marvellous, ravishing, it is *the most beautiful thing that has ever been written.*' And then he proceeded to recite it *in its entirety*, exclaiming all the time, 'What genius! What genius!', and on leaving with me he stopped suddenly in the street, raised his arms to heaven and murmured: 'The sky this evening is of a colour that no words can describe,' making the passers-by stop and stare, and making me catch a terrible cold, and then, electrified by the impact of this sentence, he stamped his feet on the ground as though his heels would snap, flinging himself backwards the while. And that is not all. Before that he had spoken to our hosts of 'La Prière devant le soleil',[2] and in an oracular tone had said: 'I shall tell you *exactly what it is: it's the most beautiful thing that's been written since* Antigone. What sublimity! *It's more beautiful than Sophocles.*' Speaking of the 'arms overflowing with tenderness', he said it was 'a myth more beautiful than the pagan myth of the Milky Way', and recited it admirably I must say, contrasting the noble 'Euxine Sea' with the touchingly rural sun-dial. Then he said that he didn't know the lines of yours I quoted in the *Bible d'Amiens*,[3] which he found more beautiful than anything, etc. etc.

Dear friend (sorry, Madame), please do not think that I find it noteworthy that people should say these things; they are the inevitable echo of your divine accents in any human ear capable of hearing them. But whenever I have the composure to silence my admiration for a while to listen to a momentarily fraternal voice take up the refrain and show me that we are in unison with regard to you, singing quite independently of one another the same canticle in your praise, it fills me with such joy that I cannot resist repeating it to you in a similar frenzy of tender prayer.

Were you told that I telephoned you the day before yesterday (the only day I was able to get up, I've forgotten which) to ask if I could come round to thank you for your sublime letter (which I shall return to you, for I feel that certain phrases applied to me outshine me infinitely, are of a beauty which does not belong to me and which you should publish).

If you see Hermant[4] would you tell him that if he could slip in, however irrelevant the context, a word or two about my *Bible d'Amiens* he would give me great pleasure, for it might sell a few copies in which I have no financial interest but I should like the generous action of my publisher in bringing out the book not to be too disastrous for him. Since I produce a book every ten years Hermant need not fear this precedent. But I imagine his articles, being too well composed to bear the introduction of a foreign element into their economy, will reject such violence even supposing, which I don't know, that he himself wished to make it possible out of friendship for me. However, if you find it a nuisance to talk to him I can perfectly well do it myself, or indeed not do it since it won't do any good!

I shall tell you some alarming things when I see you. I'm probably going to ask Hermant to do a more disagreeable chore, but this is one that I hope he won't dodge. I'm going to have my publisher[5] to dinner and at the moment I can't really invite more than two people. And I should like the second person to be Hermant. (But don't tell him this – for I'm not yet certain about the other, who is incidentally charming, a friend of Régnier's.[6])

I'm going to tell you the alarming thing here and now, but *silence tomb*. I told Montesquiou that Horatio[7] wasn't me. But what do you think Montesquiou has done: he told me that, not having been able to find the author, *he had had the article privately printed in a slim volume*, making a few corrections to it, merely to the punctuation, he said. I didn't dare say a word, for fear of giving myself away if I protested, but what do you think of this coup? *Tomb, tomb, tomb.*

Your respectful
Marcel

Have you any news of your silent and nomadic cousin, my inconstant friend, or friend that was, Antoine Bibesco?

1. Her short story (see letter 6, n. 7); it was only six pages long.

2. A recent poem of hers, also published in *La Renaissance latine* (later collected in *Les Eblouissements*, 1907), from which Montesquiou goes on to quote.

3. In his preface, Proust quoted from her poem 'Déchirement' (see letter 14, n. 1).

4. Abel Hermant (1862–1950), novelist and critic who had recently rejoined *Le Figaro*. He seems not to have obliged (cf. letter 36).

5. Alfred Vallette (1859–1936), editor-in-chief of Mercure de France.

6. Henri de Régnier (1864–1936), symbolist poet and novelist, later an Academician.

7. Horatio was Proust's pseudonym for his society articles in *Le Figaro* (see letter 10, n. 7). That of 16 January, entitled 'Fête chez Montesquiou à Neuilly (Extraits des Mémoires du duc de Saint-Simon)', was a parody of the great chronicler of Versailles under Louis XIV and, by implication, a satire on Montesquiou's entertainments at his sumptuous Pavillon des Muses in the Paris suburb of Neuilly. Montesquiou, at first ignorant of its true author, had replied in a letter to 'Horatio' (see letter 127, n. 6). This early Proust pastiche was later superseded by a much longer Saint-Simon parody in his Lemoine series (see letter 265, n. 2).

26   *From Maurice Barrès*[1]

100 boulevard Maillot
Neuilly-sur-Seine
[13 March 1904]

Dear friend,

I liked what I saw of Ruskin through Milsand years ago.[2] I never saw him in the flesh. I mistrust his verbiage, but I must also mistrust my prejudice against him. You put me thoroughly at ease with this translation, through your erudition and the book you chose. At last I shall have a sure opinion for I shall read you in Amiens itself as soon as we can make a motor-car trip. – I have already glanced through you. There is a printer's error in note 2 on page 150.[3] And I must say that I'm very touched by this foreigner's friendship for France. I've been told that Pater has described some travels in France. Who will introduce them to us?

Thank you for your appreciation, so amicably expressed and underlined.[4]

I have a great esteem for Maeterlinck, but if he has any taste he will not forgive you for having put him, or appeared to put him, on the same level as Jean Racine. Proust! Proust! Marcel! There are ways of admiring genius that are more unseemly than the brutal coarseness of the barbarians. When three centuries of noble minds have united their sincere sentiments around the pages of Maeterlinck, then one may be able to compare him to Racine. But what am I saying? When that time comes, there will be six centuries of veneration of *Esther* and *Andromache* and the rest. Let each one assume his proper rank and, to ensure his own, respect that of the masters.[5]

Forgive me, my dear friend, because I am right and because I have a great deal of sympathy with your avid passion for artistic matters.

Yours, with a handshake,

Barrès

I ought to emphasize more than I have done the richness of your preface and your notes. One senses there not only love, but saturation in the subject.

1. Maurice Barrès (1862–1923). Novelist, essayist, nationalist politician (cf. letter 113, n. 2) and leading anti-Dreyfusard, for several years the *grand ami* of Mme de Noailles.

2. Joseph Milsand (1817–66), French author of a Ruskin study also admired by Proust.

3. The note in question reads thus, in its entirety, in the original: 'On inquiry I find in the plain between Paris and Sèvres.' Proust translated it correctly but failed to supply the missing direct object.

4. Proust cites Barrès twice in his preface to *La Bible d'Amiens*, although in a postscript he reproaches him for having omitted to mention Ruskin in his latest book, a study of Venice.

5. Proust's remarks were in his preface. For his reply, see letter 28.

27 *To Robert de Montesquiou*

[Sunday, 13 March 1904]

> Rosencrantz: Who is that young lord?
> Polonius: His name's Horatio. Naught
> else I know of him.   (*Hamlet*[1])

Dear Sir,

Many thanks for the pretty booklet.[2] I should like to know where the dedication comes from and what the last line applies to. I was concerned to hear that your secretary had made a fruitless journey. But since I did not think he would come today I hadn't prepared the books.[3] I shall leave them ready since I'm told he is coming back, which I deplore as an excess of kindness on your part, but I'm afraid of complicating everything for you by not leaving them and addressing them to you myself.

I was sad to hear last night from what you said that Monsieur de Yturri was very ill. I had no idea. But I must confess that I have the impression that he is one of those condemned men who are never executed and that he will soon get his reprieve and live to spend many unhoped-for and happy days. I remember that several years ago now my poor Papa found him very ill. And you see how he recovered. I believe he would be wrong to worry, and you too.[4]

I was delighted to see you again and to hear what you had to say last night and I telegraphed at once to Madame de Noailles the splendid encomium you gave her.[5] She will, I am sure, be especially proud of it.

> For only three or four
> Among whom you are numbered,
> Have power to bestow
> Praise that lasts forever.[6]

I fear I may have given you to believe last night that my translation wasn't mine alone. On the contrary I did it entirely on my own. I asked d'Humières[7] for advice here and there, but the whole thing is mine, and rewritten twenty times. If I told you that I could not have done it without Madame H. that is because she taught me the rudiments of English. But when she left France I had never opened a book by Ruskin, so she could not have translated a single line of him for me.

Good-bye, dear Sir, and do not cultivate too bellicose a spirit at Artagnan.[8] Follow the lessons of Gautier rather than those of d'Artagnan.

Your respectful admirer
Marcel Proust

1. Proust's witty allusion to his *Figaro* pseudonym and also, as the Shakespeare quotation is itself a pastiche, to his Saint-Simon parody. Cf. letter 25, n. 7.

2. Dedicated to Proust in verse, this was one of the copies of the *Figaro* pastiche (cf. letter 25, n. 7) which Montesquiou had had privately printed and handsomely bound.

3. Complimentary copies of *La Bible d'Amiens*.

4. Montesquiou's friend Gabriel de Yturri, a diabetic, had less than two years to live.

5. See letter 25.

6. Proust had quoted the same four lines to Georges Goyau. See letter 24.

7. Vicomte Robert d'Humières. See letter 19, n. 7.

8. Montesquiou's family home was the Château d'Artagnan, in Gascony, and Dumas's hero in *The Three Musketeers* was his ancestor.

28   *To Maurice Barrès*

[13, 14 or 15 March 1904]

Dear Sir,

It is extremely kind of you to write to me at such length[1] and I thank you wholeheartedly. But what strange reproaches you level at me. I do not know M. Maeterlinck,[2] consequently I have not sent him my book, consequently he will not be aware of it, and so I don't know whether he would be pleased or sorry to be compared to Racine. But what I do know is that it was a comparison that I did not make. I said that the lives of Racine, Pascal, Tolstoy and Maeterlinck divided themselves into two parts. It's an idea that pleases me. And as soon as a newspaper agrees to take articles from me, if it ever does, I shall write an article which I shall call: 'How beautiful is a life that begins with art and ends with morality', and it will be about Maeterlinck and I shall bring in Racine, Pascal, Tolstoy and I hope to find others between now and then. But this phenomenon can occur even with inferior people and I have never sought to put them on the same plane. I may add that if I were writing an article on Maeterlinck I should also compare him to Virgil. From the point of view of talent (which I hadn't thought about) of course I don't think he is at all like Racine. But I still find him a very considerable thinker, and I read him more often. But once again, when I refer to Barrès and Chateaubriand it is the works that I'm thinking of. Whereas here it was purely in terms of divided lives, and for the sake of symmetry and to appear clever, and to make myself understood. You say: When three centuries of noble minds have united round the pages of Maeterlinck there will be six around Racine. But in that little game no one would ever catch up with anyone and it's a way of settling (and reviving) the quarrel between the ancients and the moderns which is really too cruel for us 'who are the people of now'.[3] And I am too much of an idealist (of the Bouteiller type[4]) not to conceive of works of art as to some extent outside time and independent of the admiration they arouse and I would be careful not to deprive them of the patina with which so many ardent admirers have invested them. However, what is extremely nice is that in order to mitigate your harshness you should have called me Marcel. And I should be infinitely happy if it outlived your reproaches and became a habit. I looked up page 150 where you say there is a printer's error in note 2. Note 2 on Sèvres and Paris seems perfectly correct to me.[5] As for Pater, I shall certainly not be the one who translates him. I still have two Ruskins to do,[6] and after that I shall try to translate my own poor soul, if it doesn't die in the meantime.

Your most grateful admirer
Marcel Proust

1. See letter 26.

2. The Belgian poet, playwright and essayist Maurice Maeterlinck (1862–1949), author of *Pelléas et Mélisande* and famous for his allegorical writings on bees and ants.

3. 'Les anciens, Monsieur, sont les anciens, et nous sommes les gens de maintenant.' Molière, *Le Malade imaginaire*, Act II, scene 2.

4. Character in Barrès's trilogy *Le Roman de l'énergie nationale*; a brilliant and ambitious philosophy teacher who 'deracinates' his pupils, he represents everything that Barrès anathematizes. Cf. letter 113, n. 2.

5. Cf. letter 26, n. 3.

6. *Sesame and Lilies* and a selection (*Pages choisies*) which he abandoned.

29   *To Antoine Bibesco*

[Around 28 March 1904]

My dear Antoine,

There is one point at least on which I can tell you [. . .[1]] on condition that you promise me never to make an 'inquiry' and never to violate the tomb.[2] It is on this point that it would have been more convenient for me to speak to you in person. Among other truly odious things said by you there is the following. You said – to several persons who do not know each other – that I [. . .]. I cite this depressing example so that you may understand why in the true sense of the word I am no longer your friend. And what astonishes me is the fact that the pleasure (quite negative incidentally and without exhilaration) of not seeing you when I don't see you can suddenly be succeeded, when I do see you as I did this evening, by a *very great* pleasure. This is because your person, your physical person itself, retains the unconscious memory of the marvellous qualities you once had and which, materialized, bidden by the wicked fairy of your self-destructive character to be now merely looks, gestures, inflections of the voice, preserve nonetheless for one who has known them the charm of a particularly touching image, that which objects leave of themselves, make of themselves, an imprint full of meaning and beauty.

From another point of view, since the core of my nature is sympathy,

I recreate more readily in myself the tendencies that unite me to people than those which separate me from them forever. Hence the memory of the rare kindnesses you have shown me (I don't mean that they weren't very numerous but on the contrary that they were of a rare and exquisite kind) often comes back to fill me with the desire, not to draw closer to a person from whom too many ineffaceable grievances have estranged me, but to be of help to him if the opportunity ever arose, which alas, given my situation in life, seems all too improbable but which for that very reason might be possible. So if ever in any given circumstance I can be momentarily of use to you I should be happy for you to remember that in this exceptional and precise sense I remain

Your devoted and grateful
Marcel Proust

I should be very pleased if M. Vuillard would agree to sell me the sketch of the dinner party at Armenonville last year,[3] a unique point of coincidence between his admirable talent which often pervades my memory and a consummately delightful moment of my life. Would you be kind enough to ask him.

1. Passages in square brackets are illegible, scratched out in black ink. The whole of the first manuscript page of this letter, and part of the second, were crossed in red ink and omitted from the original edition of Proust's correspondence.

2. i.e. the silence of the tomb. Cf. letter 3, n. 4, and elsewhere.

3. The painter Edouard Vuillard (1868–1940) was a guest at a dinner which took place on Thursday 18 June 1903 at the Pavillon d'Armenonville in the Bois de Boulogne.

30   *From Daniel Halévy*[1]

26 place Dauphine, 1 arr.
Thursday, 31 March [1904]

My dear friend,

Your Ruskin[2] has reached me after a detour and a stop at my father's house in the rue de Douai.

I knew it already, having glanced through it on Dreyfus's table,[3] and we talked about how clever it was.

I must confess that I haven't yet read it. I hate reading books because they have been sent to me and as soon as they are sent to me. When a book is worth reading, there will always come a moment when one wants to read it, and then one reads it properly. I am sure that will happen with your Ruskin, and I shall wait.

I entirely agree with you about the scant initiatory value of those books on the Man, the Range of his Influence and the Errors of the Doctrine. Great men have to be probed, and it's a difficult exercise, which requires tact, patience, affinity and, above all, time, that precious substance. You offer us at the foot of your pages, in the most modest fashion, the fruit of your tact, your affinity and your time; the offering is worthy of the cultivated scholar that you are.

Your

D. Halévy

1. Daniel Halévy (1872–1962), historian and art critic. A schoolfellow of Proust at the Lycée Condorcet and the object of his adolescent passion. Aged nineteen, Halévy and Proust, together with friends who included Robert Dreyfus (see n. 3), Fernand Gregh (see letter 58, n. 1) and Jacques Bizet (see letter 123, n. 1) founded *Le Banquet* (The Symposium), a short-lived periodical which published Proust's stories later collected as *Les Plaisirs et les jours*. See vol. I.

2. *La Bible d'Amiens*.

3. Robert Dreyfus (1873–1937), historian and journalist, on the staff of *Le Figaro*. See vol. I.

31   *To François de Carbonnel*[1]

45 rue de Courcelles
Wednesday [6 April? 1904]

Dear friend,

Could you tell me on what day Bertrand de Fénelon left St Petersburg. I sent him a translation of *The Bible of Amiens* and not having heard from him I fear that it may have arrived after his departure, in which case (since his books are probably not forwarded) I shall have another copy sent to him wherever he is. I ask you this strictly between ourselves, for if by any chance he has received the book there is nothing more idiotic than to appear to be clamouring for thanks by wondering at somebody's silence. Which is why I am making this discreet and indirect approach. However if you prefer to tell him that I've written to ask you this that would also be fine. It's a question of tact which I leave to yours which is bound to be as exquisite as ever. I hope that when you come to Paris on leave you will not fail to visit me; it would give me great pleasure and enable me to express to you better than in this brief note my warmest and most affectionate regards.

Marcel Proust

1. A young diplomat *en poste* at St Petersburg (cf. letter 3, n. 3).

32 *To Louisa de Mornand*

[April 1904]

FOR LOUISA
(SKY-COLOURED CANOPY
ROSE-COLOURED ANGEL)

Colour of canopy, colour of skies,
Cloud-flecked azure overhead
Floats above Louisa, where she lies
Reading on her side in bed.

Her head droops upon the pillow,
Drowsily she turns to stare
With vacant eyes towards a willow
Painted by Madeleine Lemaire.[1]

In this heaven, with angel guarding,
On the pretext that it's Sunday
Marcel Proust, all unregarding,
Stays so long he finds it's Monday!

From the calyx of her sleeve
(Fabric of the finest weave)
Languidly and half unknowing
Louisa's arm, a pure white stem,
Emerges, rises, softly glowing.

If I've said an angel bends
From this disturbing Paradise
And from the blue-white sky descends
To where the rosy virgin lies,

Doubtless I was confused by bliss,
It is two Sèvres cupids[2] who,
Delicious and surprised, espy
Two mouths united in a kiss
Two hearts in perfect harmony.[3]

And through the half-open salon door
The crimson hangings, partly seen,
Turn Tunisian blue to green

... ...

The bed is blue, the salon red
Nothing moves and nothing's said.
Dear azure casket, you enclose
A pearl the colour of a rose.

... ...

I'm mad about my photograph. I feel that what you have written to please me[4] will become a reality for me and will have the power to bind my fickle heart. Meanwhile no child that has just been given its first doll was ever as happy as I am.

Your grateful
Marcel Proust

---

1. Mme Madeleine Lemaire, *née* Jeanne-Magdeleine Coll (1845–1928). Hostess well known for her lively salon, her musical evenings – cf. Mme Verdurin in *RTP* – and her flower painting (cf. Mme de Villeparisis in *RTP*). She illustrated the original de luxe edition of Proust's *Les Plaisirs et les jours*. See vol. I.

2. i.e. porcelain cupids, presumably part of the baroque-sounding decorations of Louisa's four-poster.

3. Cf. *RTP*, where Saint-Loup's mistress Rachel (modelled on Louisa), who liked to excite his jealousy, claims that the narrator had made sly advances to her in his absence (II, 318, 360–1).

4. Louisa had sent Proust a simpering photograph of herself on which she had written, 'The Original who is so fond of little Marcel. Louisa. April 1904.'

33 *To Marie Nordlinger*

[Sunday, 17 or 24 April 1904]

Dear friend,

Thank you for the miraculous hidden flowers which have enabled me this evening to 'make a spring of my own', as Madame de Sévigné says, a fluvial and inoffensive spring.[1] Thanks to you my dark electric room has had its Far Eastern spring. Thank you too for the splendid translation[2] which I shall go through carefully and with your permission alter, though diffidently, with affectionate respect. You speak French not only better than a Frenchwoman, but like a Frenchwoman.[3] You write French not only better than a Frenchwoman but like a French-woman. But when you translate English all the original characteristics reappear: the words revert to their own kind, their affinities, their meanings, their native rules. And whatever charm there may be in this English disguise of French words, or rather in this apparition of English forms and English faces breaking through their French accoutrements and masks, all this life will have to be cooled down, gallicized, distanced from the original, and the originality extinguished.

No, I shall certainly not go and seek the *coup de grâce* from Manet and Fantin when Clouet and Fouquet have already half-killed me.[4]

Your friend

Marcel

What is 'the pearl of the Rosary of Venice'?

P.S. (Monday). Seeing that my letter has been forgotten I add a brief postscript. I'm afraid you will be angry about *Sesame* as I've turned everything upside down; however I can go back to your terms if you prefer. But I think the text you have translated cannot be the same as in the edition I'm working from as they conflict all the time. Or else you've skipped words on several occasions. It will be better for me to revise what you've done by myself and afterwards we can discuss the whole thing together. The other day I was sleeping when I heard the door-bell ring. In the first stupor of wakening I didn't ask immediately who had rung. When I did ask and learnt that it was you it was too late, you were already gone. However even if you had still been there I wouldn't have been in any state to see you. I still haven't been out again but will risk it shortly (only in the evening, not in the daytime any more).

Affectionate regards

Marcel Proust

1. April was a poignant month for Proust, the beginning of a long daytime incarceration (pollen aggravated his asthma), and she had had the charming notion of sending him a symbol of spring, a packet of the Japanese paper pellets which open in water to become exquisite flowers. Proust was to use them in a key passage in *Swann's Way* as a simile for the revelation of unconscious memory (*RTP*, I, 51).

2. Her work on *Sésame et les lys* (see letter 15, n. 3).

3. Marie Nordlinger was, of course, an Englishwoman living in France.

4. Two exhibitions had just opened: one of late nineteenth-century French painting at the Luxembourg Museum, and the other, which Proust had evidently visited, of French Primitives at the Louvre.

34 *To Robert de Flers*[1]

[Second fortnight of April 1904]

My dearest Robert,

As a result of illnesses, that is to say a worsening of my habitual state, I was unable to go to *La Liberté*. Then you told me that your article[2] had appeared two or three days before the note in the *Figaro*.[3] So I asked for the previous week's, and it wasn't there, then the following one, neither. I sent round but it couldn't be found and it's only now that I've managed to get hold of the article which dates back further than you had said, so that it's even more touching in its kind-hearted haste than I had thought. And I read those charming things; in the transparent, mocking flow of your restless and profound dialectic that grain of hackneyed common sense in Alphonse Karr[4] takes on a freshness, a brilliance that I had never found in him. You are more than a setter of pearls. You are more like the character in the *Arabian Nights* who filled my adolescence with wonder, who changed old lamps into new.[5] This modest miracle becomes much greater when applied to the things of the mind. And you have succeeded in restoring to that old, worn lamp a new light, almost a radiance which illuminates the present situation very clearly. Dare I confess to you that this is not what I was most sensible to. And that for once what I've most liked in you was not, as Musset says, yourself, but myself, or rather yourself in relation to me, the charm and sweetness of your praise, the graciousness of the intention which brought *La Bible d'Amiens* from the depths of those bright, tumultuous waters, like the cathedral of Is whose steeple could be seen in the depths of the Breton seas. You are an exquisite friend whom I adore.

Marcel

Before I fell ill again I had gone round to the *Figaro* to say good-bye to you before your departure. You were not at the *Figaro*, but neither had you left, having been held up, I was told, by a sore throat. I hope it didn't last and that you and your wife have been able to enjoy the sun and the flowers which do me so much harm but which I love so much.

1. Robert Pellevé de la Motte-Ango, Marquis de Flers (1872–1927), author, journalist, playwright, and schoolfriend of Proust.

2. In an article in *La Liberté* on the journalist and Romantic novelist Alphonse Karr (1809–90), referring to Ruskin's admiration for Karr, Flers had commended the 'elegance, strength, literary quality and taste' of Proust's translation of *La Bible d'Amicus* (sic: Flers had not corrected his proofs).

3. A *Figaro* diarist had welcomed 'a young writer of great talent, M. Marcel Proust'.

4. See note 2.

5. Aladdin: *The Arabian Nights* is a recurring theme in *RTP*.

35   *To Robert de Montesquiou*

Friday [22 April 1904]

Dear Sir,

Had I not been bedridden for some days I should very much have liked to call round to tell you how deeply and respectfully I share in your grief.[1] I scarcely knew your father, but the fact that he was your father was enough to endow him in my eyes with a sort of miraculous singularity and added the sense of mystery that invariably surrounds the engendering of a great mind to the great respect which his universal reputation everywhere aroused. When I saw him in the street some years ago, in the days when I still went out for walks, this tall old man already shadowed by the approach of the darkness but proud of the lustre which your achievements shed upon his name, I said to myself over and over again: 'M. de Montesquiou's father', with a strange feeling in which there was a mixture of astonishment and veneration. I have the literary proof, quite apart from all those others which now so clearly confirm it, of the gentle sensibility which you have bequeathed to persons who did not always have the same intellectual aspirations as you did. Hence I can imagine all too sadly the painful scission which must now have occurred between yourself and that part of your sensibility embodied in the man whom you will see no more, and the melancholy of all those memories which in this cruelly solemn moment must pitilessly assail you.

Believe me, no one is more grieved by your sorrows, more admiring of your work, more respectfully devoted to your person than myself.

Marcel Proust

1. Montesquiou's father, Comte Thierry de Montesquiou-Fezensac, had died aged eighty-four on 16 April.

36   *From Madame de Noailles*

Monday [25 April 1904]

My dear Marcel,

How can you believe that I failed to do something that you asked me to do. I told Hermant of your wish on the very evening that you expressed it, and several times since.[1] I don't know what he can be thinking of. But I'm glad that your book has aroused such passionate interest. The preface, brimful of quiet learning, and all the paths of gold which your thought, with its sublime versatility, opens up throughout the length of Ruskin's essay, filled me with fond enchantment. Dear Marcel, I am tired, and you, how are you?

With all my infinitely grateful friendship,

Anna de Noailles

1. Proust had asked her to persuade Hermant to mention his translation in *Le Figaro* (cf. letter 25).

37   *To Anatole France*[1]

Tuesday evening [10? May 1904]

Mon cher maître,

I have just this minute received your letter. What wonderful evenings I can look forward to spending with Crainquebille, Dean Malorey, General Decuir, Putois, Riquet, now brought together,[2] so freshly born of the miraculous foam of your genius, though already august because of the irresistible influence they have exercised over men's minds in the past few years during which they have changed the world so profoundly that they have taken on the majesty of the centuries, over the greatest and most original minds, which they have magisterially constrained to imitate and reproduce them. Witness only yesterday Maeterlinck's *Pelléas*,[3] in which I cannot help but see the last-born of *Riquet*, full-blooded and somehow intermixed with Flanders. Thus the great writer is obliged to make use of your symbols, and he has borrowed all your fetishes and all your divinities. I shall meet again those sublime *Upright Judges* and the alternating and successive dialogues of the judges on horseback and the horses harnessed together. And the marvellous conception of the Heavenly Horse (though the fact remains that the happiness of sacrificing oneself to comply with the wishes of the Heavenly Horse can make the recalcitrant horses more noble and more inventive than the happy ones). The 'Manoeuvres at Montil' has of course that wonderful scene of the general searching for his brigade – as brilliantly ironical as the Battle of Waterloo in the *Chartreuse*,[4] with those conversations equalled only by Balzac's though finer – the general seeing the Van Orley tapestries and saying: 'You have a big place here!' 'The General could have brought his brigade.' 'I should have been happy to welcome it.' The three lines have remained engraved in my memory as the finest comic triptych ever painted by a master hand, quite perfect of its kind, the most surprising with its strokes of inspired invention, the most satisfying in its unexpected but disconcerting truthfulness. I also seem to remember the editors of a newspaper who ask for an article with a whiff of aristocracy. The only piece I've re-read so far – the book only arrived ten minutes ago – is 'Le Christ de l'Océan', which moved me deeply. The one I should like best perhaps – I loved it so much – is 'Putois'. But then I know the story of it, having learnt it from your lips in that happy time when I could still see the little flower,[5] still living, which had suggested the form of the sculpted stone in your sublime cathedral.

Thank you again, *mon cher maître*, for not forgetting an invalid whose existence you alone have remembered, because the greatest are also the best.

Your respectful and grateful admirer
Marcel Proust

1. Anatole France (real name Thibault: 1844–1924). Novelist and critic, Dreyfus supporter and one of Proust's earliest mentors, he wrote the preface to Proust's *Les Plaisirs et les jours* (1896). See vol. I. He was one of the principal models for Bergotte in *RTP*.

2. Characters (some eponymous) in short stories by France just republished in a collection which he sent to Proust, and referred to throughout this letter.

3. Proust is presumably referring to Debussy's opera, which was first performed in April 1902. Maeterlinck's play goes back to 1892, eight years before *Riquet* first appeared, in the *Figaro* of 26 September 1900.

4. Stendhal's *La Chartreuse de Parme*.

5. Possibly a reference to France's novel *The Red Lily* (1894). See vol. I.

38   *To Henri Bergson*[1]

Wednesday [25? May 1904]

Dear Sir,

Allow me to thank you with all my heart for your great kindness. You can imagine how much store I set by a few words from the philosopher I most admire, and how delighted I am at the thought that he might be prepared to say them.[2] Had I not been so unwell recently I should simply have gone round to ask you whether you didn't find the whole thing excessive, and hence out of place, inopportune, perhaps ridiculous. If by any chance you did find it so, please don't take the trouble to write to me. I shall understand that your opinion was that it would be better to do nothing, on seeing that you had in fact done nothing. If on the other hand you find the idea natural and possible, don't bother to tell me either, and it will be a great joy to me, a just cause for gratitude and pride, that you should have done it.

Your most respectful admirer
Marcel Proust

1. Henri Bergson (1859–1941). The most celebrated French philosopher of his day and a strong influence on Proust, with his stress on memory, time and intuition. Proust was apparently his 'best man' when he married Louise Neuburger, whose mother was Mme Proust's first cousin.

2. On 28 May Bergson was to give a paper to the Académie des Sciences Morales et Politiques commenting on Proust's translation, *La Bible d'Amiens*, in particular on his preface – 'an important contribution to the psychology of Ruskin' – and on his style: 'so lively, so original, that one can hardly believe one is reading a translation'.

39 To Marie Nordlinger

[About 25 or 26 May 1904]

Dear friend,

The urn[1] is marvellous! The four harpocratic heads[2] (in identical quadruple sleep) are exquisite, the symbolic serpent closes the urn beautifully and the petals of the flowers, the leaves of the foliage and knots of the snakes are admirable. But more than anything else I like Sorrow, so intimately linked to the beloved Ashes to which she clings, her cloak as in *Middlemarch*,[3] her reverie absorbed and bitter; and above all she isn't superimposed; she isn't an additional figure tacked on; her legs, so beautifully positioned, exclude the decorative motif of the other sides, for she takes up some space, poor thing; she is not a symbol, she is there alive and real, her body excludes anything else and isn't added on top, she is there against her urn in the position she must inevitably occupy, new and yet natural, contrived and yet necessary, with no suggestion of conventional grief either, a real creation of yours, but in the context of truth, and hence revealing new truths about the meaning of grief, and the bodily inflexions and facial expressions we adopt when it takes hold of us, all of which you have recreated, making us feel the truth of it in the obscure memory of our aching muscles, in the droop of our grief-chilled flesh. And similarly with the mysterious expression of the sleeping heads, or rather Sleep (I haven't been able to work out the precise symbolism of the floral decorations).

I'm writing all this to you from my bed, where I've been lying ill for the past week, unable either to speak or to eat or to sleep, etc. etc. and from which I glimpsed for barely five minutes that beautiful human urn whose very shape is a creation of your mind. The problem now is to decide whether it is superior or inferior to the dish[4] which was at once more personal and more impersonal but less potent and soul-searching. I'm too exhausted to develop this. All I have the strength to do is sign

Your friend
Marcel

I've worked really hard!

1. Marie Nordlinger described her sculpture as 'a bronze funerary urn'.

2. Harpocrates, Greek god of silence, is traditionally portrayed with his finger to his lips.

3. George Eliot was one of Proust's favourite writers, but the allusion here to *Middlemarch* is obscure.

4. Marie Nordlinger described it as an 'incised dish sculpted in relief'.

40   *To Marie Nordlinger*

Friday [27 May 1904]

Dear friend,

I'm sorry that the latest interruption in our meetings is being further extended. And it isn't my fault. So I'm obliged to write to you since I'm unable to see you, to put to you an idea I have had thanks to which, since you seemed sad to be leaving Paris,[1] you might perhaps be able to prolong your stay here. This is it – see how it strikes you. Mama would like – however painful she might find the comparison between the precise and gentle image she retains of my father and a work of art, necessarily rather unfaithful – Mama would like, for those who come after us and who may wonder what my father looked like, to have a bust in the cemetery[2] which would answer them as simply and accurately as possible. And she proposes to ask some gifted and amenable young sculptor to be good enough to try, on the basis of photographs, to reproduce in plaster or bronze or marble the shape of my father's features, with the maximum of fidelity which will still be very far removed from our memories and may even, perhaps, make them more painful, but which will nevertheless give those who never knew him a better idea of him than his mere name carved in stone.

Would you care to be this sculptor? I should like to talk to you about it, but being unable to see you I wanted at least to write to you about it. Tell me yourself what you think of this proposal.

Our *Sesame*[3] will appear in *Les Arts de la vie* as soon as it is ready, but when will it be ready? *I'm re-doing it from top to bottom!* I haven't asked how much we shall be paid but I don't think it will be too bad for this review. Can you believe that I've found all my notebooks[4] (six of them!) for *Sesame and Lilies*? By pure chance. What had prevented me from finding them up to now was this. I told everyone to look for *green* notebooks. The first three, those that we had, were indeed green. But the next three were *yellow*! Seeing some notebooks that weren't green, no one even glanced at them. My fatigue, and above all my scrupulous insistence on starting everything over again, are delaying me to such an extent that in order to speed things up I should like you to try a rough draft of any *verses*[5] wherever they appear in *Sesame*. As for the *Preface*, since it's the one for *Sesame and Lilies* and not just for *Kings'*, I think it could be dispensed with at a pinch. If we wanted to be conscientious and include it, we would have to choose between the one in your copy and the one in mine.[6] The one in mine has thirty pages, the one in yours six.

That would perhaps mean giving preference to yours, though it is I think less good.

Your Bing is inundating me with the most magnificent catalogues. *There's no need to thank him, is there?* I suspect that I still owe this largesse to the 'blond man' who took my name when I bought the big Gillot catalogue.[7]

Your friend
Marcel

1. She was to organize Siegfried Bing's exhibition of Japanese prints in the United States (see letter 16, n. 1).

2. Dr Proust was buried in the family grave in Père Lachaise cemetery: see vol. I.

3. The translation of Part I, eventually published in 1905.

4. He had reported the loss in a letter to Marie Nordlinger in January.

5. Ruskin quotes from Patmore, Wordsworth, Tennyson and Milton.

6. In the event, the Ruskin prefaces were dropped, possibly because Proust's own preface, 'Sur la lecture', was fifty pages long. See letter 15, n. 3.

7. The blond man was presumably René Haas (see letter 17). The Gillot catalogue was to an exhibition of oriental works of art at the Durand Ruel gallery.

41 *To Lucien Daudet*

[10 or 11 June 1904]

My dearest Lucien,

I'm alarmed by your long, sudden and unexplained silence, because your silences are always either augural or zoological. Forgive these words which you dislike, you who know how to say everything in the purest French – what I mean is that your sibylline silence is always related to some imminent and anticipated reality, like the silence of augurs, or the silence of giraffes when it is about to rain. Your silence indicates sometimes the habits of augurs and sometimes the instincts of the puma. And when I see one of these tragic or solemn silences descend upon you, I'm always afraid that you may be on the point of having an eruptive fever or a marriage in your family. I very much hope it's the latter. But the most likely explanation is that quite simply, in the brilliant concatenation of your days, Madame de Mun has had you to dinner one day, Madame de Fitz-James the next, Madame de Pracontal another and Madame de Kersaint the one after that, 'grandeurs'

(These constant 'grandeurs' whose splendour importunes me[1])

which you wrongly believe must arouse my melancholy envy either because they were refused me or because they were withdrawn from me, and for the brilliance of which, out of fruitless charity, you endeavour to compensate by speaking of my 'smart friends', which is nice of you but doesn't wash.

In any case, a brief note to let me know if the cause of the disappearing act is good or bad, and if Pray[2] is postponed to an unspecified date for the celebration of an exchange of rings.

Affectionately yours,
Marcel

I haven't received *Le Visage émerveillé.*[3] Perhaps through an unavoidable delay, perhaps the beginnings of an estrangement.

1. Cf. Corneille's *Cinna*: 'C'est l'amour des grandeurs qui vous rend importune.'
2. Pray was the Daudet family house in the Loire valley, near Amboise.
3. Anna de Noailles's second novel, publication of which had been announced in the previous evening's paper: Proust received his copy the following day.

42   *To Madame de Noailles*

Saturday evening [11 June 1904]

Madame,

I had not received *Le Visage émerveillé*;[1] I thought I would never receive it and I was delighted. Because for some time now I've been extremely (but very respectfully) annoyed with you and when I have a grievance against people I like them to be guilty of wrongs against me which sound warlike fanfares in my heart and stiffen my rancour or my spleen into bellicose postures within me. But I have just this instant received *Le Visage émerveillé* and have therefore been disappointed. I read twenty pages of it and was at once transported into a region where only the greatest geniuses can lead and from which complacency and resentment are equally absent. Each step I've taken in this supernatural landscape has filled me with such ecstasy that I instantly wanted, before going on, to come and hang up my offering and lay down my wreath and bless the great and wise and powerful mind, the marvellous genius who opens up for us the secret essence of all things, before whom all appearances fall away and for whom the flower of the pansy has, as it must have had for God, 'on its violet velvet petals a beautiful yellow stain, smooth, vibrant and glossy, as though a wren's egg had fallen there from a tree and broken'.[2] Thus you remove from all things the grey mist which is nothing but the emanation of our mediocrity, and reveal to us an unknown world whose existence we had sensed in our more divinatory moments, where all is truth and beauty.

No doubt it is impious of me to repudiate your past books for the sake of the book of today. But it would also be absurd not to recognize your growth which is as wild and as astonishing as that of the shrubs that grow in hot, damp countries. Twenty pages of *La Nouvelle Espérance*[3] compared to twenty pages of *Le Visage émerveillé* appear roughly as follows. In *La Nouvelle Espérance* a fiercely innovatory genius shattered all the old ways of composing, narrating, thinking. Everything was in turmoil. In *Le Visage émerveillé*, on a divine plane, everything is ideally reconstructed. In the one there is the linguistic freedom and revolt of Saint-Simon; in the other the order and the purity of the Gospel (I speak almost exclusively in terms of grammar). Moreover in *La Nouvelle Espérance* poetry appeared here and there to add a heavenly ornament to the human story. Here and there, with a sort of coquettishness, you let fall one or two of your marvellous secrets, like Chateaubriand (in accordance with the advice which Joubert gave and which was repeated to him by Mathieu Molé, your husband's grandfather[4]) or like Barrès.

Now it's quite different. The poetry is no longer interpolated. It is perpetual, integral, as though some wonderful quality of eye enabled you to see everything in terms of beauty, truth, novelty, genius. It is impossible to choose between the daisies thickly bunched like camomiles, or the joined fingers, pointed, passive and soft, like little candles burning, or the water of the fountain which is like frozen silver (oh! the splendour of that thirst and the glory of that coolness!) or the little potted fir-trees that give out a pungent, crackling scent at noon, or the light, delicate, green spring-time, or the impulse to laugh in church when the weather was fine, or the soft, violet twilight like a shower of anemones, or the abbess who gently withdraws her habit when a sister passes too close to her.[5] It is your thought that is like that, it is your style that is like that; it isn't so much a sequence of marvellous things as a sort of inspired vision which creates continuously. There are in *Atala*[6] perhaps two or three images of perfect beauty. In each one of your pages there are as many as there are ways of saying things. And so, if *La Nouvelle Espérance* appears like the Revolution in language, and this one like the golden age in the reconquered language, the poems now appear like sketches, as it were, which here have the value of a background, or of such and such a detail. I shall be clearer about all this when I've read more of it. For you must resign yourself to hearing from me often. The sense of the absolute perfection of these pages gave me a momentary feeling of sadness. It struck me that your literary life was, if not finished, at least accomplished in the sense that you could renew the subject matter of your books indefinitely, but that your talent would never be more marvellous or more powerful and that you would never surpass what you had already achieved. I thought about this for some time, and I now no longer believe it. No, I believe that you will be able to write still more beautiful books, books that we cannot even conceive, that are enclosed in the secret recesses of your instinct, as spring and summer were hidden in the depths of the year when they had never occurred before and people did not know what they were, books so beautiful that thereafter we can burn all the libraries, close all the pianos, demolish all observatories. And if I cannot conceive what they will be, I think I can imagine how you will succeed in writing them, with what discipline, what deliberate excisions, what remodellings. But who would dare to probe that sacred future? Parenthetically, one final remark. Since *The Bible of Amiens*, the 'singularity of images' is something that exasperates me because it isn't true genius and doesn't emerge directly from the language and doesn't merge with it. But the miracle has happened a second time: in *Le Visage émerveillé* (incidentally, how you love that word *visage*, which is so soft and gentle when you pronounce it, and which at the end of your verses 'Au lac de nos visages' reproduces the sweetness of your pronunciation[7]). Thus 'as though a wren's egg had fallen there etc.' (for the petals of the pansy) is absolutely an image from the *Bible*, as inspired, as anterior to all literature, as superior to anything else.

Good-bye, Madame.

Marcel Proust

P.S. Having finished writing to you I took up *Le Visage émerveillé* again. I read it right through. And now it's morning, but a morning still so grey that it's impossible to tell from the pallor of the sky whether it's the beginning of a blue or a sombre day. Dazed by my long read, I can scarcely talk to you about the book. Besides, I shall have to begin all over again, because everything I found so beautiful, and spoke to you about, was nothing. I had stopped just before it all began. I remember vaguely among the most tremendous things: 'She is of a rustic plainness, reminiscent of those autumnal birds which smell of berberis.' 'Don't you understand that he *dotes on you.*' 'The silver church-bell garlanded with roses.' 'You see that yellow streak: it is a little gold knife pricking my heart.' Or the white bowl left behind in the garden (not a single Claude Monet, not a single Manet, is as beautiful: it's the very genius of impressionist landscape), or the sublime and hilarious prayers of Sister Catherine. The Abbess and God, mother superior erect before her superior. 'He sees the world as I see it, *blue, yellow and violet*', and '*Do not be imprudent, Sister*'. The pomade on the red roses. The chaplain who is in his soutane 'once and for all'. Sister Catherine consoling the priest: 'Will we have the strength, the two of us?' Summer, a golden road that advances, the lily a bell cracked at its four corners, the sun with its myriad tiny windows of happiness, and all the little suns appearing at the windows. 'My soft body, my supple legs.' 'Are you not put off by my being a nun?' (divine!).

I don't like your not liking *old women*. It's unkind, it's like M. Forain.[8]

I find sublimely funny: 'I know everything about his family, he loves me, etc.' and the hands plunged in water, the chaplain's eucalyptus leaves and Sister Colette's bird-like, hedgerow face. The restlessness of a beautiful woman who tells herself at the same time I could be loved by a poet etc., who is in a continual state of turmoil and insomnia feeling her ravishing face against her pillow, under her hands (this being not you but the interpretation of what you say).

I don't like your being jealous of Colette's love affair. La Plata is wonderful. The thief, a poor man with a tired look, is sublime. Marvellous litany[9] of the cherries, the plums, the medlars, the pears, the apricots, each of them a triumph of inspiration. 'When one puts one's hand on one's heart, it is perhaps because the desire is in ourselves.' But Madame, I'm too tired, I cannot go on remembering. You are more cruel than the Mother Abbess, than God, to have had Sister Saint Sophie obey, let her lover go, return the letter. It hurts us to think that other women will love him. And you didn't do it out of faith, or obedience to the rule, or secret envy like the Abbess, but out of perversity, to cause pain to your own heart, to wallow as you wrote in romantic sadness, to inflict on all of us an irremediable despair.

Good-bye, Madame, I'm too tired.

Marcel Proust

P.P.S. *The last twenty pages are the most beautiful in the book and in everything you have ever written.*

1. See letter 41, n. 3.

2. 'Sur les pétales de velours violet une tache d'un beau jaune lisse vive et luisante, comme si, tombé de l'arbre, un oeuf de roitelet se fût cassé là.' The first of many quotations in this letter from Anna de Noailles's novel *Le Visage émerveillé*, some of

which Proust has italicized. This one recalls a simile to which Proust had drawn her attention: see letter 14, n. 2.

3. Anna de Noailles's first novel, published in 1903.

4. This story seems to have been pure invention, in order to introduce the name of this illustrious ancestor, the statesman Comte Louis-Mathieu Molé (1781–1855).

5. *Le Visage émerveillé* is about a nun in love, Sister Sophie.

6. Novel by François-René de Chateaubriand (1768–1848).

7. Cf. the passage on Bergotte's style, in speech as well as writing, in *RTP* (I, 592–4).

8. Jean-Louis Forain (1852–1931), painter and cartoonist, notorious for the cruelty of his satire, and incidentally a vitriolic anti-Dreyfusard.

9. Cf. the litany of the Paris street cries in *The Captive*, and especially Albertine's rhapsody on ice-creams at the end of that passage (*RTP*, III, 122–6).

43   *To Madame de Noailles*

Wednesday [evening of 15 June 1904]

Madame,

Apologies! It's me again, but for a minute only because I'm shaking with fever and fatigue. I saw Constantin this evening and told him that he was publishing some pretty bad verse. And then I come back home and find *La Renaissance* with some sublime poems in it.[1] And then for the next hour I'm unfair to *Le Visage émerveillé*, because after all, the sun springing up from the earth in its hazy effulgence, the houses as white as Algiers, the bamboo shoes of the morning, the little bridges and the green Chinese ladies, the wicker blinds – *Ah! see how my head is thrown back* . . . – the bridges a tangle of azure and metal (Claude Monet) and all of the last two pieces (to get it over with!) – there's none of all that in *Le Visage émerveillé*. And if the novel has the inimitable accent of your speech, where else but in your poems do we find the inner music of your soul? You say more things in the one, but in the others you hint at deeper things. I shall stop this Plutarchian parallel. And in bidding you good-bye I repeat that I like *Le Visage émerveillé* more each day. Each day I find in it, as in a miraculous countryside, a new 'view'. And without wishing to force or vulgarize a nuance to the point of turning the book into a Rougon Macquart (!)[2] I like incidentally that vision of a family in which a trait is handed down: 'the absence of scruple in enjoying life' – the family, or *Sister Saint Sophia*, who deserves her beautiful name of '*Sister Saint Wisdom*'.[3]

I trust that all this incense which a truly burning heart offers up to one of the two Sisters will not make the other (the Princesse de Chimay) think that she is on this account any less admired than before, or less loved, I dare not say less doted on, in my ardent uncertainty as to her feelings, so broad and soft and bright and unconstrained is that inner sun of hers.[4]

Your respectful
Marcel Proust

1. Mme de Noailles's brother had just published four of her poems – 'L'Enivre-ment', 'Le Jour', 'Un Matin', 'La Nostalgie et la destinée' – from which Proust proceeds to quote. They were later collected in *Les Eblouissements* (cf. letter 25, n. 2).

2. Zola's family saga.

3. See letter 42, n. 5.

4. Proust's metaphor is taken from Anna de Noailles's poem 'L'Enivrement': '. . . le soleil solitaire / est si large, si mol, si vif et desserré.'

44   To Douglas Ainslie[1]

45 rue de Courcelles
[June 1904]

My dear friend,

In asking you to allow my friend Prince Antoine Bibesco,[2] who will be in London for three months as Secretary to the Embassy of the King of Roumania, to call on you, my object is not only to obtain news of you through him and to seize an opportunity of expressing anew my feelings of sincere friendship for you. I am especially delighted to enable you to enjoy the exhilarating acquaintance of one of the most seductively intelligent and warm-hearted people I have ever met, and to instigate conversations which your mutual friend, far from London, chained to his rock of solitude and silence by a misfortune commensurate with Andromeda's, will be reduced to imagining with envy. Thanks to you, if you can guide him a little, Bibesco will be able to get a glimpse of everything that is most admirable in your country, in the spheres of intellect and beauty, from George Meredith to Mrs Clayton.[3] When he has acquired your address he will pass it on to me, and I shall thus be able to send you the translation I have made of *The Bible of Amiens*. No doubt you will take a perverse pleasure in seeing thus disfigured by an always clumsy and often grotesque travesty a work whose original graces you so cordially hate.

Believe me, my dear Ainslie, yours sincerely
Marcel Proust

1. Douglas Ainslie (1865–1948), English poet, literary critic and philosopher whom Proust had met in 1897. See vol. I.

2. Cf. letter 29 where Proust offers to be of service to Bibesco despite their quarrel.

3. Presumably the wife of Joseph Clayton, novelist and publicist, author of *Father Dowling* (1902), *Grace Marlow* (1903) and *John Blankset's Business* (1904).

45 *To Louisa de Mornand*

Sunday evening on returning home
[26? June 1904]

My dear little Louisa,

I am very sad. Because yesterday, when you so sweetly came round, I had a stronger impression than ever of your gentle, trusting, devoted friendship. And tonight I feel that I'm going to go down considerably in your affections by interfering, for the first time since I've known you, in things that are not my concern. I'll explain myself more clearly when I see you. But between now and then, I beg of you, even when, as this evening, Louis[1] has an unjust suspicion in his mind, or an undeserved reproach on his lips, instead of – which is so natural and which I understand so well – letting him think what he likes, not deigning to justify yourself, or worse still, amusing yourself by anchoring him in his misconceptions, show him clearly, gently, kindly, tenderly, that he is absurdly imagining things, that he is unjust, absolutely on the wrong tack. You will tell me that that is not your character. Change your character then for forty-eight hours, if someone who loves you tenderly entreats you to; you won't in any case lose very much thereby. And if what I'm saying doesn't seem reasonable to you, you will be free to return afterwards to what I'm advising you against. I remember one evening, at the Mathurins,[2] I think, seeing Louis and you already needling each other. I was very careful not to speak to either of you about it. In the first place it wasn't my business, and besides, lovers' quarrels are of no consequence, and your relationship with Louis was in no danger of being affected by it. Now things are entirely different. I learned this very evening from Louis of an event which you too know about[3] and which has had an influence on his character and his views that I had noticed for some time without knowing the cause, so much so that twice I've been on the point of asking you for a talk. Under the circumstances, minor quarrels, seemingly insignificant, and in fact insignificant up to now, may no longer be anything of the kind. Perhaps I'm wrong, you can tell me what you think, or indeed let's not talk about it any more if that's what you would prefer. At all events, between now and then, follow my advice which is dictated by a deep and dual affection, the affection I have for Louis and the affection I have for you.

Your

Marcel

I'm writing to you on different paper so that my letter will not be noticed by Louis.

1. Her lover, Louis Marquis d'Albufera.

2. The Théâtre des Mathurins where she had had a walk-on part. In *The Guermantes Way* this quarrel is echoed in the scenes between Saint-Loup and Rachel on the day of Mme de Villeparisis's reception (*RTP*, II, 166–75; 177–84).

3. Albufera was about to announce his engagement (see following letter), an event which was not supposed to alter his role as Louisa's protector.

46   *To Louisa de Mornand*

Sunday [3 July 1904]

My dear little Louisa,

I haven't been able to thank you sooner for your charming telegram because I hadn't seen Louis and consequently didn't know your address.[1] But my heart was full of the memory of your sweetness and your charm the other morning, and I poured it all out to Louis before writing to tell you how fond of you I am. I talked to Louis about his travel plans. He told me he would of course write to you to tell you *everything* he knew. But that *everything* is very little, as far as I could gather. He knows that he will be going to Germany with the Prince d'Essling,[2] but doesn't know precisely when or how. It all seemed very vague to me. He told me that the news of his engagement would be officially announced any day now. That will be a day when I shan't be able to read my newspaper with the same indifference and absent-mindedness as usual, a day I should very much like to be with you. I've heard his fiancée very highly spoken of and that gives me great pleasure for his sake. I dined alone with him this evening. It was the first time I had dined with him without you. We were both preoccupied by the same thought. As it chanced, I was guilty of all the little quirks you always make fun of. Continually on the point of sneezing, I couldn't leave my nose or my eyes alone for a minute, with that gesture you mimic so well, and, what took me even further back, for it was quite a long time since I had done it, I said to the waiter without thinking, 'Waiter, what sort of beer have you got?' I immediately remembered your friendly teasing and I wanted to laugh and cry at the same time.

I hope your journey went off well. Please give my respectful regards to your mother and your sister. As for you, dear little Louisa, allow me to embrace you most tenderly.

Marcel Proust

1. She had gone to Vichy with her family.
2. Prince Victor d'Essling was the father of Louis d'Albufera's fiancée, Anna Masséna d'Essling de Rivoli; their engagement was announced on 5 July.

47   *To Albert Sorel*[1]

[Sunday evening 10 July 1904]

Monsieur et cher maître,

Life, doubtless in order to show us that it encompasses everything (everything, even that which seems most positively to exclude the 'philosophy' we extract from it little by little, everything, even the miraculous realization of arbitrary hopes, the 'romantic'), sometimes has surprises in store for us, fairy-tale strokes of good fortune. Certain it is that if anyone had told me that the great historian I so admired, the master whom I so often heard 'demonstrating' History, illuminating our fervent minds with his passionate geometries, the man about whose slightest literary predilections, alerted by his essay on *Une Ténébreuse Affaire* and *L'Envers de l'histoire contemporaine*,[2] I have made so many inquiries, collected and tried to verify so many anecdotes – if anyone had told me that the great Balzacian, who is also in his way a Balzac himself, if 'Le Baron Bidard',[3] as I firmly believe, is not inferior to *Le Colonel Chabert*,[4] would one day write in *Le Temps* a whole big article about a modest translation[5] by the most obscure of his pupils, and in this article, with the indulgence and generosity of a master, would go out of his way at the slightest opportunity to bestow marks of flattering attention on the translator, on the former pupil recognized and over-generously complimented[6] – if anyone had told me this, I should have been truly astonished since, acutely alive as I am to the 'logic' of life, I should have been very slow to believe in such an improbable joy.

So when I was informed of your project recently by someone at *Le Temps*, since I do not find it easy to believe in even the most deserved and natural joys, it seemed to me too wonderful ever to be true. Or rather it was true already. For I savoured with delight the infinite kindness of your intention. I knew that it would always remain no more than an intention, but that intention was delightful in itself.

This evening I was looking through an album of *The Rivers of France* by Turner, which someone had lent me, and had it open at 'Honfleur'.[7] I was wondering what you would think of it, not daring to decide myself, as I was strongly inclined to do, that Turner had entirely failed to render the infinite charm of the old Norman town, when a friend came in, bringing *Le Temps*. Sir, I am proud indeed to have been the occasion, at moments the object, of this essay, this magnificent essay, proud indeed that my name should have been uttered by you; but I am perhaps even more touched by your remembering me, and when you mention my name I seem to hear you utter it, and that fills me with the sweetest pleasure.

Your article contains only one error: Ruskin did not cure me.[8] I have been very ill for several years. But no invalid was ever so 'spoiled', was ever so overwhelmed, by virtue of Emerson's mysterious law of 'compensation', by such a magical 'surprise' as I have been this evening in reading your article. I'm sorry that you have been so discreet about the 'history' in Ruskin. I am well aware that it's flimsy, but I should have been interested to know what you thought about it. The poet in him was harmful to the historian, the economist, the philosopher. Perhaps one should not expect from his *Genius of Christianity* anything more than incomparable images. But in your case, the fact of having written the 'ethics' of historical events has not dampened your passionate devotion to Schubert and Schumann, whose rhythms and accompaniments have permeated your style, or your love of Balzac, who never managed to drill, manoeuvre, bring together in a final parade, after having made them grapple and skirmish in extended order, all the 'forces' of the French language as well as you have done in that long, gradually soaring comparison (at the beginning of this evening's article) between the different interpretations of things with the help of books and works of art. There is in this passage, which I consider to be perhaps the most remarkable you have ever written, a sort of symphonic progression, a long Wagnerian crescendo which I find marvellous, with all the sonorities of the French words, in response to the principal theme continuously transformed, jostling and exploding in their varying weight and charm (a note on the horn 'transmutes' them in passing). Thank you again, dear Sir; you have given great pleasure to your respectful and grateful admirer

Marcel Proust

However, in this incomparable piece it is perhaps the delightful passage about Leipzig and *Faust* that is the most evocative and thought-provoking and the most beautifully realized.

1. Albert Sorel (1842–1906). Historian, specialist in diplomatic history. Proust had followed one of his lecture courses at the École Libre des Sciences Politiques in the spring of 1892.

2. Novels by Balzac.

3. An 'anecdote' of the Napoleonic wars by Sorel published in *La Renaissance latine* in January 1904.

4. Novel by Balzac.

5. Sorel reviewed *La Bible d'Amiens* at length in *Le Temps* of 11 July.

6. Sorel had some penetrating things to say about Proust's style: 'He writes, when he is cogitating or musing, a prose that is flexible, floating, enveloping, opening on to infinite vistas of colours and tones, but always translucent, and reminiscent at times of those artefacts in glass in which Gallé encloses the intertwining tendrils of his lianas.'

7. Honfleur was Sorel's birthplace.

8. Sorel cited Proust's preface to *La Bible d'Amiens* to suggest that he was converted by Ruskin to a love of medieval architecture.

48   To Madame de Noailles

[Thursday evening, 14 July 1904]

Madame,

I mentioned to Guiche[1] that I very much wanted to see you and he said that he would like to as well and that I should ask you if you would care to have us to dinner. He would be free on Sunday. So would I. Between now and then I have to rest as I killed myself dining at Vallière[2] today. From Sunday onwards I don't know whether he'll be free, but I shall.

I needn't tell you, need I, what joy Barrès's article gave me.[3] I would have preferred him not to include the quotation about Fouquet[4] but apart from that it was all marvellous and it's nice to have such sensible things said by such a great writer who will make them even more widely accepted if that is still necessary.

Your respectful friend
Marcel Proust

If I dine with you some day, with or without Guiche, I should be even happier were the Princesse de Chimay dining too.

1. Armand, Duc de Guiche (1879–1962), painter, later a distinguished scientist, lifelong friend of Proust; in *RTP*, one of the models for Saint-Loup.

2. Vallière, a vast nineteenth-century château, was Guiche's family home at Mortefontaine, north of Paris. See letter 50.

3. An article by Barrès in *Le Figaro* of 9 July entitled: 'Un grand poète: la Comtesse Mathieu de Noailles'.

4. Barrès had misrepresented her when, in the context of Dreyfus, she had quoted La Fontaine on Fouquet's disgrace: 'Misfortune is a kind of innocence.' 'Il est assez puni par son sort rigoureux; / Et c'est être innocent que d'être malheureux' (*Aux Nymphes de Vaux*). Cf. Proust's reproach to Barrès in the following letter.

49   *To Antoine Bibesco*

[Shortly after 17 July 1904]

Dear Antoine,

Yes, 136 boulevard St-Germain is indeed my brother's address.[1] I shall tell him of your great kindness and he will be, as I am, very touched by it.

But yes, 'she knew'[2] the evening when we had supper together. But we know how to keep tombs,[3] old fellow.

I dined at your 'cousin Noailles's' and I broke her finest Tanagra figurine. I dined at Vallière with the Aimerys,[4] the Chevreaus (but it occurs to me that the *Herald*[5] and the *Gaulois* have described this party for you, à propos of which I offer you the following resemblance: isn't the young Lasteyrie (not the one who plays the fool but the other, known as Lolotte) exactly like Léon Blum – physically, I mean?).

I dined in the Bois – another Guiche party – with your friend Tristan Bernard.[6] But this was a very small dinner; whereas at Vallière there were thirty people. I gathered together (this was earlier) Bertrand's friends (not all of them!) and Bertrand,[7] at home, on the eve of his departure, and it ended up with supper, also at home. You know that Albu is not the only friend of mine who's getting married. Guiche is also engaged and I've been seeing quite a lot of him too (though less). As a matter of fact Albu comes round even more often than before if that's possible and is adorable to me. But I'm well aware that whatever he may say it can't stay the same afterwards. Still I'm too fond of him to consider his marriage purely from my point of view.

Since you're interested in medical matters and also like to think I'm a bit mad, I must tell you that I've consulted the doctor who along with Faisans[8] is considered the best in the field, a Dr Merklen,[9] who told me that my asthma had become a nervous habit and that the only way to cure it was to go to an anti-asthmatic establishment somewhere in Germany where they would (for I probably won't go) 'get me out of the habit' of my asthma just as they 'demorphinize' morphine addicts. I shall be delighted to see your brother[10] when he passes through, but must I really wait until then to hear what's been happening? You know how discreet I am and you could perfectly well write and tell me. Anyway, why not try it and see. I should like your brother to tell me which is more interesting, La Charité or Nevers.[11]

I've described my life since your departure exclusively in frivolous terms but you are well aware of course that '*this is the apparent life*' and that '*the real life is underneath all this*'.[12] But now I must leave you. What else can I tell you? I rebuked Barrès (without mentioning you in support of my view, having felt as I do from the very first) about the end of his article.[13] I told him that 'the Comtesse de Noailles' had never believed, à propos of Dreyfus, that to be unfortunate was to be innocent, but that it was unfortunate to be innocent when one had been convicted, etc. He simply laughed and said: 'What is the meaning of this sudden outburst of Dryfousism?'[14] There was a M. Vaschide[15] there who expounded some absurd medical theories (silence tomb) but who was charming nevertheless, a Parsifal one couldn't very well engage for the Opéra because he has too strong an accent and talks too fast, and is so anxious to enlarge the field of his speciality that to everything one mentions he invariably says, 'It's *nelvous*.' What a calamity, old man, that broken Tanagra.[16]

Good-bye, my dear Antoine. Much love to Emmanuel and yourself.

Marcel Proust

1. Dr Robert Proust (1873–1931), Proust's younger brother. See vol. I.

2. Bibesco had evidently asked whether Louisa 'knew' about Albu's marriage.

3. i.e. to keep a secret (see letter 3, n. 4).

4. Comte Aimery de La Rochefoucauld (1843–1923) and his wife, *née* Henriette de Mailly de Nesle (1852–1913), parents of Proust's friend Gabriel. Comte Aimery was well known in society for his remark when apprised of the death of a close relative as he was impatient to leave for a ball: 'These things get exaggerated' (cf. the Duc de Guermantes's remark in similar circumstances, *RTP*, II, 751–2).

5. The Paris edition of the *New York Herald* which, like *Le Gaulois*, had reported the dinner at Vallière to celebrate Guiche's engagement (cf. letter 50).

6. Tristan Bernard (1866–1947), novelist and celebrated playwright and boulevardier.

7. Bertrand de Fénelon.

8. Dr Michel Faisans (1851–1922), specialist in respiratory disorders.

9. Dr Pierre Merklen (1852–1906), heart and lung specialist.

10. Prince Emmanuel Bibesco (1877–1917), Antoine's elder brother.

11. Proust, contemplating a visit to Louisa de Mornand at Vichy (cf. letter 46), hesitates between these intermediate towns. Emmanuel Bibesco was the leader of their former cultural excursions (see letter 3, n. 1).

12. In English in the original: from Ruskin's *The Bible of Amiens*.

13. Cf. letter 48 to Mme de Noailles.

14. Barrès pronounced the word Dreyfusism with a German accent.

15. Nicolas Vaschide (1874–1907), Romanian doctor and experimental psychologist; he couldn't pronounce his r's.

16. Proust was mortified by his clumsiness (he avoided Mme de Noailles for a year: cf. letter 140) but she evidently treasured the broken figurine, for when her relics were shown at the Bibliothèque Nationale in 1953 it turned up as exhibit No. 84: 'Tanagra broken by Marcel Proust'. The incident is mirrored in *The Guermantes Way* when Bloch knocks over a flower vase at Mme de Villeparisis's (*RTP*, II, 221–2).

50   *To Bertrand de Fénelon*

[Soon after 17 July 1904]

I heard of Guiche's marriage one evening when I went to dine at Vallière. When I arrived the Duc de Gramont[1] asked me to sign the visitors' book which had already been signed by the other guests of that evening, and I was about to append my signature underneath a tiny Gutman followed by an enormous Fitz-James and an immense Cholet[2] followed by a tiny Chevreau and an equally small Mailly Nesle-La Rochefoucauld, when the Duc de Gramont, filled with anxiety by my humble and confused demeanour (in addition to the fact that he knew I wrote), addressed me in a tone at once imploring and peremptory these lapidary words: 'Your name, Monsieur Proust, but *no thoughts*!' The desire to have the name and the fear of having the 'thoughts' would have been more justifiable if I had had him to dinner and had asked him to sign: 'Your name, Monsieur le Duc, but no thoughts'.

Guiche's sister, Mme de Noailles,[3] told me she was like me and loved chatting with 'inferiors', her housemaid, her concierge. One day when she was speaking about this to her cousin Mme Léonino (Jeanne de Rothschild) she added: 'I feel I have the soul of a concierge. I don't know where I get it from. I must have had a skivvy ancestor.' And Jeanne Léonino, slightly hurt, replied: 'On the Gramont side, perhaps; but on the Rothschild side I can assure you no!'

The engagement[4] got off to an extraordinary start, M. de Gramont and M. Greffulhe[5] having similar characters and neither being cut out to tolerate a rival. M. Greffulhe to Guiche: 'It's quite nice, Vallière. Not at all unattractive. There's a nice little lake. It's rather sweet. Of course I don't suppose your father would pretend to compare it to Bois Boudran,[6] would he – that would be a real joke, etc. etc.' But gradually Guiche had become a thing for M. Greffulhe and hence an admirable thing, superior to what others possess, a sort of human Bois Boudran. M. Greffulhe to the steward of Bois Boudran: 'Let me introduce you to the Duc de Guiche who got both his baccalaureates at fourteen. The Duc de Guiche is a doctor . . .' (to Guiche: 'Doctor of what?') Guiche: 'Not a doctor, a Bachelor of Science') '. . . Bachelor of Science, doctor, the whole shooting-match. He won a medal at the Salon, he's a first-class shot, he can skewer an opponent with his first thrust etc. etc.'

A propos of the Noailles, dear Bertrand, I wanted after all to go just once to the avenue Henri Martin[7] and almost before a word had been said with a too sweeping gesture I shattered a Tanagra figurine to smithereens.[8] What am I to do?[9]

---

1. Agénor, Duc de Gramont (1879–1962), Guiche's father, one of the models for the Duc de Guermantes in *RTP*. See note 6 below.

2. Among this list of guests is the Comte de Cholet, Proust's superior officer during his military service in 1889 (see vol. I).

3. Comtesse Hélie de Noailles, *née* Corisande de Gramont.

4. Guiche had become engaged to the twenty-two-year-old Elaine Greffulhe (1882–1950), a precocious poet whose verse, written at the age of five and a half, was published with a preface by Robert de Montesquiou (cf. letter 82). The marriage took place on 14 November (cf. letter 79).

5. Comte Henri Greffulhe (1844–1952), rich banker, father of Elaine. Elaine's mother, *née* Elisabeth de Caraman-Chimay (1860–1952), famous beauty, hostess

and patron of the arts, was a model for both Guermantes ladies in *RTP*, the Duchess for style and dress and the Princess for her beauty and social position.

6. The Château de Bois Boudran (or Boisboudran) was the Greffulhes' vast sporting estate in the Brie; its huge house was considered as ugly as the Gramonts' enormous mansion, Vallière (see letter 48, n. 2), built in 1878 after the impecunious, widowed Duc Agénor (see note 1) had married a Rothschild (he was to marry for the third time in 1907: see letter 238).

7. i.e. to Mme de Noailles's house.

8. Cf. letter 49, n. 8.

9. The beginning and end of this letter are missing.

51   *From Bertrand de Fénelon*

Ambassade de France en Russie
11/24 July 1904

My dear Marcel,

I write very seldom these days; but I don't want to remain silent any longer. My letter is in any case self-interested: my object is not so much to give you news as to ask you for news. I live a peaceful existence here: the departure of most people for the green outskirts or the distant countryside has turned the city[1] into a quiet little resort to which the length of the days and the persistent soft light give a peculiar charm. I like it very much, and summer life suits me here in every way. The news from the Far East is still vague, and although they have nearly 200,000 troops concentrated in the area, the Russians do not seem anxious to engage in any serious offensive. They wear down the enemy by contesting the ground foot by foot.[2] So we have very little work. In summer, everything comes to a standstill.

I've been told that you had recently been a devoted and loyal friend in difficult times,[3] and it was particularly pleasing for me to hear your praises being sung. I know how warm-hearted and affectionate you are! – Write to me. It will give me great pleasure, and although I am too tactful to offer you advice, let me hope that you will try hard to get well so as to be able to give me better news of yourself.

Yours ever
Bertrand Fénelon

1. St Petersburg.

2. Fénelon is referring to the Russo-Japanese War in Manchuria (1904–5). He

seems not to have been very well informed. On 26 July, two days after he wrote his letter, the Japanese captured the important Manchurian port of Niu-Chuang.

3. An allusion to Proust's efforts at mediation between Albufera and Louisa de Mornand (see letters 45, 46).

52   *To Marie Nordlinger*

Thursday evening [4 August 1904]

Dear friend,

I went out today and found your letter only when I got back! I am very unwell. If I feel well tomorrow, that is today when you receive this note, I shall telephone to ask you to come, but it will depend on my condition. Perhaps one day I shall come if I may and visit you in

> your garden at Auteuil
> where the yew and the honeysuckle grow[1]

No, Reynaldo didn't come. – I'm sad to gather from Madame Hinriksen's reply[2] that she is vexed with you. Why? Don't you think that you would please her by joining her in Sweden,

> Silvery Sweden with its twin seasons[3]

You yourself would find enjoyments there, I think, a little weird but nevertheless profound. I feel that it would be even better than Auteuil. You will come back to Paris with greater pleasure if you deprive yourself of it than if you exhaust it. –

We haven't talked about the medallion of my father.[4] I asked Reynaldo to have a look at your sketch which I haven't seen, and he told me it was too late as you had already destroyed it. Don't you think my filial memory might have been of some use to you? I suppose you are often between Paris and wonderful Versailles.[5]

Till tomorrow perhaps.

Your respectful friend
Marcel Proust

1. She had a summer house in Auteuil, the prosperous residential quarter of Paris where Proust was born. Proust paraphrases the last two lines of a verse by Boileau: '. . . gouverneur de mon jardin d'Auteuil, / qui dirige chez moi l'if et le chevreuil' (from 'Epître XI, à mon jardinier', 1695), but perhaps also intends an oblique reference to the first line: 'Hardworking servant of an indulgent master' ('Laborieux valet du plus commode maître').

2. Sic. Marie Nordlinger's aunt, Mme Caroline Hindrichsen, who lived in Hamburg and was a passionate art-lover. Proust had met her with Marie Nordlinger in Venice in 1900 (cf. letter 13, n. 1).

3. 'La Suède d'argent avec ses deux saisons . . .' Anna de Noailles, 'Les Voyages', *L'Ombre des jours*.

4. See letter 40.

5. Her cousin Reynaldo lived in Versailles.

53   *To Marie Nordlinger*

Sunday evening [7 August 1904]

Dear friend,

What can I say? I caught cold on Friday and have been in bed since with a temperature. And I'm leaving on Tuesday to spend two days in a boat (don't say that it will be only two days). Could you come tomorrow Monday at about half past nine with the medallion. (Or is it too difficult?) Mama will see you back to Auteuil and I'll get up for five minutes to shake hands with you. Would that suit you?

Your friend

Marcel

54   *To Marie Nordlinger*

[August 1904?]

Dear friend,

The prettiest thing I have ever seen was once in the country, in a mirror fitted to a window, a patch of sky and landscape with a carefully chosen clump of fraternal trees.[1] And it seems to me that that necessarily fleeting enchantment of a now distant hour is the very thing you have just made me a present of forever. The actual glass of the mirror still covers with its mysterious, lucid protection the clump of trees which seem to have come to live there from choice, because they liked being together and form an imposing group, and the ephemeral hour of the mournful sky. How grateful I am to you for this marvellous present,[2] this gift of a piece of nature, an hour of time, a momentary glimpse of your inmost soul, so alive to the natural world and so full of secrets which belong to it alone. I thank you truly with all my heart.

Your

Marcel Proust

1. Proust was to reproduce this image – inspired by his visit to Holland in 1902 (see letter 3, n. 3) in his preface to *Sésame et les lys*.

2. A small watercolour she had painted at Senlis (cf. letter 15). Proust hung it by his bed and left it to Reynaldo Hahn in his will.

55 *To his Mother*[1]

Yacht *Hélène*
Thursday [11 August 1904]

My dear little Mama,

I'm going to reply in minute detail to all the questions you might ask me. Train journey to Le Havre without any attacks or even oppression thanks to a through-draught maintained by me for the purpose. Found M. Mirabaud[2] at the station. Went with him by cab to the harbour. At the very moment of arriving there, intense asthma (no suffocation, no fit, just asthma). Shown round the boat, increasing asthma, still without suffocation. Then settled into my cabin at about 1 or 2 o'clock in the morning. Fumigations, which do no good. Impossible to undress, because everything cold and damp etc. Trional about 3 o'clock and then attempt to sleep. Lie down from 3.30 to 5, still dressed. At 5 (in the morning) go up on deck. At 7 we sail. My breathing not very clear but still quite bearable. I have breakfast, then my asthma eases; calm sea, fine weather. Big lunch at 12.30, asthma subsiding more and more. I have a very good afternoon (still without having slept or undressed). At 7 o'clock we arrive off Cherbourg. We have dinner (a light one – still on board of course) which doesn't cause me any breathing trouble afterwards but prevents me from going to bed. I remain on deck *without the least sign of breathlessness* (note that for 3 hours it was very cold, everyone was freezing except me), really exceedingly clear, completely well (as far as asthma is concerned). So I decide not to leave tomorrow morning (today). At 2.30 or 3 o'clock in the morning I go to bed (undressed, properly in bed for the first time since Paris). I sleep well but am woken up at 7 o'clock by severe asthma. The weather has turned wet, and it's drizzling. My asthma seems largely to depend on how damp it is on board. I want to leave but when it's announced that we'll be going to Dinard this evening I give up the idea and decide to spend two days at Dinard. But it turns out that in the meantime M. Mirabaud hasn't been able to sleep. So he stays in bed and has given orders for the yacht not to sail so that he can sleep in peace. We are to spend the day in port. Since it is eleven a.m. by then it's too late to change my mind and leave. I don't have breakfast because last night's dinner, light as it was, has taken away my appetite for it, and I get dressed. At 12.30 I have lunch, but too frugally. After lunch a *steam-launch* takes us to Cherbourg. I ask to be left alone to write and rest, but being unable to find a suitable spot or a closed cab, I walk, I kick my heels, I tire myself out till I'm pouring with sweat. Then it's time to go back in the launch, there's a stiff breeze blowing, I begin coughing again, etc. We return to the yacht which definitely won't be moving, M. Mirabaud having lunched in bed and not proposing to get up till tea-time. A little yawl is rigged up for a trip round the harbour, and I go for a hundred yards or so in it but make them bring me back at once as I was a bit scared. Now I'm back on board the yacht (which has been stuck at anchor since yesterday evening) and writing to you. Although I thoroughly enjoyed yesterday's delightful crossing in that lovely weather (it turned fine again this afternoon, as it happens, but there was no longer the charm of sailing since we're stationary) and not having asthma and the magic of solitude yesterday off Cherbourg, I think I shall take the train to Paris tomorrow morning, stopping at Bayeux and Caen. But I haven't quite decided. On no account send me any telegrams (except in case of emergency) as the one I had from you yesterday caused enormous complications. M. Mirabaud is absolutely charming to me and Billy[3] I would call brotherly if I didn't hate sacrilege. We also have on board Mme Fourtoul *née* de Bourgoing[4] who is extremely nice to me, Mme Jacques Faure who is very pretty, Mlle Oberkampf and Madame de Billy[5] who is charming to me and charming in general. The yacht is a marvel. However, I only wanted to keep you informed about my health and am reserving my stories for when we're alone together. I haven't of course touched the syrup of ether as I haven't had a single choking fit since I left you. As for sea-sickness, I haven't even thought about it. It must be said that although the sea wasn't exactly a mill-pond, it was so beautiful and unagitated that there couldn't have been any possibility of being sea-sick. Mme de Billy was, it's true, but I think it was the steamboat that caused it. I have a slight tendency to asthma and I regret my rushing about this afternoon but still I haven't really any asthma at all. I don't know when you'll get this letter; it will probably have to go to Paris before coming back to Etretat.[6] It seems to me so odd, when I'm tucked away in your heart and have you perpetually within reach of my thoughts and my affection, when I speak to you in my imagination a hundred times a minute and hug you no less often, to think that my letter must go such a long way to find you and will reach you long after I've written it. Our hearts are more advanced; they have the telephone. I was captivated yesterday by this yachting life[7] and asked Billy about the cost of hiring a boat. You have no idea! This one (not counting of course the enormous sums it cost to buy) costs 25,000 (twenty-five thousand) francs for maintenance per month (the months when it's in commission, naturally).[8] M. Mirabaud is magnificent, a huge and powerful Saxon god with exactly the same face as Madame de Billy, her nose, her complexion, her eyes (only bluer), her hair, her way of talking. Merklen[9] has resuscitated him and he is in the best of health. I think it was an infringement of his diet yesterday (too much melon) that prevented him from sleeping last night and consequently us from sailing today. Give Robert and Marthe[10] my fondest love. I'm so glad they are such nice children for you, nicer than me. I'm sure Diaz[11] would be pleased to know that his studio was occupied by Marthe who is so exactly the type of woman he always painted with her lovely eyes, her lovely hair, her lovely face. He would surely have wanted her to sit for him if Robert hadn't objected because he usually painted goddesses with very little on.

A thousand loving kisses,
Marcel

I can't tell you how much I miss you here. The sea that you love so much, colours that would enchant you, air that is very different from that of our dining-room, a temperature that keeps everyone huddled up in shawls (not in the cabins, however, compared to which the dining-room is an ice-box – I'm obliged to leave my porthole half-open which upsets me in the morning because of the damp, but one can't open it when one likes and one has to decide in the evening if one wants it open). Having seen you suffer from the heat, always trying to escape it, or, if you stay in it, sleeping heavily, I should love to see you here admiring everything and breathing freely. Do talk to Robert about your tiredness the other day and about the minutest details of your health. I don't talk to him about it because my interference would irritate you if you knew about it (although you wouldn't know about it). But remember that I live in close proximity to you with my eyes closed, that you might have pains, kidney troubles, that you could actually have had that illness last winter, without my knowing anything about it. So do at least put my mind at rest by always talking to Robert in detail about these things. Happiness and sorrow have ripened his character like a fruit which sweetens after having been a little acid. So that his intelligence and his kindness will combine to advise you. – I'm sorry about Waldeck-Rousseau's death,[12] and doubly so because I feel that you must be saddened by it. One really feels that these last few months must have been so silently cruel for him, so bitter, even more disappointed and disillusioned by the life that was over than anxious about the death that was to come. Perhaps I'd have preferred his conversion to have been a little less sensational. 'Brother, the Christian sentiments you display.'[13] But it's very useful to show the world that religion and politics have nothing to do with each other and that one can be hard on the clergy and nonetheless devout. Only I don't think he was.

A thousand loving kisses, soon to be actual ones.

Since writing to you I've warmed up and no longer have any asthma. As in an opera, you were bending over me as I wrote, and the soothing effect of our conversation removed the last traces of oppression. I think I shall leave tomorrow morning. But I shall have to start early. And having eaten scarcely anything at lunchtime I shall have to have some dinner and that will prevent me from going to bed early. Complicated!

A thousand loving kisses,
Marcel

1. *Née* Jeanne Weil (1849–1905), the beautiful, intelligent daughter of kind, cultivated, wealthy Jewish parents. In the last days of the Second Empire, aged twenty-one, she married the successful thirty-year-old Dr Adrien Proust. A year later, on 10 July 1871, at her uncle Louis Weil's house in Auteuil where she had taken refuge during the Commune, Marcel was born. Until her death, when he was thirty-four, she organized her household around the extraordinary routine dictated by his delicate health, his work and social life, a routine which she was scarcely to allow even her last mortal illness to disturb. Not surprisingly, he was to miss her daily for the rest of his life.

2. Paul Mirabaud, Governor of the Bank of France, father of Mme Robert de Billy, Proust's hostess on the yacht.

3. Comte Robert de Billy (1869–1953), close friend of Proust since the days of their military service at Orleans in 1889, later a diplomat. See vol. I.

4. A widow, *née* Inez-Marie de Bourgoing, Mme Fourtoul subsequently married General (later Marshal) Lyautey.

5. *Née* Jeanne Mirabaud.

6. Mme Proust was staying in a hotel at Etretat, in Normandy.

7. Cf. the painter Elstir on the charms of yachting in *Within a Budding Grove* (*RTP*, I, 959–63).

8. Cf. the passage in *The Fugitive* where the narrator thinks of buying a yacht for Albertine (*RTP*, III, 463, 1115).

9. Dr Pierre Merklen. See letter 49, n. 9.

10. Proust's brother and sister-in-law, *née* Marthe Dubois-Amiot (1878–1953).

11. Narcisse-Virgile Diaz de la Peña (1807–76), painter of classical landscapes and nudes.

12. René Waldeck-Rousseau (1846–1904), Dreyfusist Prime Minister 1899–1902 and moderate anti-clerical who had died on 10 August. The socialist press noted with disapproval that he had received the last rites.

13. Cf. 'Les sentiments humains, mon frère, que voilà!' Molière, *Tartuffe*, Act I, scene 5.

56   *To his Mother*

[Paris] Monday evening [15 August 1904]

My dear little Mama,

I shan't write to you at length and certainly not very well as I've just written a long letter to M. Mirabaud which otherwise would never have reached him, and I'm tired. I arrived in Paris at 7 o'clock this morning. I was annoyed to find my bedroom window wide open, and because of the rain which had poured down all night it was terribly damp, so that it would have been impossible for me to stay in it. I had to have a fire lit to dry it out and wasn't able to go to bed till eleven o'clock. I shall explain at greater length when we meet why and how much I was put out by all this. In front of the open window my furniture was taking the air but in fact infecting the room. The house smells awful, especially the dining-room, the door of which had been left permanently open 'to let some light into the corridor' (I'm telling you all this between ourselves). I'll also give you an account when we meet of my last few days, which from the health point of view weren't quite up to the others, though the sea wasn't in the least degree responsible for this. The fact is that, going to bed very late because I had dinner, and getting up early because I had lunch (I didn't have lunch on Friday because I felt slightly sick, but I was nevertheless on deck at ten o'clock in the morning – the only day I didn't get up (until 3 o'clock) was yesterday, I'll explain why), I had nowhere near enough bed or enough sleep and that always upsets my breathing a bit. As long as I remained in the pure sea air it was no more than a faint discomfort. But when I went into the country on Saturday, with all the leaves and dust, I had an attack which in itself was not unbearable, but which was prolonged by having a fumigation on the spot in Dinan. We got back late, though I didn't want to hold up dinner by fumigating too long. My attack continued throughout dinner, and on top of it a storm blew up just then so that I couldn't get back to bed till 7 o'clock in the morning. So I stayed in bed till 3. In these circumstances I dreaded the thought of the eight hours in the train. Thanks to the most skilful precautions and thanks especially to a heavy shower of rain (nature is cleverer than the cleverest man in the world) it passed off wonderfully without a moment's discomfort. I was delighted, for I was expecting something quite different and saw myself being obliged to get off the train at three o'clock in the morning. Naturally, having gone to bed this morning at eleven, I still don't feel properly rested and without the pure sea air I'm more aware of my accumulated ailments in the form of a slight wheezing. Even on board the yacht the lack of sleep meant that I hardly knew what I was doing. I only wanted to send you one of my usual bulletins in which I never tell you anything but the worst so that you know what curve I never went below. As for all the beautiful sights to do with nature or people, that can wait for one of our cosy heart to heart chats, punctuated with kisses. Impatient though I am to be able to give you actual ones, I think the sensible thing would be for you to stay a bit longer, even for my sake. Although I'm not as mobile as I expected to be on my return, a little burnt out by all these strains, a couple of days' rest will perhaps put me in a different frame of mind, and if you come back there'll be no more absences. Let me know your plans; I'll write to you tomorrow. Give Robert and Marthe and Adrienne[1] a good hug from me. I'm so tired that I don't know whether you can sense between these incoherent lines the joy it gives me to feel you nearer to me and to think that soon we shall be one person as we are now one heart.

A thousand loving kisses.
Marcel

I've had the electricity reconnected.

---

1. Suzanne-Adrienne ('Suzy') Proust, aged nine months, daughter of Proust's brother Robert and his wife Marthe who were evidently staying at Etretat with Mme Proust.

57   *To Madame Robert de Billy*

[Wednesday, 17 August 1904]

Madame,

I was absolutely delighted to receive your nice post-card representing the bay where I was so happy and dated from the one to which I should so much have liked to be able to accompany you, and I hasten to reply so that my thanks may still reach you at Morlaix. The wind is blowing a little as I write, but in Paris it is ugly because it raises nothing – except dust – and isn't a jeweller like the sea breeze which turns sapphires into emeralds and embroiders them with bright silver. But I'd like to think that it isn't blowing at the moment on the coasts of Brittany and that the sea is calm and you feel well and are not sadly stretched out on the deck having your hands warmed by the boatswain.

Tell Robert that my article[1] appeared in yesterday's *Figaro* (dated Tuesday), but as I didn't correct the proofs, as I write very illegibly, and as in the agitation of my imminent departure I didn't have all my wits about me, it is so full of misprints as to be totally incomprehensible.

Thank you again with all my heart for your kindness to me during those wonderful days, to which, with charming coquetry, you have decided to add a final seal by sending me this post-card. I hope that all is well on board and that the indisposition that prevented Madame Fourtoul from accompanying you has disappeared. Please thank your father again from me as I expressed very badly everything I feel in my letter to him yesterday, and accept, Madame, my most sincere respects.

Marcel Proust

1. Proust's article entitled 'La Mort des cathédrales' (The Death of Cathedrals). For his second thoughts, see letter 67. For the background to the article – anti-clericalism, separation of Church and State, etc. – cf. letter 97.

58   *To Fernand Gregh*[1]

[22 August 1904]

My dear Fernand,

Returning to Paris after spending a few days on a boat (did *your* yachting trip do you good or harm?) I found the *Renaissance latine* and was painfully upset by the article in it about you.[2] It would be more tactful on my part not to mention it to you, and indeed it would also be wiser, so unimportant the thing is from every point of view. But I know how delicate and highly strung is the harmonic integument of your imagination, your sensibility and your heart, and I'm afraid that these absurd remarks may have upset you; and it struck me that perhaps an affectionate word from someone who was irritated by it, and then recognized its utter insignificance, would be of some comfort to you. The article in itself will be a good advertisement, because people don't read, and in this case moreover would have to read 'between the lines', which would be asking too much of them. As for those who do read, an article in which the beauty of the Parnassians[3] (that is to say essentially a privileged beauty, the beauty of certain things and not of others, and hence the non-beauty of things in themselves, of life in itself) is assimilated to the beauty of living, to your conception of beauty, such an article can have no importance except to show what a prominent position you have achieved, and the feebleness of the arguments to which those who want to contest it are obliged to resort.

This article gave me only one regret. Some time ago Constantin promised me the post of literary critic on his review,[4] not only promised but spontaneously offered. Without warning me, he gave it to someone else, and we fell out completely as a result, since I couldn't tolerate such a way of going about things without kicking up a fuss. We have made it up completely since then, for he has behaved very thoughtfully ever since, and I've forgotten all about it. But having seen the absurd way in which my 'usurper' talks about you, I regretted for the first time not having the right to hold forth in his place.

Don't be upset, will you, by such an absurd business; don't get worked up; I assure you that it's as though it never happened, and when all's said and done if a dissonant note had to be struck in the concert of praise, in the 'murmur of love rising in your wake',[5] this one is on the whole agreeable, has real advantages for you (what a style!). I am too unwell to write to you at greater length, but I imagine you'll have understood my thoughts.

Your

Marcel Proust

---

1. Fernand Gregh (1873–1960), poet, author of a book (published 1958) about his lifelong friendship with Proust, and one of the models for Bloch in *RTP*.

2. A patronizingly disparaging review of Gregh's poetry.

3. A group of poets led by Leconte de Lisle (1818–94), translator of Homer.

4. Cf. letter 6, n. 4, and elsewhere.

5. 'Ce murmure d'amour élevé sur tes pas.' From 'Mes heures perdues', a sonnet by Félix Arvers (1806–51).

59   *To Marie Nordlinger*

Monday evening [22 August 1904]

Dear friend,

Your letter dated Saturday, arriving here Monday evening (this evening) only (why?), brings me dreams and regrets. That lunch, as I shall explain to you, would be impossible from every point of view. But the worst of it is that I'm very ill and I'm afraid I may not be able to accompany my brother on Wednesday. And you tell me that it would be necessary to see the medallion[1] again beforehand. What's to be done? I can't after all ask you to bring it back here a second time tomorrow evening (this evening) Tuesday. That would be too complicated for you. So I shall look after myself and try and come on Wednesday. But I'm not at all well tonight. I should like to make a rendezvous with you in the near future to see a bit of you. Not, though, to tell you about my journey which was so short and since I didn't sleep is not very clear in my memory. I feel as though I've indulged in the fantasy of the Verlaine character (*Beams*):

> He wanted to walk through the waves of the sea
> And as a gentle zephyr lulled the storm . . .[2]

Till Wednesday, perhaps, though not in any case for lunch, but thank you for that charming invitation which I'd have been *thrilled* to accept. I'm sure the occasion would have been 'exquisite and beautifully arranged'[3] and I'd love to taste Jeanne's cooking which must be divine, while from rose to rose buzzes

> The wasp drunk with its mad flight.[4]

Your

Marcel

1. The commemorative medallion of his father which she had sculpted at Proust's request. It was for many years on Dr Proust's grave in the Père Lachaise cemetery, but has since been removed to Illiers and placed on the façade of the house in which he was born.

2. Cf. 'Il voulut aller sur les flots de la mer / Et comme un vent léger soufflait une embellie . . .' Verlaine, *Romances sans paroles*.

3. 'Exquise et fort bien ordonnée.' Hugo, *La Fête chez Thérèse*.

4. 'La guêpe ivre de son vol fou.' Verlaine, *Sagesse*.

60 *To Léon Yeatman*[1]

Wednesday [24? August 1904]

My dearest Léon,

You are wonderful and sweet. And your letter is marvellous. I thank you for it with all my heart. But what is bad of you is never to tell me your plans. I've been very unhappy recently, and since bed, though more and more necessary, has become intolerable to me in my misery, I lead the most active life. I'm beginning to be exhausted by it. But had I known you were going to Brittany I would have done several things with you. What can I advise? I know nothing of Brittany. I was so ill at Belle-Ile (on the sea) that I'm unfair to that famous and indeed splendid spot. But I think it needs storms. On the other hand I love Beg-Meil,[2] which in fact is no more than an apple-orchard sloping down to a sleepy bay. It isn't something to go and see, but a delightful place to live in. And if you go by boat from Beg-Meil to Concarneau some evening, your oars will scatter all the colours of the setting sun on the dazzling, sleepy waters. Douarnenez and its delightful Ile Tristan are quite beautiful, but the smell of sardines would make them unbearable for me. At the Pointe du Raz one really feels what the word *Finistère* means. But to this grandiose but familiar landscape of typically Breton cliffs I prefer, and prefer to everything else (but you need a storm), the beach, somehow Indian, American, reminiscent of Florida, of desolate Penn March which rounds off its Netherlandish hinterland. With a storm there you'll be wild with joy. And you'll see gentle, wounded little beaches clinging to the rocks like so many Andromedas. You're in the middle of the Baie des Trépassés there. It's an unforgettable part of the world. But all this is more for going to see. Whereas Beg-Meil is a delightful place to stay in (if it isn't swarming with visitors). If you go to Beg-Meil I'll give you a note for Harrison and for M. Bénac.[3]

I think of you both constantly and send to the one my admiring regards, and to the other my affectionate friendship.

Marcel Proust

And thank you also for the post-cards which enable me to travel with you, not only with my heart which I would do anyway, but also with my imagination.

1. Léon Yeatman (1873–1930) studied law at the Sorbonne with Proust; both he and his intelligent wife, *née* Madeleine Adam (1873–1955), were his loyal friends. See vol. I.

2. A small Breton resort where Proust wrote Part I of *Jean Santeuil* and which was also one of the origins of Balbec in *RTP*. Cf. Proust's letters of September 1895 reproduced in vol. I.

3. The American painter Thomas Alexander Harrison (1853–1930), whom Proust and Reynaldo Hahn had met at Beg-Meil; it was he who made the comparison with Florida, like Elstir in *RTP*, for whom he was one of the principal models (cf. Elstir on Carquethuit: I, 913). Pierre Bénac, a civil servant and friend of Proust's parents, had a château nearby.

61   *To Maurice Le Blond*[1]

[About 27 or 28 August 1904]

Sir,

I have received the questionnaire you were good enough to send me.[2] On the pretext of clarifying the meaning of the questions you pose, you hasten to point out the spirit in which the answers should be framed. And after two pages of highly interesting and as you say necessary explanations, you rightly consider the reader to be sufficiently 'prepared' so that you no longer need all that circumspection to express your thought and reveal your aim which is not at all, is it, to 'conduct an inquiry', but to get a particular opinion endorsed. Thus after the 'necessary explanations' you give your questionnaire a new and entirely unequivocal form: 'Do you accept the *age-old tyranny* of Rome, etc.' 'Do you think the State has the right to *subjugate* artistic personality?' Faced with a question posed in this way, would anyone dare to reply that he is in favour of the tyranny of Rome or the subjugation of personality? Naturally you think not, but with what heavy irony, just in case, you denounce this improbable recreant: 'If yes, then the present state of affairs is excellent.'

And yet, Monsieur! Whether or not the State has 'the *right*' to subjugate artistic personality, do you think that so important since in no circumstances will it ever have the *power* to do so. What *can* subjugate the personality of an artist is, first of all, the beneficent force of a more powerful personality – and that is a servitude which is not far from being the beginning of liberty – and secondly, the pernicious effect of sloth, sickness or snobbery. But the 'State', Monsieur, how do you imagine the State can subjugate a personality? Take for example any one of the official painters whose work you do not like any more than I do. Do you really believe that a mental utopia, skilfully managed, would find in him a Claude Monet or even a Vuillard who only wanted to live but whom the State has stifled? Do you think that M. Claude Monet, 'subjugated by the State', would have painted like Monsieur Z.? I believe that we are indeed dying, but for lack of discipline not of freedom. I don't believe that freedom is very useful to the artist and I think that, especially for the artist of today, discipline would be as entirely beneficial as it is for the neuropath. And discipline is something fruitful in itself, whatever the value of what it prescribes. If a choice has to be made, it is perhaps on the whole slightly preferable that those responsible for teaching should in fact be 'masters'. And this prompts me to put forward a not very radical but perhaps fairly sensible solution to the problem you pose.

Why, instead of demanding the suppression of the Ecole des Beaux-Arts, do you not suggest that M. Claude Monet, M. Fantin-Latour, M. Degas and M. Rodin be asked to give classes there? That would be interesting. Actually I don't know the names of any of the current professors but it must not be forgotten that Gustave Moreau and Puvis de Chavannes taught at the rue Bonaparte.[3] Moreover there are, I suppose, plenty of ancillary classes where the greatest would not necessarily be the most instructive. M. Gaston Boissier is obviously a much less good writer than M. Pierre Loti. Yet there would probably be more to learn from the former than from the latter. I recognize however that official painting, like official music, is much further removed from the real thing than official literature or philosophy. On the whole the majority of our best writers belong or will belong (if they wish) to the Académie Française. M. Lachelier, M. Darlu, M. Boutroux, M. Bergson and M. Brunschvicg belong to the Université.[4] On the other hand many of our greatest painters and our greatest musicians do not belong to the Institut[5] and do not appear to have any chance of ever doing so.

In short, Monsieur, while confessing, as will have been only too obvious to you, that I am dreadfully incompetent in the matter, on this first point I favour a mixed solution: the improvement of teaching by a freer choice of teachers, and in particular the introduction of classes by men like Messrs Claude Monet, Degas, Fantin-Latour, Rodin (and others whom you know as well as I do). I have cited these names as being particularly significant and also as being those of artists who have achieved a degree of mastery that no one today would any longer dispute.

As for the second question, I have no solution to offer because the present state of affairs seems to me excellent. Although I do not subscribe to the view that certain places on earth have a monopoly of beauty, Rome, if we are to believe the recent comments of M. Maeterlinck, for example,[6] seems after all to be one of those places which can exert the most stimulating and durable influence on the imagination of an artist. Moreover I know a number of young men without the slightest hint of an official connection who have gone to spend several years in Rome after having tried and failed to work. Of course I believe that one can find beauty elsewhere, that one can find beauty everywhere. But since one has to make a choice, I see no reason why any other beautiful place, Honfleur, Vollendam, Quimperlé or wherever, should be preferred to Rome.

As for the 'tyranny' that the 'Roman' ideal exercises over us, do you not think that it is by trying to obey others that we gradually become aware of our own identity. No power was ever more tyrannical than the power which Byzantine hieraticism exercised over the Romanesque artists. And yet, could anything be more delightful than their sculpture? The freer works which followed still submit to this yoke with an acquiescence which for my part I find incomparably attractive. 'The most beautiful sculpture in the world', said Huysmans – 'the most beautiful sculpture of the Middle Ages', said Ruskin – is the west porch of Chartres Cathedral. There is no more original, more spontaneous, more Gallic masterpiece than those admirable statues of the Queens. And yet how subservient the artist still is to the precepts of the Byzantine style. We do not feel, however, that this 'tyrannical' influence has inordinately subjugated his personality. We find there a mixture of freedom and obeisance which is exquisite. Do you not think the influence of the Impressionists has been infinitely more tyrannical than that of Rome? It is a great mistake to believe that an artistic influence needs to free itself from official constraint in order to exert itself. Love is the great tyrant, and one imitates slavishly what one loves when one isn't original. The truth is that there is only one real freedom for the artist: originality. The slaves are those who are not original, whether the State interferes with them or not. Do not try to break their chains; they would immediately forge new ones. Instead of imitating M. Jean-Paul Laurens they would imitate M. Valloton.[7] They might just as well be left as they are.

Yours sincerely,
Marcel Proust

1. Publicist who campaigned against the State funding of two institutions for teaching fine art, the Ecole des Beaux-Arts and the French Academy in Rome.

2. A questionnaire sent to 'personalities in the field of art, literature and politics' under the aegis of the magazine *Les Arts de la vie* as part of an inquiry into 'the separation of the State and the Fine Arts'.

3. Headquarters of the Ecole des Beaux-Arts.

4. Contemporary French philosophers; Alphonse Darlu (1849–1921) was Proust's former teacher. See vol. I.

5. The Institut de France is the overall body incorporating the five academies: Académie Française, Académie des Inscriptions et des Belles-Lettres, Académie des Sciences, Académie des Beaux-Arts, Académie des Sciences Morales et Politiques.

6. There is a chapter on Rome in Maeterlinck's *Le Double Jardin* which had recently been published.

7. Jean-Paul Laurens (1838–1921), professor of drawing at the Beaux-Arts, was a traditionalist, Félix Valloton (1865–1925) was more 'advanced'.

62   *To Marie Nordlinger*

Wednesday evening [31 August 1904]

Dear friend,

What a pretty letter you wrote me, what a pretty page! However I shall tell you so by word of mouth, for the first evening I feel well, perhaps even this, Thursday, evening, I'll send you a message, not to come and work, but to come round for a little chat, which may not be as sensible but much more agreeable – for me. What gave me less pleasure in your letter than the pretty thoughts and pretty phrases was a sort of vague allusion to some deep sadness which you don't express, which you weren't perhaps even thinking of as you wrote, but which seemed to circumscribe and darken your letter like an inescapable horizon. I know so little of your life that it may well contain sorrows of which I am ignorant. But ignorant or not, they grieve me.

You have an incredible memory. Since I myself because of these horrible anti-asthmatic medicaments cannot remember what happened the day before, I envy you for having retained such a precise memory of the days in Venice.[1] But to what expression (an expression of speech – or of face) do you refer? I feel that your letter itself is a charming trellis woven over something dark that I cannot see. Dear friend, I want you to feast on life, your plate loaded with cherries and loud with the cock-crow, and not weep sadly beside an urn containing nothing but regrets.

Your

Marcel

1. See letter 13, n. 1.

63 *To Antoine Bibesco*

[Friday evening, 9 September 1904]

My dear Antoine,

You would be doing me a great favour if you could pass on the following to M. Ollendorff.[1] He has had for a very long time a dialogue of mine, delivered to him by Picard,[2] which he had promised to publish. I don't mind whether he publishes it or not but since it is set in the month of September[3] if he doesn't intend to publish it immediately will he return it to me without delay so that I can take advantage of this vague topicality to place it elsewhere. But he must be told this clearly and sternly because he's an impossible man. As a publisher he had my Ruskin[4] in his hands and it took me a year to get it back from him. If he hadn't left the publishing house and been forced, or rather his successor, to go into liquidation, I don't know whether I should ever have recovered my *Bible*, at once cruelly spurned and jealously hung on to.

As for the 'snapshot',[5] Louisa took seriously what you may have said in an idle moment. You're the one who ought to do it. However, if you'd rather I did, nothing would hire me to sign it since it must be something utterly idiotic, a sort of pastiche of Delilia:[6] 'Remember this name, it will hit the headlines one day' – or 'She's fanatically devoted to her art, a real slogger.' I think you'd do it marvellously. I shall be at home tomorrow evening (that's to say tonight, Saturday, by the time this reaches you). But don't come in the daytime, however late, because I shan't be visible.

Yours ever

Marcel

I think the gentleman I said looked like Clemenceau was Leygues.[7]

I'm asking you all this because you suggested it. If it's the slightest bother don't do it, as I know quite a lot of people who could ask for this manuscript back, myself among others. But your intervention would simplify everything.

1. Paul Ollendorff (1851–1920), co-editor of the daily newspaper *Gil Blas*. Cf. letter 24, n. 4.

2. Presumably Bibesco's friend the playwright André Picard.

3. Rejected by the *Gil Blas*, this dialogue between a young man and his putative mistress over a dinner table in the Bois on a September evening, entitled 'Vacances', was not published until spring 1949, in the Harvard Library Bulletin. Cf. letter 105, n. 2.

4. *La Bible d'Amiens* (see letter 11, n. 2).

5. *Instantané*: a brief 'profile' of Louisa de Mornand for the *Gil Blas* (under the paper's rubric 'Medallion') as publicity for her coming appearance at the Vaudeville Theatre, which Bibesco seems to have agreed to write but which, in the end, Proust wrote himself.

6. Alfred Delilia, *Figaro* editor who wrote vaudeville sketches and light comedies.

7. Georges Leygues (1858–1933), politician, future Prime Minister.

64 To Reynaldo Hahn

[Friday evening, 9 September 1904]

Dear Mossieur de Binibuls,[1]

I didn't come this evening because you said. So went out, so had attack. So say if want me *Sunday evening* (while leaving me the option of coming Monday if not well enough). But since letters don't arrive Sunday daytime, send little telegram so that know. – Weepsie on reading sufferings of my Buncht.[2] How should like to be able to punish moschant mal that tortures you.[3] When Clovis heard the story of the passion of Christ, he rose, seized his axe and cried: 'If I'd been there with my brave Franks, it wouldn't have happened or I'd have avenged your sufferings.' Which is pretty pony[4] on the part of such a moschant king who and who. But I can't even say that because I wouldn't have been able to prevent and wouldn't know who to punish, not even old mother P[otocka][5] 'who wears on her head plumes that would look better on a hearse and decorations that are half way between the cross of the hundred guards and the blue ribbon of the Children of Mary' (R. de Montesquiou). You know he's finishing a book which is divided into two parts: (1) the beauty that doesn't reveal its noble old age: the Comtesse de Castiglione,[6] (2) the ugliness that exhibits its decrepitude: old P[otocka] – a book of which he gave a foretaste the other day in a letter to *Le Siècle*. At St Moritz when someone lost a superb pair of opera-glasses he said they must belong to Mme de Rothschild – or Mme Lambert – or Mme Ephrussi – or Mme Fould.[7] And next day he said: 'I was still aiming too high, they belonged to M. Untermayer, below Mayer, less than Mayer, imagine what that can be.' – Dear Binibuls, would that the casket of my memory were richer so that I could divert darlings little sick one. But I know nothing. Perhaps this though. Meyer[8] is as though intoxicated. He 'notifies' his marriage to all the sovereigns or at least dukes. His witnesses will be the Dukes of Luynes and Uzès. He ran into Barrès and said to him: 'I'm off to Versailles. Would you like me to greet my cousin Louis XIV on your behalf?' He wrote to old Mme Brancovan[9] to say that he would be supporting all the Fitz-Jameses and would leave the *Gaulois* to his fiancée's brother whom he would attach to the paper during his lifetime. And lots of other extravagances of the sort. He of all people, who was once so pitiable and described his duel[10] thus, excusing his notorious display: 'What was I to do? I thought I was going to meet a gentleman and found myself face to face *with a kind of madman who might have killed me!*' (a remark worthy of M. Jourdain, the other *bourgeois gentilhomme* who also feared for his skin). It was about the same duel that he once said: 'It will take ten years for it to be forgotten – or else a war.' But you know all that and kissikins and greetings from Birnuls. They played at Larue's[11] (whence attack after attack) such pretty musics and everyone and everything and it was 'Since every soul here below'.[12] And Antoine Bourbesco said it was reminiscent of Mozart's *trios*(?).

[Unsigned]

1. Proust and Hahn often corresponded in a private jargon, as throughout this letter. This weird idiom, a kind of baby-talk, apparently originated with Reynaldo. Cf. letter 18.

2. Proust's principal nickname for Hahn, though he uses a variety of others.

3. 'Mosch' or 'moschant' (a distortion of *méchant*) is a word that often appears in these letters, sometimes with homosexual overtones. By 'mal', Proust is referring to Hahn's tonsilitis.

4. See letter 18, n. 5.

5. Comtesse Nicolas Potocka, *née* Emmanuela Pignatelli (1852–1930), recent subject of one of Proust's pseudonymous 'salon' articles in *Le Figaro* (cf. letter 10, n. 7). A pupil of Liszt and passionately fond of music, she was an imperious patron of writers and artists and a notorious *femme fatale*. Forgotten in her old age, she died alone in her Auteuil house where her corpse was discovered devoured by rats.

6. Comtesse de Castiglione (1835–99), former mistress of Napoleon III.

7. All wives of prominent bankers.

8. Arthur Meyer (1844–1924), sixty-year-old director of *Le Gaulois*, was to marry the twenty-four-year-old Marguerite de Turenne, to the disapproval of members of her circle and family (cf. letter 130); she was the granddaughter of the Duc de Fitz-James who, like the anti-semitic Dukes of Luynes and Uzès, and Meyer himself (a converted Jew), were rabid anti-Dreyfusards. Cf. *RTP*, III, 784, 884.

9. Princesse Rachel Bassaraba de Brancovan (1847–1903), mother of Prince Constantin, Mme de Noailles and Princesse Alexandre de Caraman-Chimay.

10. In 1886, Meyer fought Edouard Drumont (1844–1917) who, in an anti-semitic tract, *La France juive*, had repeated an accusation of card-sharping; on that occasion Meyer cheated by grasping Drumont's sword in one hand while wounding him in the groin with the other. (The *bourgeois gentilhomme* in Proust's following parenthesis is, of course, a reference to Molière's play.)

11. Conventionally luxurious restaurant at the corner of the place de la Madeleine and the rue Royale frequented by Proust and his friends.

12. 'Puisque ici-bas toute âme', from 'Rêverie', a song composed by Reynaldo Hahn (when aged fifteen) to words from Hugo's *Les Voix intérieures*.

65   *To Georges Linossier*[1]

45 rue de Courcelles
[September 1904]

Dear Sir,

I don't know whether you remember that in happier days my father, Dr Proust, introduced me to you and I took the liberty of expressing to you my great admiration for your book, *The Treatment of Dyspepsia*.

This admiration is the cause of (and will also be my excuse to you for) the presumptuous approach I am now making. If I knew of another school of thought as brilliant and as profound as yours, I would spare you the boredom of being asked for advice. This advice will moreover be necessarily very limited, since I realize that from a distance it will be impossible for you to pronounce on questions of fact. And as for trying to see you, I lead too odd a life, leaving me free only during the evening hours, to dream of disturbing you.

Here however are the few questions which you might perhaps be able to answer.

I am, it appears from the medical point of view, many different things, though in fact no one has ever known exactly what. But I am above all, and indisputably, an asthmatic. Starting as hay fever, my asthma fairly soon became summer asthma and then a more or less all-the-year-round ailment. And as a result of too heavy meals, it became complicated by a condition which, though apparently asthmatic, was, I was told, intestinal and gastric in origin. This condition has long since been brought under control, but is ready to recur at the slightest imprudence. I eat one meal every 24 hours (and incidentally may I venture to ask you whether from the point of view of daily ration you consider this meal sufficient for 24 hours: two creamed eggs, a wing of roast chicken, three croissants, a dish of fried potatoes, grapes, coffee, a bottle of beer) and in between the only thing I take is a quarter of a glass of Vichy water[2] before going to bed (nine or ten hours after my meal). If I take a whole glass I am woken up by congestion; *a fortiori* if instead of Vichy it's solid food.

I may add that as regards the stomach, so long as my abdomen is properly held in by underpants, I never have a stomach-ache or any discomfort there. Asthmatic breathlessness is my only form of trouble. That is the advantage of my strange diet. Because previously, when I used to eat several meals, and drink between meals, I constantly suffered distension, wind, and other discomforts which no longer exist.

Now what I want to ask your advice about is this: I have been recommended, in order to modify my pernicious living habits, to follow one of those psychotherapeutic treatments which you certainly know of, and which consist of isolating the patient, immobilizing him, feeding him up and curing him by persuasion.[3]

I simply wanted to ask you whether, in theory and in the most general way, you believe that such a cure would have no harmful effects on me, or whether you consider that over-feeding might aggravate my condition. If from this point of view you make a distinction between the various establishments of the kind that exist either in France or abroad, if you know for instance that in this one as opposed to that the feeding up will be done more circumspectly and that everything will not be sacrificed *a priori* to the idea that all gastric troubles are nervous in origin, I should be grateful if you could also give me your opinion on this point. Needless to say, any information you give me will be absolutely confidential. I would also ask you not to mention that I asked you for it (at least on the specific point of the relative value of the different establishments) so that I should not seem to have suspected in advance the one where it is possible that I may eventually go.

Perhaps the following particulars will give you a clearer idea of my arthritic condition. My urine shows a marked excess of urea, of uric acid, and a diminution of chlorides. The analysis I had done added imponderable traces of albumin and sugar, but I believe this quite temporary. I have been urinating very little for several years. After twelve days on a milk diet[4] I did not produce half a litre in twenty-four hours. It is true that I took the milk in the form of boiling *café au lait*, which of course greatly increased my habitual perspiration, and that I could scarcely manage to exceed a litre and a half to two litres of milk per twenty-four hours.

When I lead a different life I can in fact manage two meals a day. But then I find it almost impossible to go to bed because of the number of hours (eight) I'm obliged to spend up (or on a chaise-longue) after a meal. And moreover whenever I lead that kind of life (without being able to say positively that the increase in the number of meals is the cause) I am always a little more breathless, I have more blood in the head, I am less well, than when I lead the bizarre life which is my habitual one and which I described to you at the beginning of this letter. I go a great deal – and unsatisfactorily – to the privy, always several times running. If you thought it might provide useful data I would be prepared to have my stool analysed. But I don't know how to go about it and moreover it varies considerably according to my different states of health. It is often accompanied by a certain amount of mucus, the evacuation of which, as far as I can tell, does me good rather than harm. Once a fortnight, roughly, I take a cascara pill in the middle of dinner, which makes me go to the privy seven or eight or even more times during the following twenty-four hours. Until the purgative has had its effect it oppresses me if anything, then relieves me and sometimes makes me a little unwell again when its effect has gone on too long. I never have enemas because they bring me out in the most unbearable sweat.

Monsieur, I began this letter in the belief that it would be only a few lines. If I had thought I would be asking you so many things I wouldn't have asked you anything at all. As I went along, such and such a detail occurred to me of a kind that might put you more precisely in the picture and help you to form an opinion. I am now overwhelmed with embarrassment at my importunateness which I feel is much greater than I at first thought.[5] And what I beg of you above all is to forgive me and to accept my most apologetic and grateful respects.

Marcel Proust

1. Dr Georges Linossier (1857–1923). His *Hygiène du dyspeptique* was published in 1900; he practised at Vichy.

2. Proust believed in the benefits of drinking Vichy water, as did Aunt Léonie (to a comic extent) in *RTP* (I, 74, 109, 116).

3. Treatment first proposed by Dr Merklen (cf. letter 49).

4. In *RTP* Dr Cottard recommends a milk diet (I, 536–7).

5. It seems that Proust never sent this letter; it was found among his belongings after his death.

66 *To Louisa de Mornand*

Wednesday evening [14 September 1904]

My dear little Louisa,

I've done the *Louisa de Mornand* 'medallion' for the *Gil Blas*.[1] I don't know which day it will appear. I'd very much have liked to thank you in person for your kindness in bringing me home the other day, but I couldn't because I've been ill all the time. Mama is with me a great deal because she is very unhappy at the moment – she always is, but perhaps even sadder.[2] Tomorrow I'm going to Versailles to say good-bye to Reynaldo who has just been very ill, and Versailles will no doubt make me unwell for a few days. Besides, you yourself are going to be very busy with your first performances which I believe are about now, aren't they? My little Louisa, you must believe me when I tell you that with all my heart I wish you a great, dazzling success and that my best wishes will accompany you throughout your life, desiring your happiness, the reward for your talent, your charm, your beauty, your kindness, the real nobility of your character. (Your paper is ravishing but smells too strong. I have only two letters of yours on the new paper, and the whole house is perfumed with it, just as a few moments spent with you are enough to perfume with happiness the days that follow.) This parenthesis is somewhat irrelevant here, but your happiness, your destiny and all that reminded me of your charming *Who knows!*[3] in which there is a touch of melancholy but even more of self-confidence – isn't that a very Louisa-like 'mixture'?

Tenderly yours
Marcel

1. Cf. letter 63, n. 5.
2. Because of her late husband's birthday, 8 September.
3. Louisa's writing-paper was stamped with a device consisting of the words *Qui sait* beneath a sort of long-tailed winged dragon.

67   *To Georges de Lauris*[1]

[September 1904]

My dear Georges,

Forgive me if I thank you briefly, but I'm so unwell and I have such a quantity of letters to catch up with. You are too kind and I thank you with all my heart. I'll try to answer quickly the various points in your letter.

1. I've been in Paris for a very long time; I was away for less than a week; I'm leaving again in the near future; I'd be delighted to see you if you warn me in time. If you don't give me any warning you'll still have a very good chance of finding me in at about ten o'clock in the evening.

2. You are wrong if you think I'm fickle. The memory of the pleasure I get from seeing you and talking to you remains as vivid and delightful as ever. It's true that at one time I had hoped for an even closer friendship. But having taken you by the hand and guided you step by step on to the path which leads to 'the road of my heart', on three or four occasions, as soon as I left you to yourself to see whether you would take this road, you retraced your steps with a dizzy and as it were instinctive haste. I know better than anyone that in the depths of our instincts there lie buried aversions like those of highly strung horses for such and such a track, fears like those of superstitious peasants for such and such a crossroads. But obviously, from the day I finally gave up hope of seeing you take that road, by the same token I gave up some part of you, which inevitably implies the withdrawal of some part of myself too. But you know that the greater and the better part (assuming there is anything good in me) remained with you (I suddenly realize that this letter could easily be interpreted in a particular sense . . . a small observation made simply to amuse you and to beg you not to show anyone my letters of which you alone can understand the precise meaning which is the precise opposite of that).

I didn't send you the *Figaro* article because I thought it very bad.[2] I happened by chance to return to Paris the day before it appeared. Not knowing this, I never corrected it and there are about three mistakes per line. But even if it were accurate I should still find it execrable. Some very intelligent people have written to me to say that they were profoundly *touched* by it. I wonder how something that didn't touch the author can touch the reader. The whole article is ingenious and not true. And it's overloaded with spurious poetry. There are two sentences which I find pretty but they are totally meaningless having been printed all askew. I don't agree with you. The nasty bigots are those who are incapable of aesthetic feeling, the true, the last Catholics those who will disappear one day, like the pithecanthropes of Java that scientists are sent to examine in their forests and bring back to the Zoo while there are still some of them left, and whom it is both fascinating and melancholy to see naïvely indulging in their characteristic exercises in those mysterious edifices where they dip their fingers in holy-water stoups and mimic the shape of the instrument of torture of Him whom they mourn, entreat and glorify. What would we not give to see the archers of Darius's palace enlivening the architecture described by Madame Dieulafoy.[3] I am not like you, I don't find life too difficult to fill, and what madness, what rapture if I were assured of immortal life! How can you really, I won't say not believe since the fact that a thing is desirable doesn't mean one believes in it – on the contrary, alas – but be satisfied with it (not the intellectual satisfaction of preferring the sad truth to the sweet lie). Would it not be sweet to find again, beneath another sky, in the valleys vainly promised and fruitlessly awaited, all those one has left or will leave behind! And realize oneself at last!

Yours ever, dear friend,
Marcel

1. Comte Georges de Lauris (1876–1963), who was to become one of Proust's closest friends, read law at university then became a novelist and critic.

2. 'La Mort des cathédrales'. Cf. letter 57, n. 1.

3. Mme Marcel Dieulafoy (*née* Jane Magre: 1851–1916), explorer and archaeologist.

68 To Marie Nordlinger

Saturday [17 September 1904]

Dear friend,

I don't know what you must think of me and yet it isn't for want of thinking of you! And not only because of all the things I have to say to you, but to answer as well. For I'm in arrears with you since a letter of yours written the day after my visit to Auteuil.[1] A very nice letter although it might – from anyone else but you – have appeared a little ironic or contemptuous. For on to the terrible state of health I was in dropped those arrogant words: 'I'm overflowing with health, with life, with strength, with the power to be happy' etc.

Dear friend, please don't think that I read this with bitterness! And please believe that on the contrary I understood the charity of these words which were like an offer of a moral blood transfusion. And I also had so much to reply to the rest, although I was so little the person who could reply to it. For I understood the melancholy of your solitude; but I was very much the last person who could dispel it, not having had two good hours a week etc. etc. I've tried for a fortnight to husband my strength in order to have those two hours and I went to Versailles to dine with poor convalescent Reynaldo and to bid him good-bye since I haven't seen him since his departure for Germany and it's still possible that at the first glimmer of health I shall go away for a few days which may be extended a bit longer. I didn't find him looking too bad, and there was no trace of his horrible suffering.

Is your journcy to America[2] (on the subject of which I may perhaps offer my opinion which may seem strange to you: I'm very much in favour of it if the idea amuses you and the financial terms are favourable; the offer itself I think very flattering, something that any art-loving person would be delighted to do) taking shape one way or the other? And has your stay in the Venice of the North[3] brought you any enlightenment on that score? Why don't you want your name to be associated with mine on the cover of *Sesame*? You answered me evasively only.[4] Answer: '*Dic nobis Maria, quid* . . . ?' (Office for Easter Day). I think our walks in Venice, *St Mark's Rest*[5] in hand, had little charm for M. Abel Hermant, for in an article in the *Gil Blas*, incidentally rather silly, he speaks of Englishwomen (I assume your sex and your nationality for the occasion, it appears) who think it their duty to pore over the capitals in St Mark's celebrated by Ruskin.

In this connection, isn't this touching: a bookseller in the Piazza San Marco writes to say that he obtained my address through M. Maurice Barrès and he'd like me to translate *St Mark's Rest* for him. I think I'll refuse, for otherwise I shall die without ever having written anything *of my own*.

I couldn't write to you before because I didn't have your address and to write to Reynaldo for it was too much of a nuisance. He gave it to me at Versailles but naturally I lost it. However Mama tells me she has it. So I've taken the opportunity and begun this stupid gossip in which your natural perspicacity *will know* that there is a great deal of friendship.

Marcel Proust

If Madame Henriksen has left

Silvery Sweden with its twin seasons[6]

tell her that I retain a memory of her that is as charming and imperishable as it is grateful and deeply respectful.

1. Cf. letter 52, n. 1.

2. To organize Bing's exhibition of Japanese prints (see letter 40, n. 1).

3. Marie Nordlinger had been visiting her aunt, Mme Hindrichsen, in Hamburg, a city criss-crossed with canals. Proust had met her on his trip to Venice. Cf. letter 13, n. 1.

4. She was being modest.

5. Ruskin's guide-book to Venice, published 1877.

6. Proust has used this quote from Mme de Noailles before à propos of Mme Henriksen (Hindrichsen); see letter 52.

69 *To Georges Goyau*

Tuesday 20 [September 1904]

Dear Sir,

You wouldn't believe me if I told you that it's only today that I've read the *Revue des Deux Mondes*. I read with such passionate interest your admirable *Allemagne catholique*[1] that I should never have dreamed of postponing my pleasure. But it's only today, while glancing through the rest of the *Revue*, that I saw the touching notice you were kind enough to devote to my poor translation[2] which would certainly never have thought that it would be so much talked about and that six months after it appeared it would receive yet another encomium, and from what a pen! I shall treasure this number. Since we're speaking of architecture and you yourself, à propos of Hermes,[3] also use an analogy borrowed from that art – 'faith is an edifice' – on seeing next to one another your great historical and philosophical construction and the delicate little piece on Ruskin, I am reminded of those great religious architects of the Middle Ages who, after having given proof of their intellectual boldness in some vast and erudite monument, were not ashamed to leave not far from there, in some touching carved initial, some affectionate personal thought that we cannot always decipher today, a token, like yours, of the kindness of their hearts.

Please give Madame Goyau my deepest respects and believe me, dear Sir, your very grateful

Marcel Proust

1. Extracts from a book later published, in 1905, as *L'Allemagne religieuse. Le Catholicisme (1800–1845)*.

2. *La Bible d'Amiens*.

3. Georg Hermes, German theologian, professor at the University of Bonn, where his ideas were in vogue around 1820.

70   *From his Mother*

Hôtel Métropole et des Bains
Dieppe
Wednesday 2.30
[21 September 1904]

My darling,

I think I did well for *myself* by coming here – but up to now I'm glad not to have you with me. I'm frozen. It isn't the sea wind forcing you into grandiose struggles to resist it from which one emerges ennobled. There's a clear sky, an appearance of superb weather through which one feels a 'homicidal steel',[1] a deceptive and penetrating cold. I may say that I had an icy bedroom last night. I had been struck when I was shown to it by its cellar temperature – and in the evening on going to bed there were no grounds for revising my impression. But I requisitioned four extra blankets – and an eiderdown (which must have been lined internally with some heavy geographical atlas) which was immovable. I put on two nightdresses and in order not to have to go and turn off the electric light which was at the other end of the room, in accordance with the chambermaid's advice I left it on all night (which shows you that the lighting is included). And this morning I put on all my things after effecting only the most minimal ablutions so afraid was I of catching cold. Then I put on a thick bodice, an overcoat etc., and went out for the whole morning. Unfortunately – although the weather is very fine – there's too much of a nip in the air for me to be able to remain seated and I have to go on walking (with little ten-minute stops whenever there's a bit of sun). But Dieppe is exceptionally well arranged for these healthy walks (for it's very good for me to be forced to walk). By the time one has covered the length of that immense esplanade – visited the pier – been to the post – bought a newspaper – everything is so far from everything that my walk is done. In half an hour I shall have a new room – this time on the first floor (no. 22) – described as very sunny. I haven't yet been able to get into it. At all events I shan't have the extreme conditions of last night. The small dining-tables are in the same room as the *table d'hôte* – big and small together scarcely add up to twenty-five people, perhaps thirty at the most, and at this hour there is even a . . .[2]

1. Quotation from Racine's *Athalie*: 'J'ai senti tout à coup un homicide acier / Que le traître en mon sein a plongé tout entier' (Act II, scene 5).
2. The rest of this letter is missing.

71   *To his Mother*

Wednesday 6 o'clock in the afternoon, or
'evening' as the common herd would say
[21 September 1904]

My dear little Mama,

A letter as 'yellowboy'[1] as could be, I warn you. But after all one must
be truthful. After having written to you last night I felt so well that it was
real bliss. When the thought of going to see Dubois[2] crossed my mind
again. Went to bed in a state of great agitation, as a result of which, I
think, an attack of asthma. I went to sleep nevertheless but with dreams
of breathlessness and finally awoke. Up to then it was nothing and I'd
have been perfectly well if I'd gone to sleep again. But I said to myself
that as I'm not going away I must reform.[3] So I asked for dinner in bed
(or tea if you like, as time had gone by and it was 3.30). At that moment
Marie came in to tell me she was feeling too ill and was going to bed (she
had been to visit her sister yesterday in an open carriage, which was pure
folly). As you don't like Félicie to stay by me, which she would have been
obliged to do if I stayed in bed, I gave up the idea and as my dinner was
ready I got up, but not having 'cooled off' at all since I wasn't expecting
to get up, I ran into the dining-room to keep warm. No fire, and I didn't
know where to go. By the time Félicie had lit one which wouldn't burn –
but I'll spare you the details. Whether because of this coldness, fatigue
from not having rested, an indescribable malaise set in, sore throat,
despair, inability to move, unbelievable pulse, etc. etc. François[4] whom
I'd asked to come had to go away again for I couldn't have taken it as he
could very well see. Perhaps I shall get better, and if so I'll go out
tomorrow. But if as I expect I'm not well I shall give up my timetable
temporarily for this reason: if I'm to stay at home feeling ill, at night I
don't mind too much, as I'm used to being there, but in broad daylight,
all alone, and seeing the sun outside, it would be too nostalgic. And then
I should have made this enormous effort for nothing. For I'm going to be
in bed by midnight tonight. So nothing could be easier than to be up by
midday tomorrow. But if I find I'm not well, even if I can get up without
going out, I shall arrange to stay in bed till 7 o'clock (thereby deliberately
destroying everything I have bought so dear) in order not to have to be in
the dining-room from noon to 7 o'clock, alone, in sight of the sun. 'Do
you see what I mean?' – My dear little Mama, I realize how boring it
must be to hear all this talk of my health. I assure you that when you're
here I shall never talk about it again. But I really must explain to you

what I'm doing, what I'm not doing and why I'm not doing it. 'The consolation of martyrs is that God, for whom they suffer, sees their wounds.'[5] Naturally not a hint of asthma in such a feverish condition.

Did you read yesterday evening's *Débats* – every word is worth reading, Filon's article on New Zealand, Welshinger's on H. Bismarck, Beaunier's on the Pole.[6] That's all I can think of for the moment but there must be some other small pieces. If I'm feeling well in an hour's time I shall be very annoyed at having moaned! Tell yourself that this letter is the expression of a fleeting reality which will no longer exist when you read it. I don't know whether you buy the *Figaro*. This morning it really goes beyond the bounds of silliness and insipidity. As I rely on you to burn this letter (and you're not supposed to know any of the domestic news I've given you, about Marie etc.) I shall tell you a secret that *no one* must know, not anyone, you understand, in the world. It was for Gabriel de La Rochefoucauld that Mme des Garets killed herself.[7]

I had your little note this morning. I wish you could have seen me last night, alert and cheerful in my 14° bedroom, ready for anything. I can't believe it's the same man. I shall become him again shortly. But this preoccupation with going out is sure to bother me. Several times last night I tried not to make a noise when passing your door, not remembering till afterwards that 'the nest was abandoned'. I shall have Albu and Bibesco this evening.

A thousand loving kisses,
Marcel

1. *'Jaunette'*. In *RTP* the narrator's mother calls her son 'mon petit jaunet' when comforting him at bedtime (I, 42).

2. Dr Paul Dubois (1848–1919), Swiss neurologist, author of *Psychonévroses et leur traitement moral*; he had a clinic in Berne.

3. i.e. to change his sleeping routine.

4. François Maigre (1864–1930), Proust's barber, formerly barber to Napoleon III.

5. 'La consolation des martyrs est que Dieu pour qui ils souffrent voit leurs plaies.' Unidentified quotation.

6. André Beaunier (1869–1925), influential critic.

7. Comtesse des Garets-Quiros, *née* Antonia Bernaldo de Quiros de Montreal Santiago (1870–1904), was in love with the young La Rochefoucauld; on 3 August, hearing rumours of his engagement, she shot herself through the heart. He was to marry in May 1905.

72   *To his Mother*

Friday evening [23 September 1904]

My dear little Mama,

The progress in my time-table continues, perseveres and intensifies, since today I was all ready (I mean dressed) for dinner at a quarter to five, which will enable me to go to bed very early. But the reverse progress from the point of view of health also continues. After a night of discomfort, and an execrable awakening, I had five minutes of good sleep just as I was about to get up. And as it was getting late I didn't have time to lower my body heat and dressed without a second's transition, then began to feel my skin burning and wanted to cry just as I did in the old days when I had my lotions etc. etc. Reynaldo who had come to see if I was dining at home this evening and made no noise so as not to wake me was surprised to find me in the middle of dinner. Afterwards I couldn't bear to stay at home so I went out, couldn't find any of the people I called on, and having developed a few aches and pains I left my cab to go for a walk to try and calm them, which it didn't, but I came back home without any asthma. It's possible that I haven't had an attack because of my slightly feverish state. If it goes on I shan't have a very wholesome look to offer you when you return. If my calculations are right I think I shall be able to get up at midday tomorrow. The worst of it is that however ill I may be (I mean however ill I may become as I'm not in the least unwell at the moment) I can't see any way of staying in bed for a single day since it would mean being there in the afternoon and it would be too gloomy. All this takes too long to explain and is anyhow boring. Just as I was coming in I heard someone shouting from a cab: 'Good evening, Marcel! Good evening, Monsieur!' It was Léon Daudet and his wife.[1] They must have been edified to see me coming home at 10 o'clock. And it will give Lucien the idea that I go out earlier when it's 'someone else' who asks me. It was no one! By the way, I told you that while Lucien was staying with his priest, Madame Daudet was in London. I had misread his letter; it was Lourdes.[2] Well, well, well. But perhaps I've already communicated this witty remark to you. I received two letters from you today which show me when I compare them that you appreciated my second letter much less than the first although the first was that of a man who had got up at 9 o'clock and the second that of one who'd got up at 4. Hence indeed their difference in well-being. But my dear little Mama, my letters have only one purpose, to give you the most accurate idea of what I'm doing. Having started off on the wrong foot, that is to say having undertaken, when I was unwell, these changes which I'd have postponed if you'd been there to advise me yourself, I can't recover my equilibrium. At any other time I'd have bucked myself up by going out to dinner with a friend, as I so enjoy doing when you're not here. And it's not for lack of invitations, of the most pressing kind. But I refuse them all, saying to myself that if I go out to dinner my reforms are down the drain since I can't get to bed etc. etc. So the fear of being obliged to rest torments me. Moreover I feel that the whole thing is profoundly useless. Since the aim is rest and fresh air, if I were to exhaust myself by moving about like this, my old way of life was a hundred times healthier. And I can see it will soon be too cold to do it here. And if I had to go away in search of milder air I should find I'd quite naturally changed my hours by continuing to feel well here – instead of losing from day to day what I'd stored up – and enjoying Paris by dining out with all and sundry. Please don't think I'm saying all this so that you'll tell me to go back to the old ways. It's because it gives me so much pleasure to complain to you and because I know it's of no importance since you wrongly interpret it as the nervous irritation inseparable from a change of habits.

My dear little Mama, what an idea of yours, knowing that I was determined to come to Dieppe (and I'd be there already if I wasn't unwell, which I couldn't have foreseen) to choose a hotel where they can't heat the rooms (in which I couldn't go to bed without a fire) except with hot-air vents (which I couldn't bear for a single hour without suffocating). It makes it impossible for me to come and join you, except to spend the afternoon with you and come back to Paris in the evening. I was about to send you just now some flannel nightdresses which I gather you thought for a moment of taking with you. But I know you're 'old enough to do what you like' and you'd have asked for them. So on reflection I await your orders. We've received an invitation (retrospective as they didn't ask anyone) to the Lalo wedding,[3] Madame Proust, M. Marcel Proust. I intend to leave three cards, one for Mme Fuchs, one each for the young couple (and for you what should I leave? – answer me in your next letter). If I've run out of cards I've a good mind to order some *different* ones. These ones are hideous. 45 rue de Courcelles shows through the black border. And Marcel Proust is so small that in a show of wedding presents no one will be able to read it! But where? And is there time? And if there isn't time, where are mine? And where's the present?[4] Can I go and see it, with my new hours, if I haven't completely stopped? No indeed, I haven't had my hair cut, but as I told you I slipped on some day clothes the very minute I got out of bed in order not to delay my time-table which was already behindhand because of the bad night (but without asthma) I'd had. Hence there wasn't a minute either for François or for cleanliness. My dear little Mama, I haven't seen a soul so I have nothing to tell you, and as for my personal thoughts I've transmitted my groans to you down to the very last and as for *thoughts* I had some rather fine ones, really rather fine as Schlesinger[5] would say, yesterday evening, but until I can give them an adequate literary form I prefer not to withdraw them from the penumbra which alone can keep them fresh, like food one keeps in a cool place until it's time to eat it.

Apparently Guiche came round while I was out – that's what comes of my no longer being able to give people precise times, being at the mercy of the next day. Your letters are nice, but short. Mine, you may observe, are extremely long and so closely written as to be twenty times longer still. Reynaldo is delighted by Voltaire's remark that it's impossible to believe that Christ would have thought of changing the water into wine at a feast where everyone was half tipsy already. I leave this pearl to Nuna[6] as it escapes me. Especially as in the legend (a good thing Lucien can't hear me) there was nothing but water, so they couldn't have been tipsy. Fortunately for Voltaire he wrote a few others. I've just read an article on d'Alembert[7] by M. Doumic in which he tries to astonish the reader by adopting the trenchant tone of Faguet[8] and finishes the article thus: 'He (d'Alembert) is a perfect specimen of what can be produced in the realm of thought by the combination of knowledge, methodicalness and silliness.' Not that I think d'Alembert was a great man. But neither does that facile way of being original make me think that Doumic is a great critic. As for Roujon's article,[9] I'm ashamed of myself for having read such rubbish to the end. In yesterday's *Figaro* there was a charming reply from Tristan Bernard[10] on the theatre page. Marie went to the Printemps[11] this afternoon to fetch some things for you and to explore Paris and its shops. She told me she had worked very hard this morning. I didn't ring until I woke up finally so that I didn't disturb anyone. And at six o'clock in the evening, or rather at twenty to six, everyone had finished with me and could go to bed if they so wished, for I couldn't even sit down for another cup of coffee. I may add to the list of my woes a stomach ache which prevents me from lying down properly.

A thousand loving kisses from a son who has been warmed up by walking and writing to you and feels well enough to be pleased to tell you.

Marcel

See if you can find somewhere in Dieppe where I could join you – rooms that can be heated by a wood fire, without hot-air vents. Do try not to come back at least till my new hours are settled, if you're comfortable in Dieppe.

1. Léon Daudet had married for the second time, to his cousin Marthe Allard.

2. Mme Alphonse Daudet had been to Lourdes, her son to a retreat elsewhere in the Pyrenees.

3. Pierre Lalo (1866–1943), music critic, son of the composer, was to marry Noémie Fuchs, the daughter of neighbours of the Prousts, on 24 September.

4. A lamp ordered as a wedding present for Louis d'Albufera and his bride.

5. Hans Schlesinger (1875–1932), painter friend of Proust, son of Henry Schlesinger, German genre and portrait painter who became naturalized French.

6. Probably Mme Proust's cousin Mme Bessière, *née* Hélène Weil.

7. In fact, the article by the critic René Doumic in *La Revue des Deux Mondes* was about another eighteenth-century philosopher-mathematician, Condorcet.

8. See letter 74, n. 9.

9. In a *Figaro* article, 'Neurasthénie', the writer Henri Roujon (1853–1914) deplores the signs of modern times: the Métro, the motor-car, the post-card . . .

10. Tristan Bernard (see letter 49, n. 6) had been asked to tell a *Figaro* journalist what he was writing while on holiday.

11. Printemps: the large department store.

73 *To Marie Nordlinger*

Saturday [24 September 1904]

Dear friend,

What can I tell you? I lead the most miserable life, a hundred times worse than when you last saw me, and am not sure of the next minute let alone the next day. However I hope to go to Versailles tomorrow, though with no certainty nor perhaps probability. I'd in any case prefer not to talk to you before you've seen Monsieur Bing.[1] For being totally incompetent in these matters I'm afraid of influencing you without any knowledge of the facts. I don't know when Mama will be back. I write to her every day but am careful not to say you're here for fear of making her come home. However if you need to see her[2] do let me know at once because then she would reproach me for not having done so. Otherwise, as she has only been away in the country for five days and will certainly not go away again once she has returned, I shall say nothing. But naturally if her presence is useful or urgent do tell me. No I didn't receive your letter from Liège. And you, did you receive mine in Hamburg? I very much look forward to seeing you again and wish I could be a little surer of being able to these days.

Yours with all my heart
Marcel

Your idea that your visit would be 'improper' in Mama's absence seems to me enchanting and made me laugh a lot. If you were the young man and I the girl it might be different. And yet I went to see you alone in your house in Auteuil. Only it would be too stupid to come when I'm in bed or to put yourself out uselessly in any way at all.

1. Proust had offered to talk to her about her forthcoming trip to America on Bing's behalf (see letter 68, n. 2).
2. On the subject of her sculpture of Dr Proust (see letter 59, n. 1).

74   *To his Mother*

Saturday evening [24 September 1904]

My dear little Mama,

It seems to me that I think of you more tenderly if possible (though it isn't) today the 24th of September.[1] Each time this day comes round, although all the thoughts accumulated hour by hour since the first day should make the time that has already gone by seem so terribly long, the habit of constantly looking back to that day and to all the happiness that preceded it, the habit of regarding everything that has happened since as a sort of mechanical nightmare, means that on the contrary it seems like yesterday and one has to work out the dates to convince oneself that it's already ten months ago, that one has already been unhappy too long, that there are still many long years of unhappiness ahead, that for ten whole months my poor little Papa has ceased to enjoy anything, no longer knows the sweetness of life. Such thoughts are less painful when we are near one another but when, as we two are, one is linked by a sort of wireless telegraphy, whether more or less near or more or less far one is always in close communion, always side by side.

I've had another analysis done; not a trace of sugar or albumin.[2] Everything else roughly the same (I should have a copy of the other to compare the two). My interpretation is as follows. Either the fatigue brought about by not sleeping or lying down much on the boat gave me a touch of albumin and sugar. But in fact I don't think so, because on the whole I had plenty of rest all the same and there was the compensation of dry air. Or else having two meals instead of one brought it on and I don't believe this either because our two meals were very light and very well 'oiled'. In a word I don't believe I was less well at that time than I am now when I'm feeling *unwell*. There remains the hypothesis (but this may be contrary to the principles of physiology – only a doctor could tell us) whereby the sugar and albumin are not eliminated at the moment because I'm getting very little fresh air, and thus making me feel all the more unwell, whereas they were being on the boat (but this may well be impossible and fanciful). Or the sea air may have made perspiration difficult and so obliged the sugar and albumin to pass through the urine whereas now they may be passing through my pores. But there could be one final explanation. The urine I sent was about 26 or 28 hours old, as I had inadvertently urinated twice in the lavatory. So I had to substitute. Those two cancelled times were after a meal. Therefore it's possible that the sugar and albumin were evacuated at that moment. But I don't suppose it happens like that.

I'm still very out of sorts. I came home last night without having an attack and went to bed more or less in the same state as if I hadn't gone out. On the other hand I had a protracted attack, though without spasms, in the morning which obliged me to go back to sleep very late and to get up rather hurriedly which brought on my discomfort again and did no good – *wird später gesagt*[3] . . . it all took a very long time. In an effort to get rid of my constant headache, my sore throat etc. etc., I took a strong dose of cascara and I hope that will set me up again. At all events I'm determined to go through to the bitter end. And as I feel it's dragging on too long like this, now that my nerves are completely restored I shall simply stay in bed one evening and get up the following morning. I couldn't do it this evening because after an attack (even a fairly mild one but which lasted quite a long time) I digest better by getting up to eat. In fact if I'd known I was going to adopt this system I wouldn't have put my hours forward like this, for when I woke up at seven or eight o'clock it was much easier for me to remain in bed than it is now when I wake up at two or three and it seems more depressing to stay in bed. Last night although I dined before five o'clock I didn't go to bed until relatively late (3.30) because I was obliged to open my window as I told you yesterday and stayed up late writing to you. As I had a fumigation in the morning, I went to sleep again until 4.30, and by the time I got up – *wird später gesagt warum*[4] – dinner wasn't ready until 6.30. The cascara hasn't yet had its effect. But I nevertheless have a bit of diarrhoea as when one eats when one isn't well.

I admired Emmanuel Arène[5] for resigning from the presidency of the Corsican General Assembly rather than vote for a bust of Napoleon as a great Corsican in the Hall of Congress. – Yesterday I read an article by Beaunier on the separation [of Church and State] – still the same

jokes, and quotations from Henri Maret, *L'Action*, Ranc[6] etc. A few bits
of society news: M. de Nedouchel staying with the Lignes in the country
etc., and one item where I thought I spotted an error: 'A daughter of M.
Legrand whose sister is engaged to M. Georges Menier' (whereas she
isn't engaged but married). Then I happened by chance to notice that it
was an old *Figaro* of 1903. Everything was the same, and I assure you one
could have read it from beginning to end without noticing anything.
Only the date was different. I'm very well at the moment of writing and
on the whole it's only the contrast with the well-being I felt only a week
ago that irritates me. But I'm not suffering at all. Perhaps I shall discover
the reason by chance. Mme Lemaire writes to Reynaldo to ask him if I've
had 'the courage to go away for treatment etc.', and her daughter to say
that her mother makes life unbearable for her.[7] That's something I can
never say of my dear little Mama! who would be so pleased to see me
un-suffocated as on the whole I am, since my attack this morning was the
only one, and as I say not at all violent; for instance, I sneezed quite a lot
as I used to when I got up in the daytime. You're wrong to attribute
Croisset's remarks to rage.[8] After all Faguet said his play was in the direct
line of descent from Marivaux, which was a hundred times too kind. No,
but it's become smart among the young to vilify Faguet as they used to
vilify Sarcey.[9] And it's very stupid to describe his ponderousness as
ignorance of French. Because it's deliberate and he knows very well what
he's doing. Indeed if there's anything defensible about him it's his style.
I'm enclosing a note from an article by a M. Alfassa in the *Revue de Paris*
not because it (the note) is of any interest but simply because it mentions
Papa. I've been all alone today but Antoine Bibesco sent word that he'll
call round for a moment late in the evening. If I felt well tomorrow
thanks to the cascara I might perhaps go into the country, but I very
much doubt it. Perhaps I catch cold when I get up after being covered up
in bed as I am now. Perhaps I've caught an attenuated version of
Reynaldo's malady (he's very well now). It you read the *Temps* in which
Lantilhac's[10] letter appeared you must have read Mirbeau[11] on Madame
de Noailles. What praise! Please specify, just in case, for the first day of
my recovery, the conditions in which I might come and stay in Dieppe. –
Félicie again went to bed very early, not so early as she could have but
still I suppose about 9.30 – though I didn't look at the time. Marie stayed
for a bit to tell me about her various plans and very sweetly refused to let
me send for Baptiste if I needed to stay in bed one evening, saying that
she'd willingly look after me herself. She told me to tell you that she's
finished a bodice for you. As the front door bell had stopped ringing
(completely stopped) (but perhaps I told you yesterday) we sent for the

electrician in the rue de Monceau, as Mme Gesland[12] didn't know the address of yours.

A thousand loving kisses,
Marcel

P.S. I feel extremely well tonight and am going to bed much earlier although I dined later.

1. His father had been taken mortally ill ten months previously, on 24 November 1903 (see vol. I).

2. Cf. letter 65 to Dr Linossier.

3. 'Will be explained later.' These German interjections were perhaps a case of 'not in front of the servants'.

4. 'I'll tell you why later.'

5. Emmanuel Arène (1856–1908), politician and novelist.

6. Henri Maret (1805–84), liberal and democratic priest and theologian. *L'Action* was a socialist, republican, anti-clerical daily, Arthur Ranc (1831–1908) an influential left-wing politician.

7. Madeleine Lemaire had an only child, Suzette Lemaire (1866?–1946), who became a painter like her mother but never married. Cf. letter 150.

8. Francis de Croisset (real name Franz Wiener: 1877–1937), Franco-Belgian playwright and, like Bloch in *RTP*, an exaggerated anglophile. Friend of Proust, son-in-law of Mme de Chevigné (see letter 134, n. 2) and collaborator of Robert de Flers (see letter 34) who, in this instance, had interviewed him for *Le Figaro*.

9. The style of Emile Faguet (1847–1916), critic and Academician, was parodied by Proust in his Lemoine pastiches (see letter 265) published in *Le Figaro* in 1908. Francisque Sarcey (1827–99), to whom Faguet is compared, was another influential drama critic.

10. Eugène Lantilhac, former schoolmaster of Proust at the Lycée Condorcet.

11. Octave Mirbeau (1848–1917), novelist, playwright and critic.

12. Concierge to the block of flats at 45 rue de Courcelles.

75   From his Mother

Hôtel des Bains, Dieppe
Sunday 25 [September 1904]

My darling,

Your letter only arrived this morning. I'm glad. There isn't another delivery on Sunday and if I had eaten your letter yesterday I would have had to fast today! Don't do anything for *me* about our neighbours.[1] I shall send a card from here. Your stationer for cards is Brou 12 boulevard

Malesherbes.[2] He can give you some in 24 hours (he's very helpful). But I think it would be better this time not to use your block (which is fine for white cards), and without going to the unnecessary expense of a new block, ask them to engrave a hundred cards (which will thus probably cost you 5 frs instead of 3 frs). You are wrong, darling, your letters are *all* charming, *all* appreciated by me as I appreciate my little pet from every point of view. Besides, 'even when the bird is on the move'[3] . . . and again this morning I've gleaned some golden corn. Forain[4] very funny, Tristan Bernard very witty and standing out above all the pretentious nonsense – such as Jules Renard!![5] Roujon's very 'Parisian' patronage of Mme de Sévigné deserved the punishment advocated by your uncle.[6] My darling don't send me anything and calm everyone's zeal and forbid all consignments of any kind. I didn't *choose* the hotel, there was no alternative when I couldn't get into the Royal. There are rooms with chimneys and without central heating. Only the lounge, the dining-room and the corridor on to which the rooms give are centrally heated.

A thousand kisses my darling boy,
J. Proust

Do look after your appearance. If you have to get dressed in the daytime make sure your clothes are immaculate. But above all no more of that looking like a Frankish king – your hair gets in my eyes when I think of you. I hope by the time I finish this you'll have had it done.

On Wednesday the 21st I sent Félicie two hundred francs. You've never mentioned it. I did it *at once* so that she wouldn't find herself short.

1. The Fuchs family, see letter 72, n. 3.
2. Cf. letter 72.
3. Quotation from a poet called Lemierre admired by Mme Proust.
4. Forain's cartoon in *Le Figaro* of 25 September.
5. Jules Renard (1864–1910), author of realist novels.
6. 'Spit in their faces' was what Proust's great-uncle, the irascible but generous businessman Louis Weil (1816–96), used to say.

76   *To his Mother*

[Monday evening, 3 October 1904]

My dear little Mama,

I came home at a quarter past 10 hoping to find you still up (my cabman had had to go for his dinner) so that I could tell you that I was amazed, enchanted, overwhelmed by the splendid column[1] which would be the admiration of everyone *everywhere*, whether at the Rothschilds', at the Murats', at the Ephrussis' or even at Versailles. The extraordinary taste which enabled you to discover it, the incredible ingenuity with which you acquired it at a price accessible to me, the exquisite art with which you restored it, and lastly and most of all the infinite kindness which was the stimulus of your investigations and your labours – all these fill me with wonder and amazement. At the Murats' – that is to say the finest house in Paris[2] – people won't believe it's a present but part of the furniture. I agree it would be better as a plant-stand than as a lamp. But we might as well stick to its declared use. The shade is pretty, especially the part without pictures, but even so it's nothing compared to the column. I'm wrong to say the part without pictures, because the pictures are all right too. The whole thing is far too beautiful, but absolutely right. My outing, which was complicated by a fairly long walk, taxed me somewhat but I think morally rather than physically. I came back sneezing, coughing and full of asthma. I was obliged to go down again, not knowing how to whistle, to tell the cabman I wouldn't be going out again, and the fresh air calmed me a bit. I came up again completely relieved. I sat down to write (please put the enclosed letter in the post with 25 centime stamps as usual) and now it seems to me that all fear of an attack is over (though naturally I can't be certain). At all events the attack that began so strongly has been entirely dispelled, without any need for fumigation. Anyhow, after a long rest in the dining-room I feel my general condition is excellent. Hardly even short of breath.

A thousand loving kisses.

Marcel

Please don't think the expression of my admiration was exaggerated by my gratitude. I didn't dare let Bourcelet[3] see it, but I thought I'd made a mistake, I couldn't believe it could be that!

1. Empire lamp bought by Mme Proust as her son's wedding present to the Albuferas (cf. letter 72, n. 4). Proust was too ill to go to the wedding on 11 October.

2. The reception given on the occasion of the signing of the marriage contract was to take place at the home of Prince Joachim Murat, whose wife, *née* Cécile Ney d'Elchingen, was the bride's aunt.

3. Antique dealer in the rue des Fontaines, where Mme Proust had found the lamp.

77   *To his Mother*

[October–mid-November 1904]

My dear little Mama,

Don't bother with the translation,[1] I've done it myself. You could if you like unravel (orally) the preface to *Sesame*.[2] What we need to decide first and foremost is whether it would be more advantageous for me (assuming I want to take all six series) to subscribe to the six now or to take vouchers for the first series now, and for the second later, and so on.[3] I've worked so desperately hard that I'm only writing you these few lines. I've been *perfectly well* all evening. I haven't yet read Mme Peigné[4] since I was working all the time but I've felt the weight of her manuscript and I can't help admiring such energy. Her little foreword isn't too bad. She has omitted, to please me, her criticism of Ruskin's mysticism. But she couldn't deny herself the pleasure of allowing the tips of her radical ears to stick out. She says 'The French do not share Ruskin's ardent convictions' (what does she know about it?) 'but are ready to admire those who – apart from their worship of the deity – have observed the cult of beauty etc.' Well . . .

A thousand loving kisses
[Unsigned]

1. Mme Proust had previously done rough translations of Ruskin for Proust to work on.

2. i.e. Ruskin's preface to *Sesame and Lilies* (not included in Proust's translation: cf. letter 40, n. 6).

3. Mme Proust's New Year present to her son was to be Ruskin's *Complete Works*, the publication of which, in a series of six sets of five volumes each, had just been announced in London.

4. Translator of Ruskin's *Stones of Venice* (published in Paris in 1906), Mathilde Peigné (*née* Crémieux) was Mme Proust's cousin.

78   *To Gabriel de La Rochefoucauld*

Friday evening [last months of 1904]

My dear Gabriel,

Thank you for your nice letter (just received), which has greatly touched me. Here goes. Everything that was still cloudy, clouded by actual shaking, has 'settled', as one says of liquids, 'cleared'; I now see more clearly what sort of substance we are dealing with and my first impression is if anything reinforced, more favourable, because my interest and emotion have persisted, and I've often thought again about this drama.[1]

I have to make to you:

1) An artistic criticism (a word very improperly chosen, I'll explain to you why). The end, contrary to what you think, is clear and compelling. Only, from the moment of the stabbing, Merrien seems a bit smug. Doesn't he say a bit too often to his mistress: 'I'm not a ninny, you know, I'm brave, you've seen the blood' etc.? Now, a failed suicide retains some grandeur only on condition that the protagonist doesn't think of himself as having been heroic – otherwise he irritates us slightly (I'm exaggerating out of all proportion, but in order to make you see the infinitesimal distinction I'm trying to make, I'm blowing it up ridiculously). Did you want to show that from that moment onwards he has all the traits of the neurasthenic, even the exasperating ones? In that case, though it may be medically profound, is it dramatically sensible to make him say with secret self-satisfaction that he sleeps like a log, that he's crotchety, weird, indolent, etc. (I may be utterly mistaken, it's for you to judge).

2) I also have an amatory objection to make; but this is in no way a criticism, for there are all kinds of psychological truths and possibilities; an author chooses one which may not be ours, and his work isn't an iota less valid for that. My objection is this: Merrien hears his mistress say to him: 'Let's just be good friends'; and this doesn't make him jealous, he acquiesces. And it's later on when he realizes that it's the doctor who has forbidden him to go on being her lover that he becomes tormented by jealousy. That could well be true, and it only has to be *possible* to be artistically *fine*. Moreover it's truer to the underlying scheme, the thematic harmony of the novel: 'The Lover and the Doctor'. But for me, with my particular temperament, my specific jealousy, what would have made me jealous at once is 'Let's just be friends', which I'd have interpreted as an admission of satiety, a mark of indifference, a sign of repulsion. I shall never forget the day when those words were said to me by the mistress I most loved. If I acquiesced, after some resistance, it was from pride. But I considered that from that day onwards her desire for my body had been replaced by desire for another. It was the beginning of our rupture. For years afterwards, years of tenderness and chaste kisses, we never once alluded to that moment, to my caresses, to anything that might have come near to the forbidden spot, the painful scar on my heart. Now, all that has been succeeded by such indifference that I

recently hazarded a jesting advance, and she said to me quite harshly: 'Don't do that, it's wrong.'

But once again your hypothesis is just as true and more in conformity with the general scheme.

Broadly speaking, you shouldn't have so many doubts about the overall design, the dramatic situations – that's where you excel; but on the other hand you should be more difficult to please as regards the 'artistic' execution, the stylistic flavour of the details. Remember that we often appreciate in a phrase what we saw at that moment and didn't put into it. The artist is often like Nebuchadnezzar, to whom the wall presented a vision that he alone could see. The others saw only the naked wall. One must remember that one's phrases often have the same effect on others. Whenever you have to express a moral, intellectual or emotional truth, a flash of observation or irony, your phrases are excellent. In the surrounding landscape the colour is less satisfying, the nuances are more banal. There are some fine big landscapes in your novel, but at times one would like them to be painted with more originality. It's quite true that the sky is on fire at sunset, but it's been said too often, and the moon that shines discreetly is a trifle dull.

Ignore these vain schoolmasterly remarks[2] and remember only my interest and emotion which persist. Bear in mind that you have written a fine and powerful novel, a superb, tragic work of complex and consummate craftsmanship, and believe in my sincere friendship.

Marcel Proust

1. Proust had been re-reading the manuscript of *L'Amant et le médecin*, a novel by Gabriel de la Rouchefoucauld (see letter 10, n. 3), published in January 1905. It seems to have been partly autobiographical, echoing his own pre-marital affair which ended in a suicide (cf. letter 71, n. 7).

2. Proust's criticisms were not ignored.

79   *To Lucien Daudet*

Tuesday [evening, 15 November 1904]

My dearest Lucien,

Disappointed not to see you this evening when you had promised to come. As for yesterday evening, I telephoned you at the Chimays[1] to ask you to come but you had already left, I was told. To tell you the state I was in from half past twelve to nine o'clock this evening would be impossible. You would have thought my last moment had arrived. I still have before my eyes (squint[2]) the vision (squint) of your sister,[3] so ravishing and in spite of what she herself thought so beautifully, elegantly and warmly dressed. She has a sort of youthful impetuosity combined with the wisdom of an old Brahmin who has been thinking for a thousand years, and such intelligence embodied in so ravishing a beauty, that she is an unbelievable person. On the other hand I suffered from seeing Madame Daudet, as I felt that I had the hurried, frivolous, garrulous air of a social caller in the second act of a Gymnase comedy[4] which must have given her an appalling idea of me which I find distressing.

I spent two hours hanging around the *Mercure* waiting for M. Vallette who didn't come back (M. Ban Veber not having been sufficiently explicit) and I ended up by coming to an agreement and catching a cold on the doorstep with M. Dumur.[5] But I forgot your Wells books.[6] I'll write to them. But you must remind me of the titles because I've forgotten (not the titles of Wells's novels but those you want). Tomorrow Wednesday unless you hear to the contrary I shall be here, but you needn't let me know whether you're coming or not.

Dear Lucien, do read in this morning's *Gaulois* the list of Guiche's wedding presents.[7] You know Mme Pourtalès[8] gave a painted ivory. See under what name it's put in the *Gaulois* and tell me what you think their object was in doing that. Was it you (aunt? among the hats of the guests, etc.) or who?[9]

Then again, what can the picture from Verdé-Delisle possibly be?[10] Is it a joke? And the one from Gabriel de L.R., and the one from Fouquières,[11] and the one from Hébert,[12] and the one from Bélugou (who the devil is the painter?). And Madame de Noailles, Fauré,[13] Alexandre Dumas, Montesquiou etc. – all 'their works'. And Coco transformed into Madame Lemaire.[14] Dear Lucien, this catalogue is exhausting and this letter must be very tedious. I'd quite like you to ask Gourgaud[15] this and that. But as I know you won't do it and as moreover if I persist I have other – less nice – outlets, it seems pointless.

See you tomorrow perhaps.

Tenderly

Marcel

1. Prince and Princesse Alexandre de Caraman-Chimay.
2. *Louchon*: private term for something that makes you squint, because of its tastelessness or vulgarity.

3. Edmée Daudet (1886–1937), then aged nineteen.

4. The Gymnase was a theatre for light comedies.

5. Proust is describing a visit to the editors of the *Mercure*, the review brought out by his publishers, Mercure de France; by M. Ban Veber he means Adolphe van Bever. Louis Dumur (1864–1933) was a novelist also on the staff.

6. Eight novels by H. G. Wells were being published in translation by Mercure de France.

7. Headed 'A Grand Wedding', the list took up two columns. Cf. letter 82.

8. Comtesse Edmond de Pourtalès (*née* Mélanie de Bussière). Second Empire beauty, former lady-in-waiting to the Empress Eugénie, she was a legendary hostess who appears thrice in *RTP* under her own name. In the *Gaulois* report, her present was attributed to 'Mlle de la Robertsau', presumably because she owned a château of that name near Strasbourg.

9. The *Gaulois* reported the fiancée's aunt, Comtesse Ghislaine de Caraman-Chimay, as being dressed in turquoise velvet with a gold lace cap.

10. See letter 82.

11. See letter 289, n. 2.

12. In its list of presents, the newspaper cited: 'M. et Mme Hébert, tableau de la Vierge'. The wife of Ernest Hébert (1817–1908) was Proust's model for Elstir's wife in *RTP*; her husband's academic painting is admired by M. de Norpois who refers to 'Hébert's Virgin' (II, 229).

13. Gabriel Fauré (1845–1924) composed a *Tantum ergo* for the nuptial mass and gave the couple the manuscript.

14. A reference to Proust's own, bizarre, choice of present: a revolver in a leather case inscribed with the bride's childish verse (see letter 50, n. 4) and decorated by 'Coco' de Madrazo – not by Mme Lemaire, as reported. Cf. letter 82, Proust to Guiche.

15. Napoléon Gourgaud, man-about-town, member of the Jockey Club, etc., great-grandson of Baron Gourgaud, one of Napoleon's companions on St Helena.

80 *To Robert de Montesquiou*

Sunday evening [20 November 1904]

Dear Sir,

I have received (not from you, I think) an invitation to a lecture of yours.[1] It is due to take place on a day that is for us a painful anniversary[2] with which the brief potential of meditation, fervour, perspicacity and detachment afforded by an hour of discourse with you could not but accord, especially one of those discourses to which you will doubtless bring all your most brilliant and prestigious qualities, although in fact I have never seen you, even on your least exceptional days, save at your maximum of intensity and your highest degree of spiritual magic. Such an occasion, by bringing me miraculously closer to the best of myself, could only make my sorrowful memories more efficacious and more fruitful. But I must not think only of myself. And I know that Mama would be distressed if I were to seek on that day a pleasure which would be all the more intense for being an intellectual one and in an assembly which will inevitably be the most dazzling since it is you who will be speaking. So I shall not go, although I'm conscious of what I shall be missing, and try in vain to imagine what you will say. I hope it will be published. But I shall not have seen 'the monster himself'; and since the Christian Sappho, 'the expectation of the ridiculous disappointed' at la Bodinière,[3] I have never heard you in public.

I wish you, dear Sir, a resounding success and I beg you to accept the respectful regards of your admirer

Marcel Proust

My best wishes to Yturri.

1. On Japanese colour prints, then much in vogue.

2. The lecture was to be given on Tuesday, 22 November; Proust's pretext was that this was too close to the anniversary of his father's death (cf. letter 74).

3. Montesquiou's first public lecture, at the Théâtre de la Bodinière in January 1894, was on the French Romantic poet Marceline Desbordes-Valmore (1786–1859) whom he called 'the Christian Sappho'. A line from one of her poems may have given Proust a vital name for *RTP*: 'Je veux aller mourir aux lieux où je suis née; / Le tombeau d'Albertine est près de mon berceau . . .' ('I want to die in the place where I was born; / Albertine's tomb is close to my cradle . . .'). Montesquiou had come attired in sober black, knowing that his audience expected him to be dressed outrageously. Cf. letter 159 to Robert Dreyfus.

81   *From Robert de Montesquiou*

Neuilly [21 or 22 November 1904]

Dear Marcel,

I do not in the least share the sentiments you express on the subject of anniversaries. Not that I mean they should be ignored; quite the contrary! But I don't think there can be a better way of celebrating them, and one more in keeping with the inclinations of those who are no more (or who, rather, are elsewhere), than not to turn one's back on a new occasion to thrill with harmony. And I suspect that your good and intelligent mother, if you had opened your heart to her on the matter, would have been of my opinion.

But there (and this is the truth of the matter), you claim to possess *all my vocal secrets*.[1] Allow me to disabuse you, or at least persuade you that a certain something will always be missing from these phonetic reproductions ... An illusion which may perhaps give me some hope of seeing you again!

R.M.

P.S. The death of Madame de Brantes grieved us *very much* in spite of everything. Whereas that of Madame d'Eyragues leaves us *absolutely* indifferent.[2]

1. Proust was a brilliant and entertaining mimic, Montesquiou being a favourite target. See vol. I.
2. In fact, Mme Roger Sauvage de Brantes (*née* Louise de Cessac: 1842–1914) and Mme d'Eyragues (*née* Henriette de Montesquiou-Fezensac: 1863–1928) were among Montesquiou's living relations; but in his eyes people with whom he had quarrelled, as in this case, were as good as dead.

82　*To Armand Duc de Guiche*

Wednesday [23 November 1904]

Dear friend,

Thank you for your post-card to which I venture to reply at somewhat greater length. What effrontery to proclaim that I do not know the church at Moret! I know its portrait by Sisley which is the finest thing he ever did.[1] There are some wonderful things to be read about Fontainebleau. I daren't advise you on what to read at the moment though these are admirable for reading together[2] – but on your return. Some very extraordinary things have happened since your departure which I shall tell you about.

Dear friend, watch out for your Ms. They're beginning to resemble Ms.[3] You must remember that their *raison d'être*, their charm, their special attraction is precisely to be Ns. I am sad for you about this cold weather that prevents you from enjoying

The mellow golden rays of the waning year

all the more so because Baudelaire says one must enjoy them with one's 'forehead resting on the beloved's knees'.[4]

Personally I find cold weather an even more poignant background for happiness. Happiness numbed by the cold, forced to draw in on itself, to go back into its shell, is for me the most intense. It's true that I only experience it through sadness. But it's still the same thing. The newspapers have been saying laudatory things about you. But wouldn't you prefer to be praised by me! And since you were going to be talked about why didn't you entrust me with your press rather than people who see you as 'pale, with a gentle smile'?[5] I've never seen you as pale or as gentle as all that. On the contrary I should have thought your brother was more gentle.[6] But unlike the sportsman reporter I've never seen you on horseback. Has Madame la Duchesse de Guiche any idea of all the beautiful things you say about her when she isn't there? Madame Lemaire wrote to me (because apparently one newspaper put in the list of wedding presents: M. Marcel Proust, revolver case painted by Madeleine Lemaire[7]): 'My dearest Marcel, it appears that I've painted for you a revolver case for the wedding of the Duc de Guiche. How little trouble it gave me!' I don't know which newspaper it was; *Le Gaulois* which I saw didn't mention this superb present! But on the other hand, what a marvellous list they gave:[8] 'M. Verdé-Delisle, picture entitled *Entry into the Stable.*' Is it supposed to be witty? And Mme de Pourtalès's painting attributed to Mlle de la Robertsau. If you write to Loche[9] don't forget to give him affectionate messages from me. I shall probably see him before you do. But I like these exchanges. He talks about you with a really touching tenderness. On your wedding day Madame Greffulhe recited to me some sublime lines of her daughter's, but I haven't got them. But I still enjoy reciting to myself

And like an ageless tapestry[10]

and those other marvels. I told Madame Greffulhe that you had envisaged your marriage (among other things) as a possible means of obtaining her photograph. She laughed so prettily that I felt tempted to repeat it ten times over. Would that my friendship with you might earn me the same privilege.[11] As for your own photograph, it wasn't very nice of you to promise it and not give it, and likewise the photograph of your portrait of Mathieu.[12] I've written to d'Albu to ask him to do the necessary about the Aeolian.[13] I hope he won't plunge out of despair into

The voluptuous sea where the sirens sang[14]

for he is there at the moment. I hope it wasn't indiscreet of me to write to you. I was missing you. And I regarded the absurd post-card as an authorization.

Yours ever
Marcel Proust

1. *L'Eglise à Moret* by Alfred Sisley (1839–99), now in the Musée d'Orsay in Paris.

2. i.e. on their honeymoon which the young Guiches were spending at a family château between Moret and Fontainebleau.

3. Proust is amusing himself at the expense of Guiche's handwriting.

4. 'Ah! laissez-moi, mon front posé sur vos genoux, / Goûtez, en regrettant l'été blanc et torride, / De l'arrière-saison le rayon jaune et doux!' Baudelaire, 'Chant d'automne'.

5. His 'profile' in the *Gil Blas* (cf. letter 63, n. 5) read: 'He's a tall young man, elegant and slim, with restrained gestures and a very gentle smile.'

6. Comte Louis-René de Gramont, described in the same article as a more prudent horseman than his brother.

7. Proust's present was listed in *Le Figaro*. Cf. letter 79, n. 14.

8. Cf. letter 79, notes 7–13.

9. Prince Léon Radziwill (1880–1927), nicknamed 'Loche', became a lifelong friend. His character and marital affairs are discussed in letter 228.

10. 'Et comme un canevas ne change jamais d'âge', from a poem written aged five by Elaine Greffulhe (now Duchesse de Guiche). Cf. letter 50, n. 4.

11. In *The Guermantes Way* the narrator uses the same ruse with Saint-Loup to obtain a photograph of the Duchesse de Guermantes, a character partly modelled on Mme Greffulhe (*RTP*, II, 77–8, 98–102).

12. Guiche's portrait of Mathieu de Noailles (Mme de Noailles's husband) was shown at the Salon des Artistes Français in May 1904.

13. The Aeolian Company in the avenue de l'Opéra was selling the new mechanical piano, the pianola. Cf. *RTP* (III, 378–80, 388–90) where Albertine plays a pianola in the narrator's room.

14. 'La mer voluptueuse où chantaient les sirènes.' Anatole France, 'Leuconöé', in *Les Noces corinthiennes*. Albufera's despair was at leaving his mistress Louisa de Mornand. Cf. letters 45, 46.

83   *To Robert de Montesquiou*

Friday [25 November 1904]

Dear Sir,

If I haven't thanked you sooner for the trouble you were kind enough to take to write me that beautiful letter and guide me in that Meditation, it's because I was hoping that my strength, which seemed to have recovered, would enable me to find you at the House of the Muses,[1] who have deserted me now that I no longer see you, and a rare letter only aroused a deeper desire to renew a delightful, invigorating and long interrupted intercourse. Unfortunately, I've never been so ill as I have been since I thought I was well. It's now a fortnight since I rose from my bed. My time-table will find itself back at its worst. And if I can venture out again it will be at an hour when you will be sleeping. However, I intend to get some treatment, and perhaps be cured. I've been postponing for many years the enjoyment of the divers pleasures of life. And there is none more precious to me than affectionate converse with a great mind. As for the hypothesis you hazard on the subject of my 'Imitations',[2] they were never more than scales, or rather vocal exercises, making no claim to convey a melody or anything of the genius of the original; better still, they were simple exercises and admiring games, *ludi solemnes, festivolia,* naïve canticles in which a youthful admiration exercised and indulged itself –

> O Robert, O my benign
> Master sublime,
> From you I am parted, etc.[3]

– of which Renan has left us the childish and charming model. I hope the lecture[4] will appear in print and meanwhile I'm happy to see, notably in the nice article in *Le Siècle*, what a success it had.

Your respectful
Marcel Proust

Unfortunately since the days I speak of my voice if not my heart has changed – and I no longer know how to *imitate* you! since that is what the presumptuous verb means.

1. Montesquiou called his house at Neuilly the Pavillon des Muses.

2. See letter 81, n. 1.

3. 'O Robert, O mon aimable / Maître affable / De toi je suis séparé . . .' Proust paraphrases and parodies a 'naïve canticle' beginning 'O Joseph . . .' quoted by Ernest Renan (1823–92) in a book about his childhood. Renan was the subject of one of Proust's 'Lemoine pastiches', see letters 265, 272.

4. Cf. letter 80.

84   *To Fernand Gregh*

Sunday [27 November 1904]

Dear friend,

I have received your admirable book[1] and I thank you most warmly
for remembering me so loyally. I knew almost all of it but I've read and
re-read it without being able to put it down. This 'sombre and limpid
tête-à-tête' with Victor Hugo is the only sort of criticism where the
writer doesn't give the impression of looking at the poet 'from below';
that is the only true criticism, the only criticism worthy not only of the
poet whom you read and underline for us, but of the poet you yourself
are. Thus when you refer back to the Victor Hugo of your *Clartés
humaines*[2] it isn't at all like when critics who have 'written a little verse' –
Sainte-Beuve, Bourget, Lemaître – 'place' in the middle of their prose a
few lines which 'one of their friends' 'scribbled one day'! It's genuinely
sincere, sincere with not the slightest affectation in the style. The
delightful article on Verlaine is a little more precious – though exquisitely so – 'passionate pilgrims', 'Gallic friendship' – there's plenty of wit
in it, always charming. But it's a trim, well-tailored article, a Parisian
package. For Victor Hugo, deliberately, like Rodin, you haven't
knocked the rough edges off your block. Not that there isn't an
abundance of wit, 'There's Homais in Homer'[3] is excellent, and 'well
roared, lion', and plenty of other examples. Your remarks on the
influence of Sainte-Beuve[4] 'with tears!' are of an exquisite literary
subtlety. I was stunned by the prototype line for

A trackless ocean through which galleys sped[5]

A palpable hit! On the other hand I'm more doubtful about the
relationship between

Clothed in simple probity and white linen

and

The restful shadow of her linen coifs[6]

I might perhaps reproach you for a few too many phrases such as
'since magus he was' or 'spurn one's own pleasure' which recur several
times. Once a thing has been well said it's perhaps as well to leave it
there. Similarly with René in London and *Faust* anglicized.[7] But where I
don't entirely agree with you is about Maeterlinck. Of course I know
those people who feel the necessity to talk about miracle and mystery. In

the Preface to *La Bible d'Amiens* I've already tried to puncture a few bladders of a slightly different kind. And I've done a study on Maeterlinck which isn't yet finished in which I say a few things of the sort but not with the intention of criticizing him for that. Anyhow you'll see.[8] But what separates us is what you say about chance or luck; at any rate your dialectic doesn't convince me.

You say (1) this explanation explains nothing since instead of its being Paul or Jean, it's the self of Paul or Jean that has luck. – Not at all, the self of Paul or Jean doesn't have luck, it has a divination of the future which isn't at all the same thing. Moreover you say it yourself in your (2) which knocks down your (1): if there is a self of this kind within us its flair, its divination, would no longer be the work of chance but the acts of a hidden reason.

And then there remains your second objection that this explanation removes the problem. Obviously, but it's the only way this kind of problem is ever resolved. One could only introduce chance into a rational conception of the universe by making it reasonable. – And that doesn't mean that I believe in chance. Only, I believe in a lot of things that are no less extraordinary and belong to the same category. And besides it seems to me to be part of a sort of system with Maeterlinck without his having allowed himself to be seduced by the words mystery etc. But I'll try and put this better in my article if I ever complete it. You are marvellous, you know everything, you understand everything, you create everything, you express everything. You mention Monticelli[9] twice with aptness and charm: Monticelli and Delacroix (Verlaine), and à propos of *La Fête chez Thérèse*.[10] There are poems of Hugo's in which the lines you quote are not always the ones I like best. But I have such a great respect for your instinct that I'm certain that I must be wrong in these cases. But nearly always you quote the ones I like and that fills me with a sterile pride. I hope you're well. I cannot believe that if you weren't extremely well you could produce so much so powerfully and scatter the seeds of future harvests. We all owe a great debt of gratitude to Dubois[11] who did you so much good.

Ever yours

Marcel Proust

From time to time I have misgivings. I wonder if the great freedom of execution of this book, its almost spoken quality, doesn't slightly deprive it of 'rarity', of textural refinement. But on the whole I feel that for that very reason it's even more a piece of free criticism of a great mind, the rough sound of something that will last.

1. A collection of Gregh's articles on Victor Hugo, Verlaine, Schumann, Massenet, Debussy, Maeterlinck etc. which had already appeared in various magazines.

2. Gregh's poem 'Rêve', in which Hugo appears to a young poet in a dream, published in the collection Proust mentions below, *Les Clartés humaines*.

3. Writing on Hugo's *Lettres à la fiancée*, Gregh says: 'Even if I perpetrate the sort of lame witticism I deplore, I am bound to say that these are the letters of a young, amorous Homais [i.e. the naïve progressive chemist in *Madame Bovary*] . . . to which Hugo would surely retort, with one of his usual monstrous puns, there's a Homais in Homer.' (*Op. cit.*)

4. The subject of Saint-Beuve's influence was to be the starting point for Proust's *Contre Sainte-Beuve*, begun in 1908. Cf. letter 315, n. 1.

5. Gregh quotes verses by Hugo which have influenced other poets, giving Hérédia's lines on Cleopatra as an example: '. . . Vit dans ses larges yeux étoilés de points d'or / Toute une mer immense où fuyaient des galères.'

6. i.e. between Hugo's 'Vêtu de probité candide et de lin blanc' and Verlaine's 'L'ombre douce et la paix de ses coiffes de lin'.

7. Gregh says of Byron's poem set to music by Schumann: 'Manfred is René [i.e. Chateaubriand's eponymous hero], René in London, with a little Shakespearean blood on his fingers perhaps . . . *Manfred* is an anglicized *Faust*.' (*Op. cit.*)

8. Proust's study of Maeterlinck has not been found. Cf. letter 28.

9. Proust greatly admired the Marseillais painter Adolphe Monticelli (1824–88). Cf. letter 311, n. 1.

10. Gregh writes: 'In his *Fêtes galantes* . . . Verlaine is to the Hugo of *La Fête chez Thérèse* what Monticelli is to Delacroix . . .' (*Op. cit.*)

11. See letter 71, n. 2.

85   *To Antoine Bibesco*

Friday evening [2 December 1904]

My dear Antoine,

At ten o'clock last evening you told me you would tell me instantly the indiscretion of which you claim I was guilty eighteen months ago – if I could find out the name of the woman to whom you had been 'giving the glad eye' during the last two months. You gave me as much time as I wanted and specifically granted me a fortnight.

Well, this evening at six o'clock, before I was even dressed, I sent you a telephone message with the name of the lady in question which I can't remember for the moment but which is something like Lang; anyway I can give it to you precisely. I therefore expected to receive a word from you telling me what the indiscretion was with as much promptitude as I set about gleaning one (not provoking, gleaning). Astonished at not having heard by nine o'clock, I telephoned you and you had gone out. Finally I went round to your house at half past ten and left a message saying that I would be at home until midnight. Still nothing. Under these circumstances, if you don't comply forthwith it's all over between us. But don't come and wake me up about it, if you please.

Yours

Marcel

86 *To Bertrand de Fénelon*

Monday [5 or 12 December 1904]

My dearest Bertrand,

I'm ashamed of myself for not having thanked you for the splendid reproductions which are doubly precious to me because of my admiration for Vermeer and my friendship for you.[1] Please don't imagine you are not often in my thoughts. I even wrote to you the other day to ask your advice. But I was afraid you might think I was indulging in puerilities which are quite incompatible with my age and which you already found ridiculous when you were the object of them. All the more reason not to bother you with matters of which you have no knowledge. So I tore up the letter. And if I allude to it now it's to show you that at the first upset, the slightest uncertainty, my thoughts turn to you and regret not having your sound and affectionate advice.

And you yourself, dear Bertrand, must at the moment be undergoing all sorts of conflicts of feeling. I fear that your desire to return to Paris may have been postponed by them. And I regret this for my own sake because I would have been very happy to see you. The year will soon be over. I hope with all my heart that you will begin the next one more happily than I shall, and I send you, my dearest Bertrand, my fondest regards.

Marcel Proust

1. In October 1902, Proust visited The Hague with Fénelon (cf. letter 4, n. 6) and discovered the works of Jan Vermeer (1623–75). In 1921, like the dying Bergotte in the famous passage in *RTP* (III, 184–6), he was to get up from his sick-bed to look again at Vermeer's *View of Delft*, the picture he considered the most beautiful in the world (it was in a visiting exhibition at the Jeu de Paume in Paris) and, also like Bergotte, was intensely moved by 'the little patch of yellow wall'.

87 To Antoine Bibesco

[Between 1 and 10 December 1904]

My dear Antoine,

Since you ask me to tell you things of this kind here is another I've just thought of. I've told you a hundred times how tactless if not unkind it was of you to talk perpetually in my presence about my articles in the *Figaro*.[1] On the first occasion you tortured me by speaking about the one on the Princesse Mathilde.[2] You promised me never to do it again. Shortly afterwards, in front of Mme Cahen,[3] you repeated the offence more gravely with the one on Montesquiou.[4] The other day in front of Lucien you talked openly about the one on Mme Greffulhe.[5] This evening you started again in circumstances which certain things you don't know about made particularly disagreeable to me. But what you did know was that you embarrassed me a great deal. It's not for want of telling you. I'm not saying this out of ill humour but because you ask me to point out what you call your defects of character. You complain about often meeting with hostility, but it's because you yourself are so often hostile. And you have certainly not met with hostility from me. But obviously I would eradicate it more wholeheartedly in others if I had more confidence in *your* heart. You are excessively touchy, you get angry about things that are of no importance and you constantly do things to people which if your standards were applied would call for murder at the very least. One doesn't hold it against you because one isn't like you but one ends up by getting discouraged and by no longer being such a good Samaritan. You realize that I'm telling you all this in a general sense because I feel that if you could mend your ways (not with me but with your future friends) it would be a great boon for you, first of all because you would become better, more admirable, more sensitive, but also from a utilitarian point of view. Don't change towards me. One doesn't correct a play. One writes a better one by learning from the faults of its predecessor. Make new friends and don't fall into the same bad habits (I shall use this development again).

Tibi

Marcel

Thank you for coming this evening. I think that if you invite Loche it would be as well to invite his mistress. But as he's at Le Havre you ought to write to him at once.

I should also like to point out the distaste one eventually has for doing anything whatsoever with someone who seriously thinks that one *shakes hands* with the head waiter *because* one hasn't any money!

Today I shall sleep excessively late. I've come home in a state of veritable intoxication at the face of Bernstein's[6] *ravishing* mistress glimpsed at Weber's.[7]

1. i.e. Proust's pseudonymous 'salon' articles.

2. See letter 2, n. 2.

3. Either Comtesse Louis Cahen d'Anvers, *née* Louise de Morpurgo, noted for her elegance, or Mme Albert Cahen d'Anvers, *née* Loulia Warshawsky, friend of Maupassant and of novelist Paul Bourget. The Cahen d'Anvers were a new dynasty of bankers.

4. See letter 25, n. 7 and letter 27.

5. The article on Mme Greffulhe's salon did not appear, at her request.

6. Henry Bernstein (1876–1953), prolific, sophisticated boulevard dramatist.

7. The *fin-de-siècle* Café Weber, meeting place of intellectuals and socialites, was in the place de la Madeleine.

88   *To Louisa de Mornand*

[Monday, 5 December? 1904]

My dear little Louisa,

I'm writing to you only a brief note to say two things. The first is that if a man comes round to measure your small settee you should give orders for him to be let in even if you are out, and tell Rachel not to take him for a burglar, as he's an upholsterer for the cushions.[1]

The second object of my letter is to say that I can't tell you how much I am wilting and languishing from not seeing my little Louisa. I have just been rather seriously unwell and for that reason am bedridden. I'm better now, but will very probably still be in bed tomorrow, Tuesday, evening because I'm going out on Wednesday. And my outings now put me in such a state each time that I have no hope of being in a fit state to get up or to utter a squeak on Thursday and perhaps Friday. But around Friday or Saturday I shall ask my dear little Louisa if there is any possibility of receiving a visit from her whenever it suits her. We have so many things to say to each other since we last met and none so important or so true as this – that I love her most tenderly with all my heart.

Marcel Proust

1. The cushions were a present from Proust. Rachel was Louisa's maid, cf. letter 21, n. 1.

89   *To Robert de Montesquiou*

Wednesday [evening, 7 December 1904]

Dear Sir,

It was very nice of you to ask me to that reception.[1] If I haven't written to you before it's because, having made a rare excursion this afternoon to go and see my doctors, I had hoped on leaving them to go and thank you personally at Neuilly. Unfortunately I didn't get away until seven o'clock. I shall have a splendid account from Reynaldo[2] one of these days which will accentuate my regrets. Ar : to think that it's perhaps the only time you'll invite me! My appointments with these doctors had already been made and as I was unwell it did me much more harm to go out than they will ever do me good. I have heard, with the vague distortion rumours undergo on their way to my sick-room, that you might be giving some lectures in Italy. In spite of what you've said about the greater efficacy of evangelizing an unaesthetic country like America, I think the 'Italian lectures' will be even more prestigious. The really unaesthetic country is not the one which was never fertilized by art, but the one which is covered with masterpieces that it doesn't know how to cherish or even preserve and which leaves its Tintorettos to fade gradually in the rain when it doesn't repaint them entirely, which destroys stone by stone its most beautiful palaces in order to sell the pieces, either very dear, out of cupidity, or for nothing, out of ignorance of their worth. The really unaesthetic country is not the virgin land in which art lives at least through a genuine desire for it, but the dead land in which art has ceased to live, through satiety, indifference and incomprehension. And I am sure your Epistle to the Romans will be no less splendid than your Message to the Church of Philadelphia.[3]

Your respectful

Marcel Proust

1. An afternoon reception at Montesquiou's Pavillon des Muses for the Italian writer Mathilde Serao.

2. According to the report in *Le Figaro* Reynaldo Hahn had delighted Montesquiou's guests by playing some of his works.

3. A reference to Montesquiou's American lecture tour in 1903. See vol. I.

90   *To Armand de Guiche*

Monday evening [12 December 1904]

Dear friend,

One is always having to thank you because you are always so nice. I received the beautiful photograph[1] – a very good likeness, very precious for correcting the memory of someone as forgetful as me. I hope you will come back soon. It's true that I may perhaps be away.

Ever yours

Marcel Proust

If you haven't given my regards to Loche it would be wiser not to do so. There's a risk that they might strike him like the rays of one of those stars which reach us when the star itself has already ceased to shine (if my astronomy is accurate, Duc de Guiche, you who were a doctor of science at fourteen, member of the Institut at nineteen, etc.[2]). I hope however that it is still only a cloud that is at the moment obscuring the aforementioned friendship, until its final extinction. But it's an extremely thick cloud. I should hate you to think that I'm putting on the coquettish airs of a fickle friend. I felt such a desire for stability as far as Loche was concerned that I'm always as dull as ditchwater with him. I wish you could see me. But he is so vile to me that it can't last much longer. And I assure you that it's with a melancholy heart that I shall say to myself: 'Another lemon squeezed' (Barrès), the more so because I had such a taste for that one.

Albu is re-doing the journey of *L'Itinéraire*.[3] Let's hope it will be Chateaubriand's impressions rather than Julien's[4] that he will bring back from it. You're going to tell me that you don't know who Julien is. But as the postscript to this letter is already as long as an ordinary letter I shall postpone those explanations until another time. And in the meantime Albu will have finished the voyage of Ulysses and Hannon[5] and[6]

1. See letter 82. The Guiches were about to return from honeymoon.

2. Proust anticipates, or exaggerates; the Duc de Guiche got his degree in science in 1902, aged twenty-three, and would get his doctorate, for a thesis on aerodynamics, in 1911.

3. The d'Albuferas were on honeymoon, travelling, like Chateaubriand in his *L'Itinéraire* (1911), from Paris to Jerusalem.

4. Chateaubriand's servant, Julien Potelin, had also written about their travels; his manuscript had just been published in facsimile.

5. Hannon the Great, Carthaginian general of the third century BC.

6. This letter, unfinished, was found amongst Proust's papers.

91   *To Robert Dreyfus*

Friday [16 December 1904]

Dear friend,

I received this morning – I don't know whether it was you who had it sent to me – a copy of *Les Essais*.[1] And I re-read those delightful pages, with their exceptional elegance, their sustained lucidity, even in the deepest waters. It's really quite remarkable. With a kind of Stendhalian curiosity and fondness for social details, you collect, ssemble, concentrate the whole diversified flavour of your subject. And in evoking his system you are prodigious. And in many other qualities. I am now a Gobinian. I think only of him. Do you know whether Prince Edmond de Polignac[2] knew him or wrote about him? Little as I know of Gobineau it seems to me impossible, if a ray from the one chanced to stray under the eyes of the other, that they weren't quite specially affected by it. I sense mysterious affinities here.

As for nationalism, especially 'integral' nationalism, I confess I'm extraordinarily ignorant on the subject. So that I don't know who you're referring to, or whether in fact in spite of your protests you aren't making an accusation of plagiarism. But against whom? Barrès? Léon de Montesquiou?[3] Maurras?[4] How awkward it must have been for you to have to pronounce so many inverted commas. I can imagine how Gobinism is pronounced, but 'gobinism'? Do you think it's good grammar to say: 'This course will be collected in book form'? In any case it's of no importance. I find the demonstration which leads you inexorably from Gobineau the historian to Gobineau annexing literature a marvel.

Fondest regards
Marcel Proust

1. The December number of the magazine *Les Essais* carried a reprint of a lecture given by Robert Dreyfus (see letter 30, n. 3) on the élitism of Gobineau (Comte Joseph-Arthur de Gobineau: 1816–82), writer, diplomat and traveller, who argued, in his 'Essay on the Inequality of Human Races' (1853), that the continuing strength of the Aryan race depended on its racial purity. The lecture was part of a series published in 1905.

2. Prince Edmond de Polignac (1834–1901) could have known Gobineau.

3. Comte Léon de Montesquiou, Robert de Montesquiou's cousin and notorious anti-semite.

4. Charles Maurras (1862–1952), poet, critic and political theorist, later a right-wing journalist and co-editor with Léon Daudet of the extremist monarchist and anti-semitic paper *L'Action française*.

92 *To Georges Goyau*

Sunday evening [18 December 1904]
I'm writing to you only this evening because, having been very ill, I
rested all day and only read *Le Gaulois* this evening.

Dear Sir,

There is something miraculous in the generosity of your gesture
towards me, which raises a delightful new crop of praise on the sparse
field already exhausted for more than a year.[1] I thank you for it with all
my heart. How pleasant for me to think that I have introduced Georges
Goyau to John Ruskin or at least made him a little better acquainted with
him; I am quite aware to what contingency of pretext or occasion my role
as intellectual go-between amounts to here. And doubtless the soldier
who held the bridle of the great man during an historic battle, or the one
who handed him the missive which was all-important even though he
had done no more than deliver it, caught as they are in a sense in the
'snapshot' of the event, have not for that reason played a very considerable role; but it is excusable that they should be thrilled by it and that
they should enjoy telling the tale. And doubtless you knew Ruskin
perfectly well before me. And there is a tender Muse, an admirable
interpreter close by you who better than anyone else could have taught
you to love him.[2] But your kindness towards me, and the desire to give
me a flattering public sign of it, may perhaps have made you dwell longer
on this *Bible of Amiens* and find in it as it were a motif and an orchestral
tonality particularly auspicious for the exposition of the fine thoughts
which are so characteristic of you and which make this morning's article
in *Le Gaulois* so superb. Whatever you speak of you dominate, grasping it
firmly and in its entirety, then quickly going beyond it, lifting it on to a
higher plane. Ruskin's theories find in your article not only their
consecration but their completion, a completion which he would probably have been incapable of giving them himself. I suspect he knew
nothing of those German churches of which the memory and the
evocation provided you with the striking image which compares and
separates (like the damned and the elect on 13th-century tympanums)
the tourists who stop and the faithful who kneel. No doubt his ideas on
the 'Genius of Christianity' were extremely vague, and if in the end his
entire work seems like the illustration of the truths revealed by M.
Mâle,[3] historically he was ignorant of them and might very well on
points of fact have sided with Viollet-le-Duc[4] for whom he had an ardent
and ill-informed admiration. There is nothing one can say about the
idea of his book that sums it up better, while clarifying it (which it badly

needs) than your definition: A Bible and a Chronicle. The whole of your introduction which heralds those two words is superb.

You know how much I admire Ruskin. And since I believe that each one of us has a responsibility for the souls he particularly loves, a responsibility to make them known and loved, to protect them from the wounds of misunderstanding and the darkness, the obscurity as we say, of oblivion, you know with what scrupulous hands – but pious too and as gentle as I was able – I handled that particular soul. I was thus profoundly gratified, even more than by the words that were so kind to me personally, to see how highly you placed (and, with the great authority and the power of contagion that adheres to your judgments, compelled the whole world to place) Ruskin – to have been able to win for him such a great mind, and have him honoured and served by so great a talent. And as for the words which are addressed to the translator, they are those most calculated to go directly to my heart, the path to which in any case you have long since found, for they say of this translator that he was tender, that he was deferential, and that he was pious. The only true joys that life has given me have come through respect, when the balefulness of circumstances, or the sterility of my heart, have prevented them from coming to me through love.

Please convey, dear Sir, my most respectful and friendly regards to Madame Goyau, and believe me, your most sincerely admiring and grateful

Marcel Proust

1. Goyau had published a long front-page article on Ruskin in which he writes of 'the distinguished translation by M. Marcel Proust, a true labour of art . . .'.

2. A reference to Mme Goyau, *née* Lucie Faure, elder daughter of President Félix Faure and sister of Proust's childhood sweetheart Antoinette (see vol. I).

3. Emile Mâle (1862–1954), authority on Christian art and architecture greatly admired by Proust.

4. Eugène Viollet-le-Duc (1814–79). Architect and writer, famed as the great conserver and restorer of medieval architecture. Many of his restorations have been severely criticized.

93   *To Robert de Montesquiou*

Monday [19 December 1904]

Dear Sir,

I shall not be able to forget that kindness, that privilege, that appeal, that election, that 'absolution'; nor the painful feeling of involuntary awkwardness that illness imposes on gestures of gratitude, turning the purest of effusions into the clumsiest refusal.[1] My only hope is in your perspicacious intelligence which instantly translates the most seemingly inadequate words into their sublimest meaning, like that Anatole France heroine who was left cold by all the homage paid to her until the day when a suitor was so agitated as to reply to a question of hers: 'Yes, *sir*.'[2]

Up to now I was ill. Now I am ill again, but in addition I am undergoing treatment. The combination of the two is more than human strength can bear. Hence unexpected outings punished by a fortnight of fever are my habitual lot, until my departure for some home for neurasthenics. So what shall I be able to do on Tuesday? Probably not come, or for an hour only. And since I must obviously say something definite at once, I shall give up my share of eternal salvation rather than leave you in doubt and risk that the pangs of uncertainty which I shall experience transmit themselves to you, in the quite different form of a fairly understandable irritation which will result in my final punishment. And yet, I should have been only too happy, as someone who, by dint of having ears and hearing not, is afraid of never being able to listen again, to apply them at last to their true purpose and to hear your wonderful voice

> Which, to strike our ears the more
> Rises to a city's roar[3]

And what a masterpiece you must be planning to read, that you should deem it worthy of your devoting to it the chant of the officiant and the powers of the psalmodist![4] What can it be: Xenophanes[5] or Novalis[6]? And to *appear*, finally, in such a select circle, so carefully planned, so distinguished, so dazzling. You tell me that you may perhaps call round to see me. I refuse to display too much naïve delight by taking that literally; otherwise I should tell you not to do so as I am always with the apothecary when I am not at the doctor's.

I shall come and bring you my good wishes for the New Year, 'for we are good folk who love the old customs' (Joubert[7]) and ask you if you will consent to come out one evening in Paris wherever you like so that I may briefly take up the magic chain once more.

Your respectful friend
Marcel Proust

1. Proust having refused an earlier invitation (see letter 89), Montesquiou had invited him again, this time to a reading.
2. In *Le Livre de mon ami*.

3. Cf. 'Et, pour mieux frapper leur oreille, / Que ta voix s'élève, pareille / A la rumeur d'une cité!' Victor Hugo, 'La Lyre et la harpe', in *Odes et Ballades*.

4. In a postscript to his invitation (see note 1) Montesquiou had written: 'Needless to say, the book isn't by me.'

5. Greek philosopher and poet (sixth century BC), author of a poem on 'The Nature of Things'.

6. German Romantic poet (1772–1801), author of *Hymns to the Night*.

7. Joseph Joubert (1754–1824), French moralist.

94   *To Madame Emile Straus*[1]

Tuesday evening [20 December 1904]

Madame,

It gave me great pleasure to hear your voice this afternoon, but also some distress. First and foremost because I found it sad and tired, physically tired but also betraying nervous tiredness, expressing the feeling (in yourself) that your voice is tired, that you feel the reserves of nervous energy which control the voice becoming exhausted. This is something that Déjcrine[2] would rid you of by isolating you for a month, but I well understand why you fight shy of this since I can't make up my mind to do it either. It's true that in my case I know that I'd keep all my ailments afterwards whereas you who don't have any others would be miraculously cured. In another sense it's more difficult for you than for me since I've been cut off from everything for so long that I'm like those people who when they take holy orders scarcely need to make any change in their lives, which were already monastic (this is only a comparison and doesn't mean either that I am chaste or even less that you . . . I break off, appalled). But at least I wouldn't have to sacrifice a whole life as you would. I won't say an agreeable life because having heard your poor adorable voice I'm well aware that the life to which it is addressed must seem to you very disenchanted and very exhausting. But when you were rested and cured life would recover its charm and would no longer tire you. All this distresses me a great deal because I didn't know you were unwell. Reynaldo only told me after seeing you at the Widals'[3] that you had trouble with your eyes (which have so troubled others). And since everything that is you, from the beautiful nails of your hands which are the most beautiful thing in the world, is close to my heart and moves me so much that if you have a pain in your finger it affects me personally, I was upset to hear that your eyes were hurting you. And I was also saddened to hear you speak because for me who can no longer enjoy any pleasure and am so cut off, that voice which evokes all that is dearest and most delightful to me, close to my ear but emanating from a woman I can neither see nor approach, was really too tantalizing, like all the music of the Oceanides or the lost paradises that one might ever have known or imagined.

> Do you hear those charming voices:
> Come with us, come, you who would taste
> The fragrant lotus! Here is where we reap
> The magic fruit that your heart hungers for.[4]

But I cannot respond to these calls. To still enjoy something of you I am left with imagination and memory.

I visit doctors from time to time and they urge me to go away; I do not go away, but each visit I make to them costs me whole weeks in bed. However, this time it seems to me that I shan't be ill for so long as a result of my outing. I shall perhaps go to the country for a day at Christmas. After that I shall ask you some day if you will have me to dinner. Madame, if there were anything in the world that I could do to please you – take a letter for you to Stockholm or Naples, or I don't know what – it would make me extremely happy.

Your respectful friend
Marcel Proust

If you see Monsieur Reinach[5] tell him that I thought his letter admirable. He is always a model of reason, nobility and justice:

> When all is change for us, his wisdom is the same.[6]

He is an idealist politician. That is really much the best sort to be.

---

1. Born Geneviève Halévy (1850–1926), of the musical family (see letter 283). Widow of the composer Georges Bizet, mother and aunt of Proust's schoolfriends Jacques Bizet and Daniel Halévy, she had married *en secondes noces* the lawyer and financier Emile Straus (1844–1929). During and after the Dreyfus case her salon was the rendezvous for prominent Dreyfusards (see vol. I). Sympathetic and intelligent, she became the close friend and confidante of Proust who is said to have portrayed her brilliant conversational style in the Duchesse de Guermantes in *RTP*.

2. Dr Jules Déjerine (1849–1917), specialist in nervous diseases. Mme Straus suffered from neurasthenia and was never to make a full recovery.

3. Either Paul Vidal (1863–1931), composer and conductor, or Dr Fernand Widal (1862–1929), leading physician admired by Proust.

4. Cf. 'Entendez-vous ces voix, charmantes et funèbres / Qui chantent: "Par ici: vous qui voulez manger / Le lotus parfumé! c'est ici qu'on vendange / Les fruits

miraculeux dont votre coeur a faim . . ."' Baudelaire, 'Le Voyage'. Proust slightly misquotes.

5. Joseph Reinach (1856–1921), politician, journalist, lawyer and leading Dreyfusard. He had written a letter which was reproduced in *Le Figaro*, in reply to one from the president of the League for the Rights of Man justifying the Combes government's policy of vetting Army officers for clerical or right-wing views (which involved spying and delation), and announced his resignation from the Central Committee of the League.

6. Cf. 'Quand tout change pour toi, la nature est la même.' Lamartine, 'Le Vallon'.

95   *To Louisa de Mornand*

Friday [30 December 1904]

My dear little Louisa,

This afternoon a parcel is brought to me[1] . . . As I always forbid people to give me New Year gifts, I never receive any. And I hadn't thought it necessary to renew my prohibition to my little Louisa, for how could I imagine that a person who is so exquisite to me all the year round would also take it into her head to think of me on New Year's Day. So I undo the parcel, though convinced there must be some mistake. Then I unfold a piece of tissue paper and find myself in the presence of a white casket so ravishing that already the thought of taking advantage of the mistake and keeping the casket or buying myself a similar one begins to germinate in my mind; the idea of being separated from this perfect white object was already unbearable, such a firm friendship had I already contracted with this charming little casket. If you feel that this is going a bit fast in friendship, let's say that I already felt for it, together with a great deal of admiration, an irresistible liking which was determined to turn very rapidly into an indissoluble friendship. And I contemplated the casket without thinking that something so pretty might enclose something even prettier still, so self-sufficient did it seem. However, a certain secret button (no improper allusion intended) seemed to invite me to press it gently. And if ever a button (still without impropriety) was more promising, none ever kept its promise so perfectly by giving access to such a paradise. I open . . . and I see something so adorable that I cannot understand how I ever managed to live without it, something that will never leave me for a single moment, I promise it. Its little golden frame encloses the most ravishing picture one could imagine. I do not know whether all the hours are as happy as this delightful person, with the optimism of beauty, proclaims and predicts, or at least wishes. But happy or not, they are all beautiful, when the needle that registers them passes over something richer and more marvellous in appearance than a peacock spreading its tail, or the sea in sunlight. The four golden fleurs-de-lys reclining in the four corners appear with good reason intoxicated by this enchanting spectacle. And certainly I know what life is and from the hours it has already given me I can perfectly well imagine that those it will give me in the future will not all be happy. Not all! Perhaps none! Happiness! Even after having endlessly interrogated people who claim to know what the word 'happiness' means, I have scarcely managed to arrive at a vague idea of its meaning. But whatever they may be, those hours, although they will wound me, until they eventually kill me (you know that they used to put a Latin motto on sun-dials, which I translate for you: *All wound, the last kills*), I shall never tire of observing their murderous progress when I see them glide over the sea-green or peacock-blue (I can't specify the colour in artificial light) enamel and reach and pass, beyond the consoling motto, the beautiful golden figures. And now, here comes the greatest pleasure of all. I did not believe it was for me . . . but then I saw that it wasn't a mistake, that I was indeed the owner of this ravishing object, and at the same time I learnt that the exquisite person to whom I owed this present was none other than my little Louisa. There is no one towards whom gratitude could be sweeter to me. Hence I have taken so long to thank her that I have no room left to scold her, which was even more necessary, and to send her my best wishes for her happiness and health in the New Year, which she knows that I offer her most tenderly with all my heart.

Marcel

1. She had sent him a gold bedside watch with a peacock-blue dial inscribed with the motto: 'May I only number happy hours'.

96　*To Louisa de Mornand*

[Around 1 January 1905]

My little Louisa,

Just a word from someone too tired to write but to whom the duty of never neglecting an opportunity to be helpful dictates these words. Louis[1] told me that he intended to go to your première.[2] I fought the idea but to no avail. My little Louisa, I haven't the strength to explain what I mean at length. But you know that it is generally alleged that you are still Louis's mistress. It's a rumour that may do much harm even to you (I'm now looking at it purely from that point of view). You know that when you came here I said to you: If you find someone who attracts you (I didn't say who is attracted by you – who could fail to be that?) and who is nice, don't hesitate to orient your life in that direction after mature reflection, talks with us, etc. Now if this occurred, can you imagine that a sensitive young man (and I don't wish you any other), hearing what's being said about your still being with Louis, wouldn't be put off by that? Louis was at your last première,[3] *and I know everything that was said about that.* If he goes again this time and finds Henraux and Bertrand there, and Lauris who will be brought by Henraux,[4] and the whole of Paris . . . who is to say that on that very evening the stranger who will perhaps be yours tomorrow and forever may not be in the auditorium? What do you suppose he will think when they point Louis out to him and tell him all sorts of dreadful things? My little Louisa, I think Louis is doing a very foolish thing and I feel that if you could write and tell him so without hurting his feelings you would be doing him as well as yourself a great service. Now I may very well be wrong. I've given you my pathetic little reasons, the reasons of an invalid but also of a passionately tender friend who sends you his very deepest most ardent most sincere wishes for the year which has just begun, which has begun with such pain for him but will be for you I hope rich in happiness and fame.

Marcel Proust

1. Louis d'Albufera, recently married. Cf. letter 46, n. 2.

2. The première on 18 February of a one-act comedy at the Vaudeville Theatre where she was playing in repertory.

3. At the same theatre on 16 September 1904.

4. Lucien Henraux (1877–1926), collector of oriental art; Bertrand de Fénelon; and Georges de Lauris.

97   To Paul Grunebaum-Ballin[1]

45 rue de Courcelles
Friday evening [6 January 1905]

My dear friend,
    By sheer chance, happening to open your book only today although

I had bought it as soon as it appeared, I opened it at the page headed 'La Mort des Cathédrales'![2] These words vaguely reminded me of something and since I couldn't believe that your book would concern itself with an article which went completely unnoticed and for whose author I thought you had no more than friendly feelings, I was amazed at the coincidence. When I saw that it was indeed about me, I was proud and delighted. 'Elegant language' didn't charm me as much as it ought to have, and 'infinite dexterity', which I nearly translated as 'Jesuitical cunning', I found almost wounding! (I'm joking), but the whole thing was enormously flattering to me and I thank you with all my heart for your kindness in having paid attention to those ephemeral and frivolous pages. I shall read the book with a great deal of interest, especially where it touches on religious buildings.

Having as yet read only the part that concerns me, I venture provisionally to reply to two of your objections. Yes, no doubt the taste of the clergy is deplorable and it has certainly introduced some incongruities into that admirable 'artistic ensemble'.[3] (But you know yourself how it is: the 'purists' cannot go to see a Wagner opera because the costumes are hideous, the chorus bad, such and such a passage is taken too slowly. People who really love Wagner are only too glad to have an opportunity of hearing one of his operas in its entirety and skate more easily over such errors of detail of which they are nonetheless perfectly well aware. I don't say that there aren't in Chartres cathedral – though in fact *are* there in Chartres? – things I should prefer not to see there.) And yet one could put even more hideous objects into it without perverting the meaning of the work, the 'elevation' of those dolorous and imploring statues, almost as immaterial as souls, towards the dim sky. The day when it becomes an archaeological museum! – dear friend, I prefer not to think about it.[4]

But it's precisely because I think priests (it looked as if I'd written 'poets'!) have in general horribly bad taste that I foresee the horror of what will happen. They have at all times preferred new, comfortable and hideous churches to the melancholy survivors of the ages of beauty.[5] They preferred them because they were comfortable. They preferred them also because they were hideous, that is certain. But now that we can rely on their bad taste, we must also acknowledge that for the first time bad taste will be excusable. Having almost no money any longer, it would be sacrilegious of them to think of beauty. They will look for premises that are economical, 'central', heatable, inviting. They will be right.

As for the fate of works of art which have lost their purpose, you say this doesn't mean death, and I recognize that your theory was not dreamed up for the necessities of the Separation, since it is extremely old, any more than mine was dreamed up for the needs of the Concordat, since although more recent it has nevertheless been widely held for the past forty years. Let's not bring them into conflict here in a letter which an invalid has scarcely the strength to write to a friend he thought no longer was one. But your examples in any case don't seem to me conclusive. I have never been to Greece but I know how eloquent a ruined monument can be. But a ruined monument isn't a desecrated monument. There is no longer a cult of Ceres. There is still, however distressing it may seem to certain people, a cult of Christ. In a well-organized state, the ceremonies of religion ought to be celebrated in buildings which were constructed, sculpted, painted, panelled, composed, conceived for that purpose, instead of which we shall see these buildings empty while the same religion will be celebrated in rooms or halls with which it has no affinity. I may add that for those who think we are seeing at the moment the last of the Christians, it must be a curious sensation to see them officiate and pray in the edifices whose form goes back to the earliest centuries, the very form of the Cross, and they must seem at least as strange as the anthropopitheci said to have been still observed in New Zealand.

The processions on the feast of Corpus Christi remain one of the most wonderful memories of my childhood. Since I'm prevented by hay fever from going to the country at that time of year, and since they are forbidden in Paris, I can no longer see them except in my mind's eye. I have no doubt that soon I shall no longer hear the sound of church-bells either, save in that inner atmosphere which still vibrates with the sounds that once moved us but have ceased to exist.

There is however a very interesting argument that could be put forward in favour of the deconsecration of cathedrals. It's expounded (although that's not at all what it's about) in Brunschwicg's very fine book *Introduction à la vie de l'Esprit*.[6] I summarized it in a note in *La Bible d'Amiens*. I don't know whether it's really an argument in favour of deconsecration. But it seems to me to be. And it's a powerful idea. We no longer think things beautiful except when they have ceased to be for us an object of faith but of disinterested contemplation. Only it's quite the opposite of Ruskin with whom I've lived for four years. And then after all the feeling we have is not faith at all, it's a disinterested contemplation which is at the same time understanding of the real meaning of the work of art. So I retract what I said; it isn't a thesis in favour of deconsecration.

I'm very glad to see the flattering and unanimous tributes with which your book has everywhere been greeted. Although I don't know the first thing about politics, I shall read it with great interest.

Thank you, and yours ever with the very agreeable satisfaction of a misunderstanding dispelled.

Marcel Proust

1. Paul Grunebaum-Ballin was the author of a new book, with a preface by Anatole France, on the separation of Church and State, examining the new measures proposed by the Minister for Education and Religion, the socialist Aristide Briand (1862–1932), later many times Prime Minister and Foreign Minister. Proust had already expressed his strong antipathy – which was cultural rather than religious – to the earlier intolerant anti-clerical laws in a remarkable letter to Georges de Lauris in July 1903 (see vol. I).

2. The author cites Proust in a résumé of the latter's ideas already published under the same title in *Le Figaro* in August 1904 (cf. letter 67, n. 2).

3. The author had questioned Proust's statement that the Catholic Church remained virtually unchanged insofar as its rites and ceremonies and aesthetic ambience were concerned.

4. Briand (see note 1) proposed that, once state subsidies were withdrawn, any cathedrals which proved uneconomic to maintain should be secularized and used as museums.

5. Cf. the views of the curé of Combray in *RTP* (I, 111–12).

6. Léon Brunschwicg (1869–1944), French philosopher, and an old acquaintance of Proust.

98 *To Gabriel Mourey*

[January or February 1905]

Dear Sir,

Thank you so much for your very friendly remarks about my essay in translation.[1] I say 'essay' because as you know one 'translates' only a little and half the graces of the original are left behind, expiring at the very beginning of the perilous journey, incapable of living in the too different atmosphere of another language. I put as much care into it as I could and yet am well aware of offering only flowerless branches. If a little perfume remains it is only a memory. I have no need to tell you that all the terms you offer me are accepted in advance; in any case they seem to me highly satisfactory and I'm pleased to see that *Les Arts de la vie* is no longer so prodigal. For it was really too much for such a young magazine which already offers so many things, and so delightfully printed, in each number. You don't mention whether in one of those numbers – the sooner the better – I might write a few lines – or more – in praise of *L'Amant et le médecin*,[2] or its author Gabriel de La Rochefoucauld. Would this be possible? I shall put a little note underneath the title of *Trésors des rois*, indicating that the complete work will be published later by Mercure de France. If you would prefer me not to announce this you have only to cross out that part of the note. It's immaterial to me. What gave me the idea was that *La Renaissance latine* made me do it when they published *La Bible d'Amiens*.[3] But I absolutely don't insist. It's of no importance.

Your devoted
Marcel Proust

1. Ruskin's *Of Kings' Treasuries* (see letter 15, n. 3). Gabriel Mourey was publisher of *Les Arts de la vie* (see letter 26, n. 6), the magazine which was planning to serialize Proust's translation.

2. Cf. letter 78. La Rochefoucauld's novel was about to be published.

3. See letter 11, n. 2.

99 *To Joseph Reinach*

Wednesday [26? January 1905]

Dear Sir,

I should like to be able to write for *Le Siècle*[1] a little article in praise of my friend Gabriel de La Rochefoucauld's novel *L'Amant et le médecin*. Does *Le Siècle* publish articles of this kind, and if so, do you think it would take one from me? –

I don't know if Madame Straus expressed to you, as I asked her to (by letter, alas, as I haven't seen her for a very long time), my admiration for the letter you wrote to the League for the Rights of Man against informing.[2] It was a fine letter and a noble act. The whole thing is greatly to your honour.

Accept, dear Sir, my most devoted respects.

Marcel Proust

1. Left-wing daily edited by Reinach (see letter 94, n. 5).

2. See letter 94, n. 5.

100 *To Joseph Reinach*

Wednesday [1 February 1905]

Dear Sir,

You are most delightfully kind and I assure you I am deeply touched. A Ruskin which I've been asked to deliver in the shortest possible time, most inopportunely for my present state of health, prevents me from writing an 'article' at the moment. If I can get by with a brief note which I could place somewhere or other, that would be simpler. If I don't find another outlet, I shall come back and bother you to put it in *Le Siècle*,[1] with some misgiving as I sense from your tone of indecision about the outcome that it may not be convenient for you.

What good news that the intolerant and calumnious ministry has fallen.[2] Did you notice that it was with your very words that the minister justified before the House the action taken against General Peigné.[3] The very adjective 'deplorable' appeared.

Gratefully yours
Marcel Proust

1. Despite having appeared to solicit *Le Siècle* (see letter 99), Proust was reluctant to appear in that journal for fear that an article sponsored by the Dreyfusist Reinach might damage his chances of joining an exclusive club (see letter 114).

2. The radically anti-clerical ministry of Emile Combes (1835–1921), Prime Minister from 1902 to 1905.

3. The general was in trouble over the delation scandal, known as *l'affaire des fiches*, which brought down the Combes government and caused Reinach to resign from the League for the Rights of Man. Cf. letter 94, n. 5.

101 *To Louisa de Mornand*

[First fortnight in February 1905]

Oh my little Louisa, how unkind of you! I got up specially to see you. So as not to risk taking up all your evening hours in advance in case you wanted to have supper, I told you that we could spend only half an hour together, but when I told Mama she went to bed in order to leave me undisturbed, and everything was arranged when my little Louisa (who doubtless in the meantime had received a more agreeable invitation and gave herself the excuse that she didn't want to disturb me – though she knows quite well that when one is expecting people the disturbing thing is not seeing them), when my naughty little Louisa, whom I love with all my heart, taking not the slightest notice of the cab and the concierge I had sent in great haste to her at the Vaudeville, leaves me in the lurch. Nevertheless I'm still dying to kiss her on both cheeks and even on the nape of her lovely neck if she'll let me (I'll tell you when I see you what I crossed out). But when? Will I be well enough to get up tomorrow? Will she be at the Vaudeville at ten o'clock if I did feel well enough to get up? But it's unlikely. Anyway, I hope to see you one of these days. Tenderly yours, my little Louisa.

Marcel

Tell me the name of the play you're going to act in and the names of the newspapers in which you would like a friendly mention.

102 *To Marie Nordlinger*

[9 or 10 February 1905]

Dear friend,

The words 'I have been so ill, I am still so ill' have been pronounced by me so often – signifying an almost habitual state, painful but not excluding from time to time the possibility of epistolaɪ y relations – that I'm very much afraid they must reach your only too accustomed (I certainly don't mean incredulous) ears somewhat faded and lacking in exculpatory and absolving power. And yet there it is, I have been horribly unwell, almost permanently bedridden, and without the strength to maintain any human relations other than the disembodied ones of friendship and memory. But those, you may rest assured, have not been interrupted and my self-accusations of forgetfulness have not been justified as far as you are concerned, for I've thought of you a great deal and what you tell me somewhat mysteriously but very forcefully about your intense happiness has delighted me. Happiness is something I have sometimes ceased to hope for myself but never stopped wishing for others. And I bless the Whistlerian magician[1] who has brought you happiness, because I don't think circumstances alone have given it to you, and I have a feeling that people have contributed. Thank you so much for the delightful little Whistler book.[2] I've searched everywhere for the article you ask for and haven't been able to find it. Tell me the name of the review and I'll go round there and have it sent to you. I shall read the little monograph on Whistler more carefully when I'm better and write to you in detail about it. I regret to think that your new friend, since you were good enough to talk to him about me, must as a good Whistlerian be full of contempt for an admirer of Ruskin. But in reality I think that although their *theories*, which are the least personal aspect of each one of us, were opposed, *at a certain depth* they agreed more often than they realized. Thus Whistler's most celebrated remark was delivered against Ruskin: 'I did indeed take only a few moments to paint that picture, but I painted it with the experience of a lifetime.'[3] And I've read somewhere that Ruskin once said to Rossetti: 'Your only good pictures are the ones you paint fast; the carefully worked ones are bad. That's because in reality the ones you work over give you trouble because you haven't ever thought about them, so you take three or four months to paint them. Whereas the impromptu ones, those you do for pleasure, realize in a moment an obscure desire which you have cherished in your heart for a long time, so that in reality you paint them with a lifetime's knowledge.' Here Ruskin abandons theory and in a precious flash of intimate experience, he is at one with Whistler.

Having a little breathing space at the moment I've begun *Queens' Gardens*,[4] which I've decided to add to *Sesame*. My old and charming English scholar of whom I've told you will act as my 'Mary'.[5]

I've received as a New Year's gift the splendid new edition of Ruskin.[6] You will enjoy reading it when you come back. And you'll see the magnificent new illustrations. I have a friend (M. Lucien Daudet) who studied painting with Whistler. If the information he might perhaps (?) provide your friend is likely to be of interest to him, I'd be only too pleased to question him. I myself did not know Whistler, except for one evening when I made him say a few kind words about Ruskin! and appropriated his handsome grey gloves which I've since lost. But I've heard a lot about him from Robert de Montesquiou and Boldini.[7] Tell your friend that in my intentionally naked room there is only one reproduction of a work of art: an excellent photograph of Whistler's *Carlyle* in a serpentine overcoat like the dress in his portrait of his mother. The more I think about the theories of Ruskin and Whistler the more I believe they are not irreconcilable. Whistler is right when he says in *Ten O'Clock* that Art is distinct from Morality. And yet Ruskin, too, utters a truth, on a different plane, when he says that all great art is morality. But that's quite enough chatter, so without further ado I send you my respectful and grateful affection.

Marcel Proust

1. Charles Land Freer (1856–1919), American railway millionaire, art collector and friend of Whistler, whom Marie Nordlinger had met with her employer Siegfried Bing.

2. *Ten O'Clock. A Lecture* by James McNeill Whistler (London 1888). She had given Proust the English original, rather than Mallarmé's translation.

3. Proust has already quoted in an earlier letter to Marie Nordlinger both Whistler's remark about his painting, *Nocturne: the Falling Rocket*, and Ruskin's remarks to Rossetti, mentioned in the next sentence. Cf. letter 16, notes 4 and 5.

4. See letter 15, n. 3.

5. i.e. to replace her as collaborator; this was apparently Charles Newton Scott, friend of Ruskin.

6. See letter 77, n. 3.

7. Jean Boldini (1842–1931), society portraitist who had painted a sensational portrait of Montesquiou in 1897.

103   *To Madame Alphonse Daudet*

[February–March 1905]

Madame,

It gave me great pleasure to see once again your beautiful hand-writing, tall, supple and deliciously sloping, reminiscent of Holy Writ (no pun intended) and seeming to reflect at once the purity of your gaze and the freshness of your voice, the handwriting of a Parisienne with its delicate imprint, of a Frenchwoman, the Lady of Pray,[1] which seems also to be the living symbol, constantly renewed, of your delightful poetic style. It is one of my regrets not to have taken better advantage of the happiness it gives me to come to visit you in an environment of living poetry and tender and profound thought. It seems to me that I spend my life close by you without seeing you any longer save in that dream of Pray in which, as the sun rose above the misty Loire, long golden sunbeams suddenly burst forth – like those flowerings in dreams, fields passing by in a momentary flash of light. But since it is perhaps your most original gift to combine what is most simple, most true, most sincerely intimate with the most profound visions of poetic thought, I feel I must be trying the patience of the marvellous hostess that you are by giving her so many futile lines to read through when all she wants is the yes or the no which will modify the dinner-table she has mentally drawn up by a disastrous over-loading or a void left open for more qualified guests. Alas, Madame (alas for me), it will be no this time, the reason being a melancholy secret which I shall confide after promises and oaths to Lucien who will not have the permission to tell you but the means of assuring you that my acceptance would be absolutely impossible.[2] I hope it won't be long before he comes to see me – it's a long time since I saw him. Tell him I haven't left my bed since he last came and I dare not say with any certainty how I shall be on any particular evening. But I hope it won't recur too often and I'd very much like him to come.

Your respectful and deeply grateful servant
Marcel Proust

1. The Daudet family home (see letter 41, n. 2) where Proust stayed in September 1902 (see vol. I).

2. Proust was suffering from chronic digestive problems, too delicate to mention except to her son Lucien Daudet.

104   *To Lucien Daudet*

[About 27 February 1905]

My dearest Lucien,

Since you are unjust; since I believe you don't realize how distressed I am when I know you have a bad cold, are obliged to stay in your room, when you find yourself driven to use the adjective deadly, I'd give you more pleasure by finding you some cough mixture than by telling you that I feel out of sorts, aching all over, without appetite, listless, because of you, that your indisposition poisons my good moments and makes the others even more unendurable, by blocking off the field of the imagination so that I can no longer constantly visualize (during the time you are unwell) an irresistible, gifted, mettlesome Lucien (d'Orsay, Rubens and Franconi[1]) galloping from the place Dauphine to the avenue Marceau (where for some reason or other I situate the de Muns, as Clemenceau calls them[2]). So it would give me great pleasure to know that you are off to Ripaille (or is it Ribaude?)[3] because that would mean that once more you have an appetite for lunch, pleasure in eating, going out, reading, etc. (etc. is sublime). I am slowly recovering from the strain of the Clermont-Tonnerre dinner where I was alone with Reynaldo and the Guiches.[4] You see, my little rat, it would have ratted you.[5] My little rat, we are both of us too nice not to be well. We really must get better and then lead an intelligent and close-knit life in which we see each other a great deal (if you don't suddenly take against me in the meantime) and have all the time in the world to show each other the splendours we produce, you in pictures and I in books. And wonderful excursions, etc. These are dreams. But dreams are the only things one can realize.

Tenderly yours, rattiest one
Marcel

Have you read an article three columns long on M. Higginson[6] (M. is very m.g.[7])? That reminds me that I still don't know whether it's the husband of Mme Higginson (or herself) and her address. Is Mme de Brantes[8] back? This comedy has been going on for a year to the very day!

1. Typical dandy, painter, circus rider.

2. Comte and Comtesse Albert de Mun; presumably Proust means the incorrect use, when referring to people by their surname, of the particle 'de'.

3. Ripaille: castle in Savoie where the fifteenth-century Duc Amédée VIII (later Pope Felix V) led a dissolute life, giving rise to the saying '*faire ripaille*'.

4. The Marquise (later Duchesse) Philibert de Clermont-Tonnerre, *née* Elisabeth de Gramont (1875–1954), was half-sister to the Duc de Guiche. Her husband was a violent anti-Dreyfusard but she became Proust's close friend.

5. 'Rat' was a favourite term of endearment between Proust and Lucien Daudet.

6. John Higginson was an Irish-born naturalized Frenchman who had made a fortune out of nickel in the French Pacific. Proust had met a Mme Higginson with her sister, the late Mary Dutton, companion to Mme de Brantes, and perhaps wanted her help in restoring his relations with the latter.

7. *mauvais genre*. See letter 20, n. 6.

8. Mme de Brantes (see letter 81, n. 2) had failed to acknowledge a complimentary copy of *La Bible d'Amiens* which Proust had sent her in February 1904.

105   *To Prince Léon Radziwill*

Tuesday morning [28 February ? 1905]

My dearest Loche,

I want to write this so amiably that you recognize that it is not dictated by any feeling of hostility or even bitterness, especially not by any resentment. Spare me the necessity of stating my grievances.[1] They are grave and they are well-founded. I no longer wish to see you, I no longer want us to write to each other, to know each other. When the friendship I felt for you is dead, since such friendships never revive, then

I shall be delighted to meet you if you too so wish. But when? That will depend on the new friendships I may form. I've no idea.

I simply wanted to put a request to you. The other day, à propos of that business about an article which is developing into a mare's nest,[2] I put to you the supposition that Guiche might have started it all. I did not really believe this. I said it to you in order to test out every possible hypothesis with you. I should be sorry if you told Guiche. I like him and respect him and this could make him doubt it.

There are some other recommendations I wanted to make to you which are not easy to put into writing. I had originally thought of going to Le Havre to talk to you about it all. But you would have been nice to me, I should have withdrawn my resignation as a friend, my friendship would have resumed, as would my worries. And after a while, it would have started all over again. Better to get it over with here and now. No one else could possibly understand it all. So I shall say nothing to anyone. As I'm going to give a few small dinner parties,[3] if anyone asks me why I don't invite you I shall say that you're very preoccupied with your family, your mistress, your country house and your regiment. (I shall need du Breuil's address.[4])

I shall send back to you in a few days' time not your beautiful ink-well, which I'm very happy to keep, but the plaque on the lid. Those ironical lines

> How pleasant to have a true friend
> Who seeks our needs in the depths of our hearts
> And spares us the shame
> Of revealing them to him ourselves[5]

would be too painful to have constantly under my nose.

I should like you to do everything that I don't do but should, and especially work. I'm sure you would do some excellent things.

I was your truly sincere friend.

Marcel Proust

1. Proust was trying to arrange another press notice for Louisa de Mornand (cf. letter 63, n. 5) and Radziwill (see letter 82, n. 9) had been unhelpful.

2. A propos of the dialogue turned down by *Gil Blas* (see letter 63, n. 3), and 'malicious articles' Proust was accused of having written. Cf. letter 114.

3. Proust was to give one large reception (which Loche in fact attended), on Monday 6 March, and, as Mme Proust was in mourning, in the afternoon rather than the evening. The guest list was published in *Le Figaro*, *Le Gaulois* and *L'Echo de Paris*; it was to be the last such gathering at 45 rue de Courcelles. The tea-party caused a

brawl among the Proust servants and the departure of their cook, Marie (cf. letter 110).

4. Pierre du Breuil de Saint-Germain, man-about-town.

5. 'Qu'un ami véritable est une douce chose . . . / Il cherche vos besoins au fond de votre coeur / Il vous épargne la pudeur / De les lui découvrir nous-mêmes.' La Fontaine, 'Les Amis'.

106 To his Mother

[End of February 1905]

My dear little Mama,

Reynaldo has announced not only that he won't sing but that as he has to be at Mme Mazza's[1] at ½ past nine he'll leave before the end of dinner if it goes on too long. I should like you to send him a note, which mustn't seem to have been inspired by me, couched in these terms:

'My dear Reynaldo,

Marcel tells me that you won't be singing. He's very upset about it and I must say that although I'm usually inclined to put the blame on him, this time he seems to me to be entirely right. I feel that you could at least write and tell Mme Mazza that you won't be arriving until a quarter to eleven and when we leave the table you could fling yourself at the piano without letting people go and smoke, until ½ past ten. I should prefer you to tell Mme Mazza that you won't be going at all. But if you do what I suggest everything won't be lost. You can't possibly do less for us than Mme de Guerne,[2] who after finishing dinner at half past ten as a result of a series of accidents nevertheless sang three pieces before going on to a very important soirée.'

You can add a few witty remarks so that he can see that it's really from you, but no argumentation. When I saw that he was determined not to come I invited Coco[3] who will always be useful if Reynaldo refuses to be swayed.

A thousand loving kisses
Marcel

We'll only make up one table since that's what you want. But it will prevent my having people like the Régniers and the Yeatmans to whom I really needed to make some show of politeness and whom I shan't be able to ask to a purely Faubourg St-Germain dinner. The same goes for Rod.[4]

1. Sic. Mme Achille Matza, *née* Colette Alexandre-Dumas (1860–1907).

2. The Comtesse de Guerne, with whom Hahn sang duets. Proust thanked her by eulogizing her singing in an article signed 'Echo' in *Le Figaro* of 7 May.

3. Frédéric de Madrazo (see letter 16, n. 6).

4. Edouard Rod (1857–1910), Swiss-born critic and prolific novelist, widely read at the time. Faubourg St-Germain = the social 'upper crust'.

107   *To Madame Straus*

[2 or 3 March 1905]

Madame,

Will you in your infinite kindness break the evil spell that keeps me apart from you and give me immense pleasure. Come *next Monday* at about four o'clock to hear Mme de Guerne and Reynaldo sing some duets, and to have tea. There will be very few people, and I'd be really overjoyed! I seem to remember that it's the hour when Monsieur Straus leaves the law-courts, so it would be extremely nice and extremely conjugal if you were to arrange to meet here at four o'clock and thus do me a great honour as well as giving me great pleasure. You can eat or not, you can drink or not, you can talk or remain silent, and if you get bored you can come with me into another room. And if I bore you I shall leave you on your own. I apologize for importuning you for so little. But I couldn't arrange a dinner party. So I resigned myself to a tea party, asking only that it should not be a Saturday so that you wouldn't excuse yourself on the grounds of its being your 'at home' day. Have I any chance? Please don't take the trouble to reply – if you take any trouble at all, take the trouble of coming – and staying.[1]

Your respectful friend
Marcel Proust

1. Mme Straus, suffering from neurasthenia, was unable to leave her house.

108   *To Anatole France*

[Saturday 4 March 1905]

Mon cher Maître,

I have to thank you for a sublime spiritual present. I have been ill. Forgive me for not having written to say why I think there is nothing finer than *Sur la pierre blanche*.[1] I shall do so in the near future. Today I simply want to say to you that I should be greatly honoured and deeply gratified if you would care to come and have a cup of tea and listen to a little music in intimate surroundings the day after tomorrow, Monday, at half past four *precisely*. I know that this kind of gathering is little to your taste and that one of the reasons why your life is not unhappy is that, like Hippolyte Dufresne, you 'do not move in social circles'.[2] So I'm asking you with some diffidence and without much hope, as the poet addressed the star

Knowing full well that it is infinitely far.[3]

Nevertheless if you came you would see the phrases of Mozart 'floating their white columns and their garlands of roses in the air',[4] and you would give me great delight.

Please accept, *mon cher Maître*, the assurance of my admiration and respect.

Marcel Proust

To tell you the whole truth and avoid the risk of displeasing you, there will probably be one or two academicians who are, not exactly nationalists, but liberals of the kind who 'claim privileges'.[5] But there is no need for you to be near them, and if the sight of them offends you I shall go and have tea with you while they listen to some music or listen to some music while they have tea. All I want is to stay by your side. As for them, they would be delighted to see you, but I know it isn't reciprocal.

1. Anatole France's most recent work, a long politico-philosophical dialogue, published in February 1905.

2. Hippolyte Dufresne in France's book dreams of a socialist society in the year 2270 and says of himself: 'I can't complain of my lot, compared to that of others. I have neither wife nor child. I'm not in love nor in bad health. I'm not very rich, and I don't move in social circles. Therefore I'm one of the lucky ones.'

3. 'Avec le sentiment qu'elle est à l'infini.' From the sonnet 'De loin', by Sully Prudhomme (1839–1907).

4. The full sentence reads: 'Depuis quelques instants, les phrases d'une sonate de Mozart suspendent dans l'air leurs colonnes blanches et leurs guirlandes de roses.' Ibid.

5. A reference to a paragraph which begins, translated: 'There can be no liberty in society, since there is none in nature.' And ends: 'Moreover, in the last days of capitalist anarchy the word liberty is so strangely misused that it has come to have but a single meaning, the claiming of privileges.' Ibid.

109 *To Robert de Montesquiou*

[Monday evening, 6 March 1905]

Dear Sir,

I come home . . . I see a letter propped against my candlestick . . . your handwriting . . . post-marked Paris! Preliminary reaction, simply the egotism (which can turn into ferocity) of the host (I was still in some degree a host some hours after the lights had been put out). 'M. de Montesquiou is in Paris!' I might, perhaps, have had him to my party! Perhaps he would have come! The gathering would have been given a unique glamour and prestige. And now it is *too late*. It's no longer possible: the party has been over for two hours, and the guests have gone home. So then, respectful mental maledictions against Madame de Clermont-Tonnerre who told me you weren't back, against Lucien[1] who told me you were at Artagnan,[2] and against yourself who told me you were going away until the spring. In spite of what you said, when I decided on this tea party I wrote to Madame de Clermont-Tonnerre to ask her if you were coming back. She answered no. I then asked Lucien who said to me: No, no, I've just had a letter from M. de Montesquiou dated yesterday and he doesn't mention coming back. But in any case I was so inured to the idea that you were staying until the spring that the opposite idea would have required proof. All these thoughts go through my head on seeing your envelope post-marked Paris, thoughts of pure rage, the predominant fury of the host cheated of his glory. No admiring, respectful or affectionate thoughts for you. Everything seen from the 'party' point of view, on seeing the big turbot one might have served up to one's guests slip away through the glass of the aquarium of 'too late'; 'My name is might have been; I am also called No-more, Too-late, Farewell.'[3] – A big turbot not only in its importance, but also fairly literally, big, fat and pink. For now there is a M. de Montesquiou of Artagnan, fat, very pink, I won't say unctuously ingratiating.[4] And naturally, since the tea party was chiefly given for the young Clermont-Tonnerre couple, both of them delightful, but neither of whom turned up, my first thought had been: what a pity M. de Montesquiou isn't here.

Still looking at the envelope, I think to myself: I wasn't feeling too bad for a week and M. de Montesquiou was at Artagnan. Now I am horribly ill and he is here, and I shan't see him. Nowadays I don't see him once in two years.

And then I open the letter and the heading makes me travel four hundred kilometres in a second, which is dizzying. And I feel more and more that you are the Aladdin whose garb you once assumed at a fancy-dress ball and whose poetically and despotically jesting soul you have retained, the imaginative feats once realized. So where are you? I sense that you are at the same time in Paris and at Artagnan – a Pan-like ubiquity.

Hark to his pipe, when far away he seemed.[5]

You are hidden behind the most seemingly inoffensive things!

Matter itself conceals a hidden god

It's true that since I do not 'put it to any impious use' . . .[6] Still, this letter perhaps written at Artagnan and posted in Paris, perhaps the opposite – who will ever know? – is somewhat disturbing.

No, not a resurrection but a farewell to social life, or rather a *mental alibi*. Since I'm probably going to go and take a rest-cure, to ensure that no one could say I had had a fit of madness and been put away, I gathered some people together to show them that I would be leaving sound in mind (insofar as I have ever been) and of my own volition (insofar as one ever does anything of one's own volition). But the farewell would have been more solemn and the alibi more impressive if the features of the poet of Versailles and Louis XIV had been silhouetted behind the microbial sunlight of my chandelier.

On reflection, in trying to decide 'if the man is invisible' or if the Martian is absent,[7] I conclude that you are at Artagnan, where you are kept informed, by the instinct which enables you to 'thought-read', of the most trifling events in Parisian life.

I have often thought about your splendid Platonic theory, a new version of the Phaedrus myth, about images avoiding promiscuity because it doesn't lead to anything.[8] I called round to thank you for it. But you had just left, on your promised journey. And now God knows when I shall see you again. You wrote to me 'kindly', ending with the words 'because I'm fond of you'. That touched me more than I can say, in spite of the subsisting uncertainty, and what I should like to send you herewith is the respectful, admiring and grateful transposition of it.

Marcel Proust

I shall put on the envelope, giving it all its equivocal and in this case mysterious meaning, 'to be forwarded in case of absence'.

1. Lucien Daudet.

2. The Château d'Artagnan, Montesquiou's family home in the Gers, south-west France. Cf. letter 27, n. 8.

3. Dante Gabriel Rossetti, *The House of Life*.

4. *Faisant sa chattemite*: an allusion to lines from La Fontaine, *Fables*, VII, 15: 'Un chat faisant la chattemite / Un saint homme de chat, bien fourré, gros et gras . . .'

5. Cf. Jose-Maria Heredia, 'Le Chevrier', in *Les Trophées*: 'Entends-tu le pipeau qui chante sur ses lèvres?'

6. The sentence is unfinished. The quotations are from a sonnet entitled 'Vers dorés' by Gérard de Nerval, in *Les Chimères*: 'Crains, dans le mur aveugle, un regard qui l'épie: / A la matière même un verbe est attaché . . . / Ne la fais pas servir à quelque usage impie! / Souvent dans l'être obscur habite un Dieu caché . . .'

7. A possible allusion to H. G. Wells (cf. letter 79, n. 6).

8. Allusion to a letter to Proust from Montesquiou written in January 1905, just before his departure for the country.

110   *To Louisa de Mornand*

[Monday evening, 6 March 1905]

My dear little Louisa,

If I'm writing to you only today during an appalling attack it's because I've seen Radziwill who came to a tea party I gave and whom you must know. I had written to him fruitlessly a dozen times to ask for news of the 'snapshot'.[1] He never answered. And I, ill as I was, with all my invitations to do, and domestic dramas which you probably know about when Henraux saved my cook's life,[2] I couldn't do more, I had a raging fever, I couldn't go out, etc. . . .

Here's what Radziwill said to me: 'Germain (the columnist on the *Echo*) told me that he was perfectly prepared to insert the piece. But as the Vaudeville is shortly going to ask him to say another word or two about the play he would prefer to insert it then as he'd have a pretext.' I replied that that was neither here nor there, that the thing was perfectly possible, that there was no need for a pretext, and I hope it will appear in three or four days' time.

My little Louisa, did you not receive the cushions for the settee that I had sent to you three days ago?[3] I don't know when I shall see you. You must know what a state the house is in at the moment. My faithful Marie[4] left yesterday as a result of those quarrels, and I'm desolate. This tea party has killed me and I fear I shall be very ill for several days. Immediately afterwards I shall ask if we can meet – especially as I have a million things to tell you. Forgive my handwriting. Having seen Loche at last I wanted to write to you at once, but I can scarcely hold a pen in my hand and I wonder if, in spite of my unparalleled efforts, you're going to be able to read me. At least don't forget me entirely during the time you don't see me. And keep a little friendship for your poor Marcel who has so much friendship for you.

Ever yours
Marcel

1. See letter 105, n. 1.
2. See letter 105, n. 3.
3. See letter 88, n. 1.
4. Marie was the cook (cf. letter 105, n. 3).

111   *To Robert de Montesquiou*

[Soon after 6 March 1905]

Dear Sir,

You are not 'kind', if it can be said that the worst harshness can offset to the point of cancelling the supreme kindness which consists in gratifying an undeserving reader with a marvellous letter.[1] But you tell me that I made you laugh loud and long, without meaning to, presumably by the absurdity of my remarks. That in itself is somewhat unkind. Then you tease me by discrediting my gathering retrospectively on the grounds of the presence of people whom I didn't invite and who *in spite of this* didn't come[2] – and finally you accuse me of being an inaccurate reader of people's ages by their faces,[3] an ill-informed observer of social rank, and a letter-writer whose epithets are so lamentably ill-chosen that they serve only to dishonour those whom I purport to praise. That makes me very well provided for. I greatly appreciate the honour of being the chosen recipient of this wonderfully witty piece and I send you the steadfast and respectful expression of my admiration and gratitude.

Marcel Proust

1. Montesquiou's letter in response to Proust's last letter to him (letter 109) is missing.
2. i.e. 'the young Clermont-Tonnerres' (see letter 109).
3. The Marquis de Clermont-Tonnerre was thirty-four, the Marquise thirty.

112 *To Louisa de Mornand*

Friday, 7 o'clock in the morning
[10 March 1905]

My dear little Louisa,

By an extraordinary chance your note must have arrived while I was asleep, and was put in the letter-pouch where it was forgotten. And it was only on returning home at four o'clock in the morning that I found it. I feel all the more sad about it because Reynaldo gave a small supper-party with Fragson[1] and told me to bring anyone I liked. And because of your strange silence, believing without knowing why that you were angry with me, I didn't dare ask you to come with me. And I think you would have enjoyed yourself. Because Fragson is charming when seen at close quarters, and sang without stopping. I left them at half past three because of a terrible attack of asthma and he was still singing at the top of his voice. This supper-party has killed me. I swallowed all the dust and smoke in the universe, and I feel I shall expiate it by several days in bed. So if you have nothing in particular to ask me, don't bother to write to me. But if you do have something to ask me, as I'm physically incapable of seeing you, write me a brief word telling me what it is so that I can do it.

I embrace you affectionately
Marcel

I won't mention the snapshot again until it has appeared.[2]

I feel you aren't satisfied with the cushions. Perhaps you find them too hard. But apparently that is how they ought to be. And they'll soften up only too soon. It's such a long time since I've seen you that I can no longer remember your face!

1. Harry Fragson (real name Victor Pot: 1869–1913), celebrated music-hall singer of homosexual bent.
2. See letter 110.

113 *To Georges de Lauris*

[? March 1905]

Dear friend,

The foreword is bad and you should alter it. Proof of this is that whereas the whole article from beginning to end is admirably written, the foreword is very badly written.[1] 'Il a écrit des pages du plus beau français d'aujourd'hui et du plus pur' – that's not the most beautiful French, even of today. And 'évoquer des problèmes' isn't the purest. And further on it's *pure gibberish*. On the other hand, once the article itself begins there is a wealth of superb phrases: instinct which must remain in the unconscious like the foetus in the mother's womb (forgive the comparison) in order to build up its strength; boredom which is the fever not the sleep of the will; the whole page on the bad influence of Barrès and the mysterious signs by which one learns to recognize the inevitable onset of disenchantment; the tactless and perhaps irritated women (the nice sentence with the quotations from Barrès and from Baudelaire); studied but genuine sincerity; love which alone can determine the quest for sensual pleasure – all these are so many delightful perceptions which spring one from the other with a sort of enchanted logic. No coldness there as one might have feared. You have filled in Barrès's formulations with your own experience of love, thought and life. So your work is personal though objective, artistic though dialectical ('the personal side of every artist', G. de Lauris). The finest thing is the passage on the usefulness of feeling which ends with the example of Rousseau whose very errors uncovered the soul and who, disdaining close friends, found the way to acquire distant ones. This, together with the passage on Mme de Bonnemains,[2] already famous in my eyes, is the best thing, and also the magnificent comparison between suffering which is bigger than we are and happiness which is made to our measure, so that one is irritated by the happiness but not the suffering of ordinary people. I find extremely apt the parallel between Montaigne/Renan and Pascal/Barrès and between clearsightedness in love and in hate.

The analysis of the theory of rootlessness is superb. I did not know the passages from Gide and Gourmont.[3]

The passage on extracted gold is copied from something of Ruskin's which I read to you, but very good.

The idea of Barrès's split personality is very profound.

'More offended by vulgarity than saddened by poverty' is charming. Sturel's ultimate solitude very good.[4] Is the idea about Loyola Barrès's or yours? I haven't the strength to scribble another word. Bravo, thank you, good-bye.

Marcel

Titles: *Barrès: Politics and Sentiment*; *Barrès and the Muses*; *Barrès: Love, Hate, Politics and Friendship*.[5]

1. Lauris had asked Proust to read, prior to publication, an article he had written on Barrès. The foreword in question was dropped when the article appeared in *La Nouvelle Revue* (August/September 1905).

2. Marguerite de Bonnemains, mistress of the failed nationalist revolutionary General Boulanger (1837–92) who committed suicide on her grave. Barrès was a Boulangist deputy, and his fictional trilogy *Le Roman de l'énergie nationale* includes an account of his disillusioning involvement with the demagogic general.

3. Lauris refers firstly to an article by André Gide entitled 'A propos des déracinés de Maurice Barrès' (in *L'Ermitage*: February 1898) in which he made his famous remark: 'Born in Paris, with a father from Uzès and a mother from Normandy, where in the world, M. Barrès, am I supposed to find my roots?'; and secondly to *Promenades littéraires* by the critic and writer Remy de Gourmont (Paris 1904).

4. Sturel is one of the central characters in *Le Roman de l'énergie nationale*.

5. Lauris chose the last of Proust's suggestions: *M. Maurice Barrès: L'Amour, la Haine, la Politique et l'Amitié*.

114 *To Gabriel de La Rochefoucauld*

[Thursday, 16 March 1905]

Dear friend,

Forgive me for bothering you incessantly with this business about an article, but it reflects a great desire on my part to talk about your book.[1] If the *Gil Blas* raises any difficulties (perhaps by the time you receive this letter I shall have had one from you in reply to the one I sent you yesterday, but never mind, naturally I'll do what you tell me to) I could perfectly well write to Reinach again. You remember he offered me *Le Siècle*, but since at that time we were hoping for something better, *Figaro* or another, we rejected his offer. And very sweetly you were afraid it might harm me as regards the Union.[2] But as I've given up any idea of ever joining a club, that would no longer be a problem.

I think I could also get the *Petit Parisien*. Would you like that? The advantage over the *Gil Blas* is that for reasons which I think I explained to you there are objections to my appearing in the latter. And half the pleasure of writing the article would have gone if it didn't carry my signature. I'm not deluding myself; I realize that, unknown as it is, it will almost weaken the effect of the article, and those who know us are aware that we're friends. But since in fact you don't in the least care about this article, and it's entirely to please me, the pleasure will be so much greater for me if it's signed, precisely because of the affirmation of that friendship which is so dear to me. So I would sign even in the *Gil Blas*, although there are powerful objections (I think you know the story – their foul behaviour towards me à propos of a dialogue,[3] and then especially the malicious articles they attributed to me[4]). However it can't be helped, but with the *Figaro*, the *Petit Parisien* or *La Presse* there would be none of these objections.

I hope you will reply to my letter of yesterday and that I shall know what you feel about the Régnier dinner. I'm extremely unwell and miss you a great deal. But it's impossible for me to go and see you in this state, and often I can't even get up to receive people.

I was enchanted by Guillaume's drawing.[5] It's the kind of publicity one would be wrong to despise, for it's by such means that a book ends up by merging completely into the consciousness of the masses. Indeed I've already picked up some unequivocal signs of the 'popularization' of yours.

Yours ever
Marcel Proust

1. *L'Amant et le médecin.* See letter 78, n. 1.

2. The Union Club. Proust's membership was rejected on anti-semitic grounds. Cf. letter 100, n. 1.

3. See letters 63 and 105.

4. See letter 105.

5. A drawing by Albert Guillaume (1873–1942) in *Le Figaro* of 16 March showing a man and a woman engaged in the following dialogue: '*The Lover and the Doctor.* Have you read it? What's the answer do you think?' 'Easy, my dear . . . in my case, they're one and the same.'

115   *To Eugène Fould*[1]

[Sunday, 19 March 1905]

My dear Eugène,

The other evening at your house, from the very first moments after my arrival (that is to say the last moments of your dinner!) I never ceased to celebrate the ravishing beauty of Mademoiselle Springer.[2] And you are engaged! I'm astonished at myself, for I'm usually such a blunderer that it's stupendous to have praised someone who was to become your wife, without my knowing it. But she really is beautiful. She also seems very intelligent – I say 'seems' only because I didn't talk to her. So it was a very solemn occasion for you. And I now understand the look of somewhat arrogant hebetude which you had assumed and which was disconcerting for a friend who was not forewarned and had for some time been dropped. If I ever have to describe the emotion of happiness in someone, I shall henceforth take care not to give him the blissful and tender mien which is traditional, but will make it consist rather of an assemblage of mannerisms stamped with a rare acerbity and an apparent stupidity (in the old sense of the word – 'stunned' or 'stupefied' – which is not in the least disobliging, for you know better than anyone that there is nothing stupid about you!).

Dear friend, all these pleasantries which are addressed to the humorist you are do not of course preclude the fervent good wishes which I cherish but hesitate to send to the friend I would like you to be. If only I could be as sure of the friend as of the humorist! I am moved by the thought that your life is now going to change course, directed towards a profound happiness which will separate you from your friends, even from those you never see. Irony is sometimes the deceptive sign of a deep-seated tenderness. That being so, you have a greater need than most of a woman's affection and the desiccating spectacle of ridicule will give you a greater thirst than most for private demonstrations of love. Your wife's task will be a very delicate and a very lofty one, and all your friends have great hopes that she will fulfil it well.

Please accept, dear friend, my profound wishes for your happiness.

Marcel Proust

1. Eugène Fould (1876–1929), son of the banker Léon Fould and Mme Fould, *née* Thérèse Prascovia Ephrussi.
2. Fould had become engaged to Marie-Caecilia Springer (1886–1978), daughter of Baron Gustav Springer of Vienna. They were married on 12 April.

116 *To Madame Straus*

[22 or 23 March 1905]

Madame,

I can't tell you what joy I felt when I heard you were going out.[1] But for the past two days I've been suffering from veritable convulsions of asthma and asphyxia during which it was as impossible for me to write as to speak. Otherwise I should have told you of my happiness at once. My greatest suffering during those two days was not to be able to express the joy I felt through my own illness at the thought that the air which I could not manage to breathe was being absorbed by you in the open, that it must have given you an appetite, replenished your blood, given you back some strength. I can feel the innate good of it, the perfect well-being. If I could think that it was my alarming advice about a cure of isolation that might, for fear of such a fate, have bolstered your desire to get better and your determination to accept treatment, I should be extremely happy.[2] But whatever the cause I'm delighted. I remember your telling me that M. Halévy[3] had recovered out of fear of being obliged to take the waters. But he had an organic disease, which could have had after-effects. You, luckily, have none; what's needed is for your body to refuse its consent to illness from now on (to paraphrase a remark of Goethe's which you know[4]). Ah, what pleasure it would have given me to see added to all your other sweetnesses the greatest sweetness of them all, the one for which we all bless you, the sweetness of the convalescent. Perhaps I shall still be able to. At this moment I'm about to have myself cupped to see if that gives me back a bit of breath! In any case it may go away within a day, but any outing during normal hours would be impossible.

The other day I telephoned to M. Straus to ask him if I might come and share his solitary dinner, but there was no reply. And I was engaged almost immediately afterwards. I read in the *Figaro* the other day that you were going to the South of France and would then resume your 'receptions'. That would be too much to ask. But perhaps the South will do you good. I don't know why I have an idea that if you don't suffer from sea-sickness, since you have friends such as the Prince de Monaco and M. Edmond de Rothschild[5] with splendid yachts, that might perhaps do you a world of good. Last summer I spent five days on a small yacht. And I thought at the time that it might perhaps be good for you because one takes the air without ever *going out*. Unfortunately I'm going to be obliged to retire to a sort of sanatorium for three or four months, but I think I'll put it off until after my hay fever. So I might be able to see you before. If I could think of some books that might amuse you I'd send them at once, but you've read everything, as in Mallarmé's line

The flesh is sad, alas, and all the books are read.[6]

I'm sure that now that you're going to slake your thirst a little more every day with the real health that basically you haven't experienced for a long time, you will want to be as robust as possible and you will consequently be all the happier, more intelligent and more beautiful (that will be something!). You're going to feel like a child out of Maeterlinck with sensations of exquisite freshness for everything natural and good. Do everything the doctors tell you but don't take too many medicines. Almost all of them are toxic, and I can assure you that nothing is more denourishing. I also think that unless you are told to you'd be wise not to 'knock back the Vichy water' as you used to say. Those health-seeking orgies seemed admirable, but really I'm not sure champagne wouldn't be less unhealthy. However there are plenty of people around you to tell you about such things. I don't want to tire you by writing at too great length but if the sight of kneeling crowds is balm to your slightly indolent convalescent's eyes I should like to give you the feeling (it isn't very original since it's like everyone else, though perhaps more keenly and more specially) that I love you to a point which makes me feel your pain like a torture and sing with joy metaphorically (since I can't open my mouth) at your recovery.

Yours most respectfully

Marcel Proust

Please thank Monsieur Straus most warmly for having been the most patient and most active correspondent – and the most charming.

1. Cf. letter 107, n. 1.

2. See letter 94.

3. Presumably her father, the composer Fromental Halévy (1799–1862).

4. 'If a man stops to think about his physical or mental condition, he usually finds that he is ill.' Goethe, *Maxims and Reflections*.

5. Prince Albert of Monaco, former husband of Princess Alice (see letter 130, n. 9); and Baron Edmond de Rothschild. M. Straus was the Rothschilds' lawyer and rumoured to be their illegitimate half-brother.

6. 'La chair est triste, hélas! et j'ai lu tous les livres.' Stéphane Mallarmé, 'Brise marine'.

117 *To Georges de Lauris*

Night of Monday/Tuesday 6 a.m.
[About March 1905]

My dear Georges,

I'm ashamed of myself for not having been able to receive you last night. But that's not the feeling that dominates as it should (I mean the shame, but I'm putting it badly because I'm ill), for I'm infinitely sadder about it, my friendship for you having recently developed a strange, new and greatly increased strength. After constant attacks and all these days trying to rest, I wanted to try going out. And since I didn't get up until after eight I was only just beginning my inhalations. And I made my poor Mama dine at a quarter to midnight. All this in order to go out afterwards (without the slightest aim of enjoyment, or indeed any aim at all) for half an hour. But this half-hour intensified my attacks to such an extent that I can't see when I shall be able to go to bed. Nor, in consequence, when I shall be able to get up. So it's only if you happened to be going to the theatre that you might stand a chance, around midnight, of finding me up, if you wanted to come this evening, Tuesday. And even then I shall probably be exhausted, though still very happy to see you.

Because of my illness I haven't been able to look at your books,[1] which I shall therefore not send back till tomorrow, but not later, as I've since remembered the sentence from *La Vie des abeilles* and it won't fit into my note, what I thought was from *La Sagesse et la Destinée* must be from *Le Temple enseveli*, which I have, and I think I've already made a note of the sentence on the passions from *Le Livre de mon ami*. There remains the charming passage on Leisure from *Adam Bede* which we once read together, which is I believe at the end of the second volume and which I shall copy out with your permission.

Thank you most warmly for sending them.

Marcel

1. Proust had asked Lauris for the loan of a number of books (as he was to do again: cf. letter 320, n. 2) including Maeterlinck's *La Vie des abeilles* and *La Sagesse et la Destinée*, Anatole France's *Le Livre de mon ami*, George Eliot's *Adam Bede* (cf. letter 39, n. 3) and John Stuart Mill's *Memoirs*. He was putting the finishing touches to his translation of Ruskin's *Sesame and Lilies* and needed these books to check quotations in his preface to that work.

118   *To Madame Straus*

Sunday evening [9 April 1905]

Madame,

I was just about to write to you beseeching you not to reply. You can imagine what a joy it is for me to receive a few lines from you, especially at the moment. I seem to feel the impact of your recovered strength, your brief notes have a convalescent charm, a sort of delicious freshness and hope (I don't mean the hope of recovery, since that is no longer even a certainty but something already achieved; what I mean is that it's as precious as a hope). Only I don't want even the tiniest quantity of this precious strength to be wasted on me. The thought that afterwards you might put down your pen with weariness, and especially that beforehand you must take it up with an effort of will, and even before that the decision to write to me (like all decisions, even when not disagreeable) puts a little black spot on your horizon, an horizon that must remain pure, almost colourless, monotonous and gentle – all that distresses and torments me. So don't ever write to me again; from time to time I shall seem to catch an affectionate and reposeful silence wafted to me from Switzerland,[1] and that will make me happier than anything. Monsieur Straus, who described to me (by letter) with touching delight your journey, your lunch in the restaurant car, etc., tells me that you've had a visit and received advice from Dr Widner.[2] It's precisely at his clinic (I'm hesitating between two) that I intend to take my cure this autumn, for I've waited too long and now I wouldn't be able to travel around countrysides which are about to burst into flower. I've heard him very well spoken of and I'm extremely glad that you know him. I'm sure he'll be able to do a lot for you. For me, unfortunately, if it's to him that I'm ordered to go, he won't be able to do much. For I'm very much afraid that these incessant attacks have destroyed something in my organism which will never be able to recover. But still, if he makes my life more bearable that will be something.

My dear little Madame Straus, do you know that you have just made a great sacrifice to the god of youth, and if you weren't a woman, which would make it look as though you were having an unnatural liaison with a lady, albeit divine, I would tell you that, like Hercules, you have espoused Eternal Youth. For you know that that is the effect of these maladies. The rebirth that follows takes you back ten or twenty years. You must sense it yourself and feel in you impressions of exquisite freshness and youth. I won't say that it makes it worth while having been ill. But since you have been ill, now that you have gone through the pain, you will henceforth enjoy only the benefits. I shall soon be sending you a new Ruskin,[3] if my publisher puts as much zeal into printing it as I have put into translating and annotating it. I don't think it's likely to please you much, supposing that you read it. I often wondered as I was translating: would Madame Straus find that good or bad? On the whole it's a very irritating book and I regret having chosen it.

I imagine you must have been pleased at M. d'Eichtal's election to the Institut,[4] for I know you're very fond of him. I was pleased, too, though with a touch of melancholy, because it reminded me of Papa's unfulfilled ambition to join that same Academy. He was a rival candidate of M. d'Eichtal's. If by any chance you had any shopping to be done for you in Paris, or things that you wanted to know, or indeed anything whatsoever, and you don't want to tire or disturb Monsieur Straus, I'd be only too pleased for you to address yourself to me. In that case alone you have permission to write to me. Otherwise, if you have nothing to ask me, I beg you *seriously* not to write. After a time when I see that you haven't replied and that consequently my letters are not a cause of fatigue and irritation for you, I shall take the liberty of writing again. In the meantime please accept the assurance of my most respectful and profound friendship.

Marcel Proust

1. Where she had gone to convalesce.
2. Proprietor of a clinic at Valmont on the Lake of Geneva near Montreux.
3. His translation of *Sesame and Lilies*.
4. Eugène d'Eichtal (1844–1936), director of railways, economist and author of works on socialism and communism, elected to the Political and Economic section of the Académie des Sciences Morales et Politiques.

119   *From Robert de Montesquiou*

Holy Saturday [22 April 1905]

Dear Marcel,

Don't prepare anything, either 'lean' or 'fat', except your regret[1] – and also ours.

In spite of all my preoccupations, I would have come round to you to share that 'commissioner's fodder'[2] if you yourself had appeared yesterday.[3]

I know that you are ill. But why do you recover when it's a question of the La Rochefoucauld orangeade[4] or the 'unique Gala'[5] (the thousand and first)? Montesquiou isn't 'small beer' either. 'He' of whom I spoke yesterday would have *laid his hands on you, through my voice*.[6] I am giving a lecture in Brussels on the 29th. Between now and then I shall have little time to spare. But on my return I shall come to see you if you are not better.

It seems to me that you should be given some other treatment. You spoke of going away. That was already something. Anything would be better than your present state and that enclosed nun's room!

A bientôt

R.M.

1. There are no clues in the correspondence to this remark.

2. *Chair de commissaire* (sometimes *chère de commissaire*) meaning a meal that is half fat and half lean – a learned pun referring to the commissions responsible for the implementation of the Edict of Nantes, which were half Catholic and half Protestant.

3. Proust had failed to turn up the previous day at a lecture given by Montesquiou to introduce excerpts from Victor Hugo's *La Fin de Satan* at the Théâtre des Bouffes-Parisiens.

4. Cf. the orangeade ritual at the Guermantes' (*RTP*, II, 532–3). For Proust's reaction to these reproaches, see letter 120.

5. Proust was listed as a subscriber to the Automobile Club's 'Unique Gala' announced in *Le Figaro* of 10 April.

6. Victor Hugo. See n. 3 above.

120 *To Robert de Montesquiou*

Monday evening [24 April 1905]

Dear Sir,

Had I not been even more unwell for the past two days, I should have thanked you sooner for the trouble you were kind enough to take in sending that letter from Neuilly without waiting for me to send for it. As I was unable to get up, it was difficult for me to write. You are, Monsieur, more cruel than the cruellest Catholic theologians, who wished us to regard our illnesses as punishment for our sins. You want us to consider them as sins in themselves and not only to suffer physically from our ills but to feel remorse for them; although unavoidable and already painful enough, they must be culpable as well. I must confess that there is something else in your letter which I try not to think about because it would embitter my rare days of health, add to other people's worries, and end – through the egotism of the invalid – by infusing a nervous resentment into my admiration, my respectful and grateful affection for you. This is the idea that when, roughly once a fortnight, I can get up, dress, and go out for an hour or two around ten o'clock at night, this single innocent relaxation can be regarded as a deliberate recovery (implying that the illness is deliberate too) with a view to vain pleasure. I know perfectly well that a number of people, because on such and such a day I am ill and can't do something, would rather I wasn't seen on a single day of the year, and, if I am, say: 'Ah! so you're well enough for the things that amuse you.' But that *you* should say more or less the same thing and not realize that if I could recover for the things I like most it would always be to go wherever I can see and hear you – that hurts me very much. When will I meet a person who truly understands my real life, my inmost feelings, who, having seen me miss some great pleasure through illness and catching sight of me an hour later (something which in any case doesn't happen at such a short interval!) at some common-place gathering, will approach me and say to me sincerely: 'How wonderful that your attack is over!' Your reproaches with regard to what you call 'the La Rochefoucauld orangeade' are also most unfair. For the last four Thursdays I've wanted to go to the rue de l'Université to thank Madame de La Rochefoucauld for coming to my house[1] since I cannot call on her as I never go out in the daytime and since she receives in the evening at an hour that is convenient for me and not tiring. But I haven't yet been able to get there in three years. The idea you seem to have that my choice of outings is dictated by snobbery (which would in fact lead me to places where you are speaking) astonishes me even more than it humiliates me. The arrangement and rearrangement of pleasures during an illness, when one is deprived of nearly all of them, is so straightfor-ward and sincere that it seems to us inevitable that everyone should understand what is so transparent and directs our actions towards such noble and disinterested ends, or would so direct them if the body could obey. If you knew all the things for which I've taken injudicious risks and those which have left me with a sense of profound regret, you would see that social considerations play little part in them. That doesn't preclude, on the evenings when I do go out, my going to a place where I can find refuge for that particular evening and see some human faces again, as long as it's a place that doesn't do me any harm. There are such places: at a time when the countryside was already fatal to me in the afternoon, I was always miraculously able to go to the 'Pavillon des Muses'[2] (if I wasn't ill that day) without any ill effects.

Forgive me, dear Sir, for boring you with explanations which must be very tedious for you. It will seem inexplicable to you that I should even think of offering them. It all stems from a subjective state induced by your letter. You have no idea of the nervous fatigue that overwhelms an invalid who feels himself misjudged by someone of whom he's fond and feels that his most innocent diversions will be interpreted against him. It can be a source of great agitation. I remember when I fought with Monsieur Lorrain,[3] at a time when I was not yet spending the whole day in bed but only the morning, my sole anxiety was that the duel might take place well before noon. When I was told that it would be in the afternoon, the whole thing became a matter of complete indifference. Which is to say that the torment of a neurasthenic is out of all proportion to the gravity of its cause. What would also torment me a great deal would be to have you come and see me since it's more or less impossible for me to receive in the daytime, and so painful! and you would be offering me such a tantalizing pleasure and honour without my being able to enjoy it – the more so because my hay fever is about to begin and I'm wondering with dread whether instead of being able to get up at nine o'clock in the evening I shall have to wait until one in the morning.

Monsieur, forgive this letter which is so stupid that I feel myself turning into a *catoblepas*[4] as I write. Having thought so much about what that admirable lecture must have been like, I haven't said a word about it, nor about you; I've talked only about myself. But you are always so kind to me that I wanted to tell you that you had hurt me. How I should like to be well and to be able to go and hear you in Brussels! Perhaps after my cure I shall be able to do such things. Give my warmest regards to Yturri who seemed well and was so kind the other day, and believe, dear Sir, in my respectful admiration and grateful affection.

Marcel Proust

And I haven't told you why I couldn't go to the lecture. If you only knew!

M.P.

1. The Comtesse Aimery de La Rochefoucauld had been a guest at Proust's tea-party on 6 March. See letter 105, n. 3.

2. Montesquiou's house at Neuilly.

3. See letter 3, n. 2.

4. A species of African bull described by Pliny the Elder. In Flaubert's *La Tentation de Saint-Antoine* the beast in question is made to say: 'I once devoured my own hoofs without noticing.'

121   *To Robert de Montesquiou*

Thursday [27 April 1905]

Dear Sir,

It would be ungracious of me to re-open the session which you at once so magisterially closed.[1] I remembered another fable in which Jupiter, anxious to put everything in its place, regulate sizes and shapes, keep things to scale, says to the audacious mortal: 'You shall be an ant' – a proportion very well illustrated by the four times' life-size Jupiter of Gustave Moreau.

The cedar smells not the rose at its feet.[2]

I imagine that the seventy[3] represent the number of tricks you think I have up my sleeve. I cannot thank you enough for that delightful letter which has made me appreciate once more, by thus *exhibiting* them on the picture-rail of contemporary application, those astonishing verses in which the tone of the fable is so marvellously observed in all its archaisms of style, with an oriental colouring as well: the *Fables* of La Fontaine illustrated by Gustave Moreau: 'Whereat Solomon rose' is wonderful. It was indeed all that had to be done and I admire the majesty with which you always give yourself what is too weakly called the 'beau rôle',[4] making the requisite gesture that raises you up, as king, sage and prophet, high above the puny efforts of the rest of us who are reduced to ants. 'You shall be an ant.' I yield to the necessary metamorphosis that places me, 'base insect, excrement of the earth',[5] at the feet of Suleiman more than ever the Magnificent. And while apologizing for having 'kept him standing' after he had risen, misunderstanding the traditional meaning of this formula of dismissal, and hurled at his legs yet another of my seventy battalions, a figure which it would be unwise to challenge me to prove is well below the reality, I send, dear Sir, my respectful regards.

Your affectionate admirer
Marcel Proust

As I've given my card for the *Figaro* reception for Duse,[6] it is probable that my name will be in the paper although I didn't go. So that won't mean I've 'resurrected' for this gala.[7] – I often think of another feast which is closer to the Resurrection and about which Flament spoke this morning in delightful terms.[8]

---

1. Montesquiou had mocked Proust's handwriting in terms of the fable of Solomon and the Ants on which he had based a poem, 'La Makédienne' (in *Le Parcours du rêve au souvenir*: 1895). For the 'session' referred to, see note 3.

2. 'Un cèdre ne sent pas une rose à sa base.' Hugo, 'Booz endormi', in *La Légende des siècles*.

3. Montesquiou had compared Proust's handwriting to the cavalcade of insects paraded in front of Solomon by Takia, Queen of the Ants. After seventy days during which she 'launched many a battalion', she announced: 'You have seen but one

species; there are seventy more to come.' The verse continues: 'Whereat Solomon rose' – i.e. declared the session closed.

4. For Montesquiou's retort, see letter 126.

5. 'Va t'en, chétif insecte, excrément de la terre.' La Fontaine, *Le Lion et le Moucheron*.

6. Eleonora Duse (1858–1924), the great Italian tragedienne.

7. Cf. letter 119, n. 5.

8. Albert Flament (1877–1956), journalist and gossip-columnist who had written in *L'Echo de Paris* about Montesquiou's lecture (see letter 119, n. 3).

122 *To Robert de Billy*

[Thursday evening, 27 April? 1905]

My dearest Robert,

I've come home with a fearsome attack of asthma, so much so that it's doubtful – though perfectly possible – that I shall be able to receive anyone. I'm telling you this just in case because you've already gone out of your way several times for nothing and I should hate to think you had called without my having the strength to receive you. I've worked a great deal since I last saw you. *I'm very fond of you.*

Marcel

If by any chance you happen to be speaking to your colleague d'Espeuilles, Duke of I don't know what,[1] tell him that I said good evening to him at Mme A. de La Rochefoucauld's because I thought he recognized me, but afterwards I wasn't sure. I had dined with him once at Mme de Saint-Paul's[2] and we had talked a great deal, whence my greeting. (This is not in the least a reproach, by the way; he was perfectly amiable and pretended to know me quite well, a flawless pretence for a lesser psychologist than me.) In any case all this is utterly trivial and I shan't in the least mind if you don't do it. I hope you are not too worried and that things are not going too badly.[3] Also at the same soirée was Radolin, your guest of the other day, who didn't reply to the questions put to him by all those society people about Germany's intentions – and indeed almost the entire Foreign Ministry including the Horrible Beaucaires, Pange etc.[4]

Affectionately yours
Marcel

1. Adrien Vicomte d'Espeuilles de Caulaincourt, Duc de Vicence. He was no longer a colleague of Robert de Billy (see letter 55, n. 3), having left the diplomatic service.

2. Marquise de Saint-Paul, *née* Diane de Feydeau de Brou, principal model for Mme de Sainte-Euverte in *RTP*. Cf. letter 123.

3. The Franco-German quarrel over Morocco was causing great anxiety in the diplomatic community.

4. The guests Proust mentions were all diplomats: the German ambassador Prince Hugo von Radolin, the Comte and Comtesse Horric de Beaucaire (hence Proust's pun) and the Comte de Pange.

123   *To Madame Straus*

Friday [28 April 1905]

Madame,

I met Jacques[1] yesterday (since you left I've only been able to leave the house twice and both times I was lucky enough to meet Jacques at Wéber's). He introduced me to his wife whom I wasn't able to talk to because I was already full of asthma, but I found her enchanting, with something so attractive, so rare, distinguished and superior about her that I thought a great deal about her eyes.[2] (I imagine Jacques won't have been offended!) He told me that if I wanted to write to you he thought I could and that it wouldn't tire you. In spite of this, he told me you were still under the weather, which greatly distressed me. As I see no one I can't 'keep you posted' about anything. I've worked so hard for the past two months, in the moments of respite my attacks allowed me, that I haven't even been able to have visits from friends at home. A very small piece of mine will appear in the *Renaissance latine* on 15 June.[3] By then you will have returned, and if it doesn't tire you I'll give it to you. One of the times I went out (it isn't very proper but then we're both invalids), at the end of a concert I heard Mme de Saint-Paul[4] make some unintentionally obscene remarks as she does every time I hear her open her mouth. She thought some things had been badly sung. 'My dear, I'm going to get Plançon to do it for me' (I suppose she meant sing it for her), 'that will be a real pleasure. Thursday next at ten o'clock. Lovers of the art who may wish to be present, please note.' An eddy of the crowd carried me out of earshot, but I was pushed back towards her again and this is what I heard: 'My dear, say what you like, Madeleine likes good cooking. I prefer to give you good music. She gives two thousand francs to her cook; I prefer to give it to my artists. It's still the same sum. Madeleine likes it in her mouth, but I prefer it in my ear. Each to her taste, my dear, it's a free country, after all.'[5] I don't know whether you've read the article in the *Echo de Paris* in which Flament, meaning to be agreeable to Primoli[6] and in order to flatter him, says how happy he had been to dine with him at Mme Calvé's.[7] And he adds: 'This entirely bald old man, with his white beard, reminds me irresistibly of some Eternal Father majestic in his indulgence.' I'm not sure whether Primoli will have been as pleased as all that; nor am I sure whether it's because my eyes are already so old that I can only see things through the deceptive veil of old images, but it seems to me that his beard is fair even if it has gone a bit white, and that he isn't a venerable old man at all.

I'm worn out by letters from Montesquiou. Every time he gives a lecture or throws a party etc. etc. he refuses to acknowledge that I'm ill, and beforehand there are summonses, threats, visits from Yturri who wakes me up, and afterwards, reproaches for not having gone. I feel it might still be possible to get well if it weren't for 'other people'. But the exhaustion they cause you, one's powerlessness to make them understand the suffering, sometimes lasting a whole month, that follows the imprudence one has committed for the sake of what they imagine to be a great pleasure – all that is death.

I read so many things every night that without tiring you I could send you countless extracts from books. But they are such boring and serious things that I don't know whether you'd enjoy it. Last night I read some letters of Mme Desbordes-Valmore,[8] full of pretentiousness in the style of Mme Daudet etc. But in fact we're wrong to make fun of pretentiousness, of irritating mannerisms. Talented people are like that. She sends little notes like this to Sainte-Beuve:

> If you were still our angel
> Lightly you would take wing
> And you would fly tomorrow
> To shake your sister's hand.
> She will sail to England,
> To rest awhile from land;
> She will embark upon a ship,
> And I shall brave the waters
> To fetch her back again.[9]

The frightening thing in these letters is to see how selfish love is. As soon as the lover is dead, since nothing more can be expected from him, it's all over (in any case it isn't true of everyone). But still she had been madly in love with Latouche.[10] He had treated her abominably. Nothing could have made her forget him. He dies. Sainte-Beuve, who wants to write an article about him, asks Mme Desbordes for some information about this man 'who had come so near to talent' (isn't that typical of Sainte-Beuve?). 'I should like to speak of him' (he adds benevolently) 'with the indulgence that one owes to a man who did not do all the harm he might have done.' I thought that Mme Desbordes, indignant, would at least beg him to do a kind article. She lets some time go by and then answers. First, protestations of grief: 'I write to you, my eyes dimmed with tears which never cease to flow, etc. etc.' Then she comes to the projected article: 'You wish to speak of him with the indulgence one owes to a man who did not do all the harm he might have done, you say. Oh, how true, how very true. He did a great deal of harm, but, etc. This man who came so close to talent. How apt that is. To tell the truth, I'm not even sure of that and can give you no information, as *I didn't dare open his last books for fear of finding them too bad*. Everyone told me they were worthless, etc.' What bitchiness! And when one thinks that she was at his feet, to the point of being mad with joy because his Christian name (Joseph) corresponded to hers (Josèphe). 'Your name! You know that heaven deigned to write it into mine!' (which made people think, since she was also called Marceline, that her letters were addressed to a M. de Marcellus). Now I may be interpreting her letter to Sainte-Beuve wrongly. One would have to ask someone who knew more about it all. I'd always been told she was an angel. I must say I was rather surprised. Madame, I hope you will get well at once and I send you my respectful affection.

Marcel Proust

1. Jacques Bizet (1872–1922), Proust's former schoolfellow (see letter 30, n. 1) and Mme Straus's son by her first husband, the composer Georges Bizet (1838–75).

2. His first wife having died, in 1904 Jacques Bizet married a divorcee (Mme Georges Sachs, *née* Alice Frankel) about whom Proust was not always to be so complimentary (cf. letter 308, n. 5).

3. 'Sur la lecture' (On Reading), his preface to *Sésame et les lys*.

4. See letter 122, n. 2.

5. Cf. the scene in *Swann's Way* where Mme Verdurin speaks in similar, unconsciously equivocal terms about her Beauvais tapestry illustrating 'The Bear and the Grapes' – 'I don't need to put them in my mouth, I get my pleasure through my eyes' – and when inviting Swann to stroke her bronzes as the young pianist is about to play – 'Now it's you who are going to be caressed, caressed in your ear' (*RTP*, I, 226–7).

6. Comte Joseph Primoli (see letter 2).

7. Mme Emma Calvé, society hostess.

8. See letter 80, n. 3.

9. 'Si vous étiez toujours notre ange / Et sans qu'un tel vol vous dérange / Vous

viendriez demain / A votre soeur serrer la main. / Pour la reposer de la terre / On nous l'envoie en Angleterrc / On la mettra sur un bateau / Où j'irai la chercher malgré ma peur de l'eau.' Proust misquotes these lines which he had read in a book by Léon Séché (biographer of Musset) published in 1904, for the centenary of Sainte-Beuve's birth.

10. Henri de Latouche, real name Alexandre Thibaud (1755–1851).

124   *To Madame Catusse*[1]

[April–May 1905]

Dear Madame,

What to read? I really have nothing on Laon. Somewhat vague memories of the cathedral itself and of Mâle's fine book on religious art[2] in which of course there is very little mention of Laon. It's mainly from the point of view of architecture that Laon is interesting, and more engaging than any other because the green shoot is not yet open. It's there, better than in the rich subsequent efflorescences, that one can see the first burgeoning of the Gothic and how 'the marvellous flower slowly emerges'. But how to show this otherwise than on the spot, under the vaults of the magical forest? The iconography would lend itself more to epistolary exchanges if my recollection were more precise. Erudite, impregnated with the scholasticism that was taught there, Laon will offer you with delightfully pedantic insistence the liberal arts in its main portal and in the stained glass of its rose window. You will recognize Philosophy from the ladder (of learning) placed in front of her chest, Astronomy gazing at the heavens, Geometry with her compass, Arithmetic counting on her fingers, Dialectics with the wily snake. Architecture is very splendid. Medicine, though, is rather banal, not as at Reims, where she is examining an invalid's urine (pardon me, Madame). On the portal too you will like the Erythraean Sibyl with a verse by St Augustine, and the Wise and Foolish Virgins. But the big thing on the portal, one of the essential pieces of medieval sculpture, is the life of the Virgin told by anticipation with the aid of Holy Scripture. I confess I'm too tired this evening to describe it to you. But would you care to refer to my *Bible d'Amiens*? I say a word about it in a note (p. 326). The stained-glass windows are interesting, especially those in the choir, because of interpretations that are still a bit Romanesque; the legend of the Midwives coming to examine the virginity of the Virgin figures there probably for the first time, a very strange fish adorns the Last Supper, the Good Thief (why?) accompanies the Flight into Egypt. If I remember rightly, the artist did not dare (this is very Romanesque) represent the risen Christ. Only the Holy Women weep at the tomb. There's a lovely vase of flowers between the Virgin and the Angel of the Annunciation.

Your respectful friend
Marcel Proust

1. Mme Anatole Catusse, *née* Marie-Marguerite Bertrin, widow of the former Préfêt of Nice and old friend of Mme Proust. Proust was later to rely on her advice in domestic matters after the latter's death (see letter 172 *et seq.*).
2. *L'Art religieux du XIIIe siècle en France* by Emile Mâle (see letter 92, n. 3).

125   *To Madame Fortoul*[1]

Saturday [April–May? 1905]

Of course I remember, Madame, the beauties of the floating Museum and the pleasures of the enchanted Isle! I think of them very often and also of the 'Lady of the Sea' who is also, I know, the Lady of kindness, courage and learning. If I hadn't led an unbelievable life since the *Hélène*, bedridden for nine days out of ten, I should have asked Robert whether there was any way of seeing you on land, but I've had to postpone all pleasures until my recovery, which is becoming less problematical as I have at last decided to undertake a serious cure.

As regards the subject which interests you and which it's high time I got round to, I haven't read M. Crépet's book[2] but here, in a Lemaître[3] which my friend Reynaldo Hahn has lent me for the purpose (forgive the lamentable state of the copy – I'm sending it to you just as it is for the sake of speed and in any case you are noble enough not to shy away from sordid things), is an analysis of it which is fairly accurate and pleasant to read. As far as I can remember, Lemaître is very unjust to Baudelaire and I think the book will help you to anticipate the criticisms rather than to refute them. My copy of *Les Fleurs du Mal*[4] contains various articles by Barbey, Asselineau etc. which represent pretty well what can be said against Baudelaire. But you must have exactly the same edition. I confess that Baudelaire, I don't know why, is not only one of the poets I like most and know best, but at the same time one of those about whose life and biography I am least well informed. So I should find little to say to you from my own resources. Moreover, in order to reinforce the defence, however feebly, I should need to know at least on what specific points the attack was targeted. You don't tell me whether it was the man or the artist who was taken to task. If it was the artist and if it was his 'satanism', a bit outmoded I grant you, that was considered rather jejune, one must I think reply that this is only a secondary aspect of Baudelaire, one of those aspects which in the lifetime of a writer can momentarily dominate and obnubilate all the rest, but which we have the right and duty to disregard. In reality this poet, who is accused of being inhuman, with his slightly silly pose of aristocracy, was one of the most tender, most warm-hearted, most humane, most 'popular' of poets. A piece like 'Le Vin du travail-leur' (if I'm not mistaken about the title of the poem[5]) evinces a delightful democratic fraternity and if it weren't so formally admirable one would realize that there's as much real tenderness for humble people in it as in the whole of François Coppée.[6] Of course it's a tear that flowed only once from a face too proud not to remain impassive. But it's enough that it should be an 'immortal tear' for us to penetrate the poet's heart. Did they say he was decadent? Nothing could be more false. Baudelaire isn't even a Romantic. He writes like Racine. I could quote you a score of examples. Moreover he's a Christian poet and this is why, like Bossuet,[7] or Massillon,[9] he talks incessantly about sin. Let's say that, like all Christians who are at the same time hysterics (don't misunderstand me, I'm not saying that Christians are hysterics, I mean 'those Christians who by chance happen also to be hysterics') he indulged in the sadism of blasphemy. I should be happy if I had chanced to hit upon the answers to the objections which have been expressed to you and I beg you, Madame, to accept my most respectful regards.

Marcel Proust

1. Mme Fortoul (see letter 55, n. 4) had been Proust's fellow guest in August 1904 aboard the yacht *Hélène* owned by Robert de Billy's father-in-law.

2. Eugène Crépet (1827–92) wrote an authoritative biographical study of Baude-laire and edited his posthumously published works and correspondence.

3. Jules Lemaître (1853–1914), literary critic, author of *Les Contemporains* (Paris, 1893) which includes a study of Crépet's book.

4. Proust's copy of Baudelaire's works (*Les Fleurs du Mal* augmented by previously unpublished poems) was either the second edition (Paris, 1869) or its reimpression (1894); both had a preface by Théophile Gautier and various appendices including articles by the novelist Barbey d'Aurevilly and the critic Charles Asselineau.

5. In fact, 'Le Vin des chiffoniers' (The Ragpickers' Wine).

6. François Coppée (1842–1908), poet and playwright, author of *Les Humbles*.

7. Jacques-Bénigne Bossuet (1627–1704), the great seventeenth-century prelate, preacher and writer, famous for his funeral orations: *oraisons funèbres*.

8. Jean-Baptiste Massillon (1663–1742), prelate and preacher who delivered the *oraison funèbre* for Louis XIV.

126   *From Robert de Montesquiou*

Neuilly, 3 May [1905]

Dear Marcel,

It seems to me that, this time, the battalions[1] have brought along a little sourness, and a great deal of inaccuracy. What makes you think I take the *beau rôle*?[2] I don't *take* it, I already have it. And I hasten to add that *everyone has his own*. That of *formica-leo*[3] would not displease me, I assure you. But since my symbolism applies to *characters* (not character) I could really only have claimed for mine a *snake kingship* which not everyone considers flattering.

Please believe that I assert no rights over the herd of the Behemoths and Leviathans, and indeed that I gladly surrender them to you, especially when they assume the shape of society people.

So, to prove that I attribute merely to *aegri somnia*[4] your reply (usually better inspired) to a friendly sally whose only aim was to beguile what Mallarmé calls 'the curtains' vapid whiteness',[5] I propose to come and give a lecture (of a quarter of an hour) in your home, at nine o'clock in the evening on the day of publication of my book,[6] if that would amuse you, in the presence of two or three of our friends, male and female, on the choice of whom we can agree together. And if you consider that it would be another case of me *taking the beau rôle*, at least this time we would be sharing it.

R.M.

1. See letter 121.
2. Ibid.
3. i.e. 'lion-ant' (ibid.).
4. *Aegri somnia*, the futile visions of a sick man. Horace, *Ars Poetica* II, 7.
5. 'Las du triste hôpital, et de l'encens fétide / Qui monte en la blancheur banale des rideaux . . .' Stéphane Mallarmé, *Les Fenêtres*.
6. Entitled *Professionelles beautés*.

127 *To Robert de Montesquiou*

[Thursday evening, 4 May 1905]

Dear Sir,

'You cannot read what lies within my heart'[1] if you thought you detected 'sourness'[2] in it, as you say. It has nothing but gratitude and admiration for you.

Your all too flattering and honorific proposal[3] delights me. But is it feasible? My attacks give no warning. I might find myself that very day incapable of getting out of bed, or even speaking, in the throes of asphyxiations as unbearable for you as for me and with a fever amounting almost to delirium. It's true that I can reduce this risk, if not to zero, at least a good deal, by remaining completely recumbent for some days beforehand, by not speaking, and by taking nothing but milk. But then you would have to tell me when it would be, and it would have to be a day when I can banish my mother, whom I couldn't persuade to see even five people. Since my last unfortunate letter, I've been iller than I've ever been before, indescribably ill. And since I have a volume to deliver to the Mercure[4] in a month (I'll explain to you what it is but I think it's really too indifferent), each attack fills me with despair by putting me back and making me fear that I shall never get to the end. Forgive me for talking about things that must be so uninteresting to you. I'm very glad to know that I shall soon be reading a new book of yours.[5] It will be a great joy. It's sad that I don't do any literary criticism in a newspaper, as I should very much like to write about it. No matter; I hope that, without waiting for the degree of notoriety which might perhaps earn me such a column, I shall be able to talk about you a little in a general way. Only I can't quite see how or when. But that must interest you even less – as I know better than anyone, having read the letter you sent to *Horatio* at the *Figaro*.[6] It didn't contain a word of thanks, and was not at all encouraging! But for my own sake, for my own pleasure, I'd like to write about you. After my cure, if I lead a normal life as I'm assured I will, I shall try to have slightly more regular relations with newspapers and reviews. I won't write to you at greater length because these nights of working by electric light spoil my eyesight to such an extent that I can no longer see my 'characters' as I write to you.

Your respectful admirer
Marcel Proust

1. 'Dans le fond de mon coeur vous ne savez pas lire!' Racine, *Phèdre*, Act II, scene 5.

2. See previous letter.

3. i.e. Montesquiou's proposed reading at Proust's flat (see previous letter). See following letters concerning this event, which eventually took place on Friday, 2 June.

4. Mercure de France were about to publish Ruskin's *Sesame and Lilies* in Proust's translation.

5. See previous letter.

6. Gaston Calmette, the editor of the *Figaro*, must have shown Proust a letter from Montesquiou about the pseudonymous pastiche, 'Fête chez Montesquiou à Neuilly', signed 'Horatio', which had appeared in January 1904 in *Le Figaro*. Cf. letters 25, n. 7, 27, n. 1, 129 *et al*.

128   *From Robert de Montesquiou*

Neuilly
Friday evening [5 May 1905]

Dear Marcel,

It would be tedious, and pointless, this evening, to revert to the *inexplicable* letter[1] . . . You will re-read it, and I think you will withdraw . . . Let us go on to other matters.

I am touched by your sentiments. I am distressed by your health.

I shall appoint a day, with no obligation, for the *reading* (half an hour at most), and if it can be done, it will be done. I should have liked to have Madame Straus, our friends the Lemaires,[2] and Flament, *with no others*. Yturri would come with me.

It would provide a preliminary documentation for your article, *which would please me very much*.

(If Horatio reads your letters, as you, I see, do his, will he be satisfied, *for two?*)

I'm sending you at the same time two fragments of a provincial frieze,[3] certain characteristics of which may, perhaps, 'beguile your idle hours'.

R.M.

1. Montesquiou is presumably referring to his letter of 22 April (119).

2. Madeleine Lemaire and her daughter Suzette.

3. See note 1 of the following letter.

129   *To Robert de Montesquiou*

Friday evening [5 May 1905]

Dear Sir,

What marvellous pages![1] And how I thank you for having sent them to me. This provincial spawn, these local blooms, enchant me, and I'm a little ashamed, on hearing myself reproved for 'the desire to have learned something', to catch myself wanting to know the name of the 'tennis-playing anglomaniac', not having hesitated, in the case of the 'cold draught' and 'the exuberant duchess', to put beside them the names Standish and Rohan, which probably applied.[2] I'm now afraid that what will be revealed in the preface to my Ruskin of my profound affection for country life, doubtless because I know so little about it, will seem to you stupid.

I find the composition of the audience for this 'lecture' 'exquisite and beautifully arranged'.[3] Unfortunately Madame Straus, after two years of nervous illness from which she nearly died, is at present in Switzerland. *Uno avulso deficit alter aureus.*[4] I don't see who could replace her. She is without equal. Perhaps we should await her return. From the point of view of documentation, as you call it, it wouldn't matter, since alas, it will be a while before I'm able to write, and perhaps not about the book. In this connection, it's too kind of you to say that it would be agreeable to you, because I know it wouldn't and that you only say it out of niceness to please me. As I told you, I have positive proof on that score, your irritated coldness towards Horatio as long as he was for you only 'the author of an article', a coldness which gave way to your habitual kindly feelings when you learned that he was none other than the person whom you always treat with benevolence.[5] But I assure you that that pretence of being pleased by praise from me, which you put on only to please me, touches me all the more because I know what real indifference it conceals. If on the other hand you would prefer not to wait for Madame Straus's return, you have only, whenever you wish, to let me know what your intentions are. As for the other guests, they are admirably chosen – if there are to be any. I say this because I find that one only listens properly to what one hears alone, or in a public place (the crowd absorbing the individualities which compose it) and the minor pre-occupation of being a 'host' lessens the perfect adaptation of the mind to receive the suggestions of the Spirit. But I can understand that, wishing to address, like St Paul or St John, certain churches which you regard with special favour, you cannot address each member of the faithful separately and are constrained to 'hearings' comprising a greater number of people, in this case flatteringly small.

Please accept, dear Sir, my most admiring and grateful respects.

Marcel Proust

1. Montesquiou had sent Proust an article, 'Encycliques mondaines' (Social Encyclicals), later collected in his book of literary portraits, *Altesses sérénissimes* (Serene Highnesses).

2. Mrs Henry Standish (*née* Hélène de Pérusse des Cars: 1840–1933), elegant style-setter, friend of the Prince of Wales; and the Duchesse de Rohan (*née* Herminie de la Brousse de Verteillac: 1853–1926), poet and watercolourist.

3. 'La chose fut exquise et fort bien ordonnée.' Hugo, *La Fête chez Thérèse*.

4. Cf. 'Primo avulso non deficit, alter aureus, et simili frondescit virga metallo' (Virgil, *Aeneid*, VI, 194): 'When the first is plucked away another gold one grows in its place with leaves of the same metal' – the golden bough without which it is forbidden to enter the underworld.

5. See letter 127, n. 6.

130   *To Madame Straus*

Sunday [7 May 1905]

Madame,

It's not 'sporting' of you to have replied to me. Don't do it again or I shall stop writing to you.[1] I don't know how I'm going to write you this letter. Working for so many nights by electric light without a lampshade has suddenly (not tonight but for the past week) strained my eyesight, which used to be excellent, to such an extraordinary extent that I can scarcely make out my characters as I write. So that I more or less don't write. If I find an oculist who is prepared to see me at eleven o'clock in the evening I shall consult him. However I hope it will pass off.

Having gone at about that hour to see Gabriel de La Rochefoucauld, I entered a salon filled with a dreadful group of people from the midst of whom rose barnyard shrieks. It was a very courtly squabble between Madame Ganderax[2] and a tall woman with yellow hair, a sort of queen of the roost, who, without listening to what Madame Ganderax said to her any more than Madame Ganderax listened to her, kept repeating: 'No, no, Madame, to have founded *Psst*'[3] (they were talking about Forain) 'when one owes everything to the Strauses, when one owes them one's bread, when without them one wouldn't know how to draw . . .' I think

Madame Ganderax was of the same opinion, but since the other one went on shouting without listening, it looked as though she was being contradicted, which was not the case. 'No, Madame,' she went on, 'I repeat, when one has behaved so infamously, when one would have starved if it hadn't been for the Strauses, etc. . . .' After these premisses, I expected some such conclusion as: 'When one has done that, one is the lowest of the low' or something like that. Not at all: 'When one has behaved so infamously, when one owes one's livelihood to the Strauses, etc., do you know what one is, Madame? Well, I'm not afraid to say it, you may find the expression a bit strong: one is a puppet!' – I felt that after such an exordium the word was somewhat feeble. But that didn't seem to be the lady's opinion. She seemed so delighted by it that she repeated three or four times in an airy way: 'He's a puppet, he's a puppet', then got up and left. When I asked Gabriel what her name was, he said Madame Uhring (?) and as he sensed that I didn't seem to admire her, he added with an air of authority as though to hypnotize me: 'One of Madame Straus's best friends.' To me she seemed more like a boring old cocotte than one of your best friends. But I said nothing. Moreover she spoke to me very obligingly as she left.

Something else concerning you (as Goncourt[4] used to say, it's the only interesting subject): I think I told you that I didn't go to Montesquiou's lecture[5] which made him furious and resulted, under the pretext of contrasting my tiny handwriting with his, in his comparing himself to Solomon and me to an ant.[6] Since this comparison rather irritated me, I replied that he always tried to give himself the *beau rôle* (I also said some better things than that), to which he replied: 'Where do you get the idea that I take the *beau rôle*. I don't have to take it, *I already have it*.'[7] Finally, to console me for not having heard his lecture, he announced that he would come and give one at my house at such and such a time; actually not for a month, or I don't know when.[8] Well, this occasion, so solemnly announced (which has prevented me from sleeping ever since I heard of it, and in fact I told him it was impossible) was to comprise an audience of only *three*! At the head of the list: Madame Straus. I told him you were away from Paris and since of course you were irreplaceable it would be best to await your return. I don't know what his answer will be. As I never have any luck with the Ganderaxes and people are always gossiping, I'd rather you didn't mention the Uhring conversation to them, although it's not in the least disobliging as far as they're concerned, but they always take what I say amiss. I'd prefer silence, which at least would avoid misconstruction. I forgot to mention that there was another man of wit involved in the conversation, to whom Gabriel, who seems to have gone over completely to his wife's family, introduced me as 'my *uncle*, M. Georges Heine'.[9]

At the concert where I heard Madame de Saint-Paul express herself so well,[10] I saw a kind of colossus dressed as a woman, with false red kiss-curls dangling over her forehead, a sort of Queen of the Spiders, a red ant played by Baron[11] in a revue with a voice of thunder: Madame Bartholoni.[12] She's what's called a splendid old trout. Perhaps from the peaks of your indifference you tower above life too loftily to have the slightest pleasure in learning that Mme de Pourtalès asked Reynaldo to tell Madame Lemaire that nothing would be more disagreeable to her than to meet the Arthur Meyers[13] in her house (your Turenne having prevailed upon Mme de Pourtalès not to receive them, contrary to her promise, she is terrified of meeting them).

I need hardly tell you that if the letters of Mme Desbordes which I have in this book on Sainte-Beuve might interest you,[14] I would hasten to send them to you. But apart from what I told you, there is nothing that you wouldn't find tedious. I'll find out whether her *Complete Correspondence* exists. In that case you could see if there was something that might interest you. I can't tell you, Madame, how enchanted I was by your little card, so charmingly, so wittily written. At the moment anything coming from you gives me a hundred times more pleasure than in the past. I am so continually grateful to you for being convalescent. How wonderfully well you write! While I was doing the thing that will appear in the *Renaissance latine* in June,[15] I kept saying to myself: Ah, if only I could write like Madame Straus! And I said it really feelingly, I can assure you, with a heartfelt longing for all that lucidity, that delightful equilibrium that makes your sentences so enchanting. But don't, I beg of you, write to me. I am too fond of you for the pleasure you give me not to be dwarfed by the pain of having tired you.

Please accept, Madame, my respectful, my profound friendship.

Marcel Proust

1. Cf. letter 118.

2. *Née* Nina Vermicati (1855–1920), of Italian origin. Her second husband, Louis Ganderax (1855–1940), was co-founder of the conservative *Revue de Paris*.

3. *Psst*: short-lived satirical anti-Dreyfusard news-sheet founded by Forain (see letter 42, n. 8) and the caricaturist Poiré, known as Caran d'Ache (1858–1909).

4. Edmond de Goncourt (1822–96), the elder Goncourt in the famous fraternal literary partnership, later the subject of one of Proust's Lemoine parodies (see letter 265, n. 2).

5. See letter 119, n. 3.

6. See letter 121, n. 1, n. 3.

7. See letter 126.

8. Ibid.

9. Brother of the Princesse Alice de Monaco (see letter 227, n. 2) whose daughter by her first marriage, Odile de Chapelle de Jumilhac de Richelieu (1876–1925), heiress to the Heine fortune, had just married Gabriel de La Rochefoucauld.

10. Cf. letter 122, n. 2.

11. Baron (stage name of Louis Bouchenez: 1838–1920), vaudeville actor and singer.

12. Mme Anatole Bartholoni, Second Empire beauty and Chateaubriand's granddaughter.

13. See letter 64, n. 8.

14. See letter 123.

15. See letter 123, n. 3.

131   *To Robert Dreyfus*

45 rue de Courcelles
·Sunday [14 May 1905]

Dear friend,

I shall certainly be writing to you again about this fascinating book,[1] not in order to please you but because I feel that I'm going to be ingobinated, at the risk of having to undergo a bit of degobination later on, and want to talk about him all the time and learn more about him. But my eyes are still tired, I have a long article for the *Renaissance latine* of 15 June and a big volume of Ruskin for October,[2] so I shall only be able to read your book slowly. That is, if I'm sensible. But will I be? For I received it not long ago and I can't extricate myself from it. The 'prefatory' portrait of Gobineau is charming (I mean the portrait by Mme de la Tour).[3] But yours ought to have been there too; it's true that 'in our minds', as you say, we have an extremely handsome portrait of you. For the extraordinary restraint with which you relate and judge, that sort of cold fervour, of unexpressed malice, of casual meticulousness, that exquisite 'measure' (pun intended) in which there are always four rests for two quavers, two demisemiquavers, those suppressed tears, that astounding brevity when the subject calls for eloquence but which is supremely eloquent itself, a conclusion of five lines and a preface of four – all this is of an extreme 'distinction', as they say, and individuality. This old madman (not you now, but Gobineau) enchants me. His historical conception of oral tradition which consists of interrogating a 'nomad horseman' on the reign of a legendary king!; his passion for collecting engraved stones and his love of parakeets; his arrogance towards society people; his irritation at always being called 'the good M. Gobineau' – all this I find delightful. What he says about Herodotus is pure Ruskin (another charming madman). 'Would to heaven' is touching and splendid, the passage about Michelangelo is really fine, Camille Doucet's[4] verdict on him killingly funny – what could it not be applied to. You don't say anything but I suspect you laughed as you transcribed it. I'm sure I've seen (at Madame Straus's) M. de Basterot,[5] who in my mind was spelt Bastro and whom I had taken for an old dotard. He was a friend of Mme Cahen[6] and of Bourget.[7] Madame Straus will tell you all about that; I can scarcely remember him. Mérimée[8] goes up in my estimation where he was already fairly high. But I don't want to bore you with all this. I *admire* you (I hasten to rewrite the word so that you see that it's 'admire' – it looked like 'I adore you' or I don't know what, I'm so illegible).

What do you mean by saying that Viollet-le-Duc[9] wrote a great deal for an artist? Is it a reproach?

Ever yours
Marcel Proust

1. *La Vie et les prophéties du comte de Gobineau* (The Life and Prophecies of Count Gobineau). Cf. letter 91, n. 1.

2. His translation of *Sesame and Lilies*, the preface to which was to appear in *La Renaissance latine*.

3. The Comtesse Victor de la Tour whose husband was a diplomatic colleague of Gobineau.

4. Doucet was Secretary to the Académie Française.

5. Comte Florimont de Basterot, friend and biographer of Gobineau.

6. Mme Albert Cahen d'Anvers, see letter 87, n. 3.

7. Paul Bourget (1852–1935), fashionable novelist.

8. Dreyfus cites letters from the writer and scholar Prosper Mérimée (1803–70) to Gobineau.

9. Reference to a footnote by Dreyfus on the architect and writer, see letter 92, n. 4. In his reminiscences of Proust, Dreyfus was to write: 'An execrable remark buried in a footnote of my book. Marcel Proust read footnotes.'

132 *To Robert Dreyfus*

[About mid-May 1905]

My dear Robert,

A friend of mine who has been in Germany and Russia has written an article on the situation in Russia about which he is pretty well informed.[1] There is nothing very wonderful about this article but it is very sensible and fair and it would give me great pleasure to be able to get it published in a newspaper or review, however modest. But it is extremely harsh about the Tsar's family and the Grand Dukes, which might make it rather difficult to place in a 'bourgeois' sheet. On the other hand he pins hopes on the benevolence of the Tsar which I don't believe are shared by the socialists and which perhaps would debar him from access to *L'Humanité*. I've had it for quite a long time and I want to do something about it but I've just been so ill that it's been impossible. Do you have any dealings with newspapers or reviews where it might be placed; at a pinch even young people's reviews like *Les Essais*, although the literary quality of that review would be a matter of indifference to my friend, who lives very much outside the world of literature. If you could do something about it you would be doubly kind in repairing my remissness by your haste. If you don't know of anything, or if you haven't the time to think about it, there's no point in my sending it to you. You would have to return it and it would be a nuisance for you. If however you can think of some outlet or other in principle, I'll send it to you at once. I'd much sooner have talked to you about all this in person, but I'm too ill at the moment to entertain anyone. I go out roughly once a month for an hour, and afterwards, quite apart from attacks of asthma, have to spend a week in bed. As for my friend, I can only put you in touch with him, if it's of any use, by correspondence, because he lives in Savoie and almost never comes to Paris, although he intends to settle here one day.

I hope life is treating you kindly and that you are happily engaged in fruitful work. As for myself, since you are kind enough to take an interest in me, I am not too unhappy at the moment. I can work a little – except however during my recent terrible crisis – and I lead a very quiet, restful life of reading and very studious intimacy with Mama.

Ever yours

Marcel Proust

1. Evidently Proust's former close friend Comte Clément de Maugny, whose family home was in Haute-Savoie. Cf. vol. I.

133 To Robert de Montesquiou

Tuesday evening [16 May 1905]

Dear Sir,

First of all, allow me to thank you with all my heart for the kindness and 'grace' (in the Jansenist sense) which prompt you to overwhelm me thus with honour and joy.

Then, as for the day,[1] here are the various considerations which I submit to you, and you will decide in the last resort however you please.

On Thursday the 25th there is a final performance of a thing Reynaldo wrote this year which he is more pleased with than anything else he has done up to now and which he specially wanted me to hear.[2] But since it has always been performed in the daytime, and only once in the evening when I was ill, I wasn't able to go. I promised him that if I was well on the 25th, since it's an evening performance, I would go. I am all the more anxious to do so as I haven't been able to go to a single one of the various recitals he has given, and as I shan't be going to the concert he is conducting tomorrow, nor in a week's time because it's in the afternoon, nor have I been able to go to any of the performances of his *Esther* at Mme de Guerne's etc.[3] Now, if on the *Tuesday* I get up, dress, and 'receive', it's extremely likely that the attacks that follow will not be over by Thursday evening. But still, especially if you could arrange it for ten o'clock or half past ten instead of nine, it wouldn't be at all impossible (I say ten or half past, if you have no objection, because my attacks of hay fever, which at the moment are an additional affliction, generally occur around four o'clock and I go back to sleep afterwards). So, if you wish, Tuesday the 23rd at ten or half past or the same evening at nine.

The following Tuesday wouldn't present the same problems and would be perfect, but perhaps too far distant and not convenient for you. However, whether next Tuesday or the one after, it could be a Tuesday only if the gathering retained the character which you fixed for it in advance, and was limited to three or four people. Because, since it was on a Tuesday that my father was struck down,[4] my mother still feels a kind of sadness and is anxious that there shouldn't be 'people' in the house on that day, even without her, and I already make her lead a rather sad life, obliging her to dine every evening at eleven o'clock, without saddening her further. But if we're only four Tuesday will be fine. If you had changed your mind on this point, I mean if we were to be more than five or six, I would ask you to make it rather Monday the 19th or Wednesday the 31st. Otherwise Tuesday the 30th or, if you prefer, Tuesday the 23rd. And if it's Tuesday the 23rd, ten o'clock or half past ten for preference but only if that suits you. I have too much 'abulia' and terror of fixed plans to decide on the day myself, even if deference didn't prevent me from doing so. So please decide for me. As for the names of the people to be invited, we won't have to 'come to an agreement' as you put it, for the very good reason that they will be those you indicated to me. And I shall write to them as soon as you have done so. You mentioned Madame Lemaire and Flament. I think if you would allow me to add Gabriel de La Rochefoucauld he would be very pleased. But I'm merely submitting his name and will naturally comply with your decision. Perhaps after all it would be wiser to stick to Tuesday the 23rd, since it seemed to suit you. I don't know what to say when I see the waves of unforeseen difficulties mounting on every side. My suggestion of a reading for me alone, which would have been so much better, didn't please you? No doubt you felt that 'it wasn't my place' to behave thus.

Thank you again with all my heart, dear Sir. Decide whatever you wish and accept once more the respectful expression of my admiring and grateful sentiments.

Marcel Proust

1. i.e. for Montesquiou's reading in Proust's house. Cf. letter 126 *et al.*

2. Reynaldo Hahn's suite for wind instruments, two harps and piano, the *Bal de Béatrice d'Este*, had had its first two performances at Mme Lemaire's during April.

3. Hahn's conducting arrangements included Lully and Rameau at the Théâtre de l'Athénée, Lully at the Opéra-Comique and a charity performance at Mme de Guerne's of his own *Choeurs d'Esther*.

4. Proust has given this excuse before, cf. letter 80, n. 2.

134 *From Robert de Montesquiou*

Neuilly
[Wednesday, 17 May 1905]

Dear Marcel,

You put so much sauce around the fish that it's in danger of disappearing. So let us retrieve it. The 23rd seems to suit you best. It suits me. Ten o'clock seems to me a bit late, but I accept. As it's to be a short reading, insist on *punctuality* – or *absence*. The person who arrives late, to see what it's all about, and stays for five minutes, is to be avoided.

Say what is the case: that I am doing you the kindness, since you don't go out, of going to your house to read a (short) chapter of my new book; and that this chapter being the portrait of Mme Aubernon,[1] the guests have been chosen among those who knew her. We've agreed on our friends the Lemaire ladies, and on Flament. I should also like Mme de Chevigné[2] and Mme Baignères (Laure)[3] if you agree, and Pozzi[4] who can perfectly well come for this, in spite of his mourning. I shall write to him myself, after I've had your reply. The *Gabriels*[5] would be fine, *if it amuses them*. I appreciate *them*. But remember, *a reading from someone else's work*[6] . . . However . . .[7]

1. Formidable in manner and looks, the late Mme Georges Aubernon (*née* Lydie de Nerville: 1825–99) had had a celebrated Paris salon of which Montesquiou had been an habitué; he was to read his satirical portrait of her, entitled 'La Sonnette' (The Handbell), at Proust's apartment on 2 June. Proust was to use her as one of his models for Mme Verdurin in *RTP*.

2. Comtesse Adhéaume de Chevigné (*née* Laure. de Sade: 1860–1936). Distinguished by birth and wit, she was a model for the Duchesse de Guermantes as the unattainable great lady whom the narrator, as Proust himself had done, waits to see pass in the street (*RTP*, II, 55–60).

3. Mme Henri Baignères (*née* Laure Boilay: 1841?–1918), prominent Parisian hostess, also noted for her wit. Asked by Mme Aubernon what she thought of love, she replied, like Blanche Leroi in *The Guermantes Way*, 'I make it, often, but I never, never talk about it' (*RTP*, II, 199). Until 1892 her family owned the villa Les Frémonts at Trouville – La Raspelière of the 'three views' near Balbec (*RTP*, II, 1030–2) – where, like the Princesse de Luxembourg, she was attended by a negro page (*RTP*, I, 750–3). Cf. Proust's Normandy tour, letters 242, 245.

4. Samuel Pozzi (1848–1918), distinguished and fashionable surgeon.

5. Gabriel de La Rochefoucauld and his wife. See letter 133.

6. An allusion to La Rochefoucauld's *L'Amant et le médecin*.

7. The rest of the letter is missing.

135   *To Maurice Duplay*[1]

[Second half of May 1905]

My dear Maurice,

It's very difficult for me to give you an hour at the moment; I'll write and explain why[2] but this evening my eyes are hurting so much that I can hardly see the words I write, and I don't want to tire myself. I shall just say one thing which I feel I must write and tell you at once (because I'm more of a writer of letters than you are – letters in the sense of correspondence – although alas less of a writer!), and that is that I've just read your book[3] and I'm amazed, stunned. Yes, stunned. I was aware of all your 'potentialities'. How could I have suspected such mastery. There's no other word. Not a sign of weakness, of clumsiness, of monotony. Intellectually elevated but always rooted in the most powerful sense of life, no abstraction. A wonderful landscape, river, houses covered with roses and vines, the sunset glow of the vineyards reflected in the sky, but all adapted, geared to the subject, blood-red images of nature refracted by the book. Delightful impressions (the flamingoes, birds of the dawn, etc. . . . sun-soaked windows, marvellous). The sergeant who makes them laugh and punishes them for laughing, superb. It really is a fine book. I must re-read it and think about it so as to compare it, rank it among others, establish its true position, make a guess at the books you'll write in the future. I haven't at the moment either the strength or the time or the inclination to do more than admire your effortless power, your stylistic gift, I repeat your real mastery. It's *stunning*. I shall put this book into the hands of all my friends. All of them will want to know you. And as I shan't easily be able to introduce you, I shall send them to you or you to them. I congratulate you once more and I shake your hand.

Marcel Proust

What form! What eloquence! What life!

1. Son of a former colleague of Dr Proust and intimate family friend. A novelist, later a theatre director.

2. See letter 141.

3. Duplay's first novel, *La Trempe: l'Ecole du Héros*. The hero, on military service, describes the countryside and the interior of a grand hotel.

136 *To Louisa de Mornand*

[Shortly before 21 May 1905]

My dear little Louisa,

You are always an even more wonderful person than one thinks. You have written me a letter full of every sort of beauty, with an additional one which is more beautiful than all the others – the fact that you are unaware of the beauties you have put into these pages that are at once so exquisite, and so simple. Indeed if you weren't so simple I should hesitate to tell you that you're so exquisite for fear that you might become conceited. But that I know you will never become. What's this I hear about your coming to my door yesterday – when I thought you were lying with your eyes half closed in the darkness? It's a case of the blind coming to inquire after the paralytic. It was a great joy for me to learn that you are recovered. Is that the reason why, after a day of raging fever, I was better last night? This afternoon it all began again. This evening I'm better once more. But for how long?

I'm told you were radiantly beautiful in Prévost's play.[1] If you happen to talk to him, you might remember me to him. I send you all my most tender and heartfelt thoughts.

Marcel

1. *Les Demi-vierges*, a three-act comedy which Marcel Prévost (1862–1941) had scripted from his novel.

137   To Louisa de Mornand

[Sunday morning, 21 May 1905]

My little Louisa,

I hear that you plan to spend the summer near Trouville. As I'm mad about that countryside, I shall take the liberty of giving you a few particulars about it. Trouville is extremely ugly, Deauville frightful, the countryside between Trouville and Villers uninteresting. But on the heights between Trouville and Honfleur is the most wonderful landscape you could possibly see, beautiful open country with superb sea views.[1] And there are houses there known only to artists in front of whom I've heard millionaires exclaim: 'What a pity I've already got a château instead of living here.' And remote paths which are perfect for riding, real nests of poetry and contentment. The most beautiful place (but is it to let?) is 'Les Allées Marguerite',[2] a stunning property with miles of rhododendrons by the sea. It belonged to a M. d'Andigné,[3] and Guitry,[4] who was mad about it, rented it for several years. Does he still? Is it still to let? I can't tell you. But I could find out for you or you yourself if you know Sacha Guitry could find out. Perhaps it would be too enormous for you. But I believe one can have it for a song. There are also some ideal houses near Honfleur. Do you want me to make inquiries?

But you'd have to give me a bit of time. Porto-Riche[5] who knows the area like the back of his hand has recently lost his son so there's no question of asking him. My friend Mme Straus, who lives there,[6] is ill in Switzerland. But her husband is here. If you like I could write to him and if he himself doesn't know of anywhere he will recommend you to people who do – he knows everyone in Trouville. I'm telling you all this because I'm in love with that idyllic region. Would you believe that I persuaded some friends to buy a property there for 200,000 francs.[7] You didn't know I was a salesman, did you? You see!

I embrace you tenderly
Marcel

I warn you that if you write to me today there's a likelihood that I shan't get your note until very late in the evening as I've just been overcome, at three o'clock in the morning (I'm writing to you at five), by a fearsome attack. In any case, I shan't be able to see you this evening nor alas any evening in the near future.

I embrace you tenderly. I was sad to hear (from your charming note) about your mother's accident.[8] If I didn't reply you know it's because my eyes are so painful.

Lovingly yours
Marcel

Don't forget that if you want me to make inquiries you must give me a little time. I need to be able to see some people and you know that isn't often easy for me.

1. Proust was to revisit this part of Normandy, Balbec country in *RTP*, in 1907. See letters 242, 245.

2. A long avenue of pines and rhododendrons above the Seine estuary visited by the 'little band' in *RTP*.

3. Comte d'Andigné owned the Château de Barneville near Honfleur.

4. Lucien Guitry (1860–1925), actor and theatre director. His son Sacha, mentioned further on, was to become famous not only as an actor but as playwright and film-maker (1885–1957).

5. Georges de Porto-Riche (1849–1930), dramatist; his son Marcel (b. 1881) had just died.

6. Mme Straus, who had taken Mme Aubernon's villa near Cabourg, Le Manoir de La Cour Brûlée, now rented a villa called Le Clos des Mûriers.

7. The property was Les Frémonts, La Raspelière in *RTP* (see letter 134, n. 3). In 1892 it was bought for Mme Hugo Finaly, mother of Proust's schoolfriend Horace Finaly, by her uncle (see vol. I) for FF152,000, not FF200,000.

8. Mme Montaud had set herself alight with a paraffin lamp.

138   To Robert de Montesquiou

[21 or 22 May 1905]

Dear Sir,

Madame Lemaire, to whom I communicated your decision, tells me: 'But I'm engaged until the 30th by others, and I have a reception in my house on the 30th and another on the 31st, presenting an opera[1] by Moret two days running! Ask M. de Montesquiou if the 2nd[2] would be convenient for him.' My chief worry is that all this will exhaust your patience and you will throw the whole thing up. If Madame Lemaire isn't necessary we could stick to the 31st or, if you put it forward only for her, go back to Wednesday the 24th (next Wednesday) which doesn't suit me very well but which I would willingly accept if it suited you better. I can't tell you how much this note from Madame Lemaire has upset me, as I'm ashamed of writing to you and having you write back about these petty problems of dates. But what can I do? I was absolutely unaware of these celebrations on the 30th and 31st, Madame Lemaire not having sent me invitations. She thought I knew about them. I had only vaguely heard. She doesn't suggest the 1st because she wonders whether she may not be obliged to present this opera (Moret and Morand) a third time, on the 1st. But after the 1st, on the 2nd, the 3rd and the 4th, she will be free. But she realizes only too well that having already decided on a date you would prefer not to change it, and since I haven't yet written to anyone else, having started with her, I shall send out the invitations at once for the date which, without taking the trouble to write to me, you will put on a piece of paper (after the 4th, I think she's no longer free, at least not every day). You haven't told me if you accept Madame Cahen. I've made a careful note of the enemies to be crossed off.[3] I am as grateful to you perhaps as much as for the event itself for the even more tiresome trouble you have given yourself over these questions of dates and people. Please believe, dear Sir, in my gratitude and respect.

Marcel Proust

1. *L'Ile heureuse* by Ernest Moret and Eugène Morand.

2. The 2nd of June was the date finally chosen.

3. In a letter dated 20 May, Montesquiou lists as 'personal enemies', among guests proposed by Proust, the Marquise de Ludre (the Ludres had attended Proust's tea-party: see letter 105), Constantin de Brancovan and the influential critic André Beaunier (1869–1925) upon whom Montesquiou makes a scarcely veiled attack in his new book, *Professionelles beautés*.

139   *From Robert de Montesquiou*

Neuilly
[Monday, 22? May 1905]

Dear Marcel,

Since no one chucks [*balance*] anything for us, we shall have to let ourselves be lulled [*bercer*] by these fluctuations of mundanity, which crush [*broient*] other people.[1]

'Other people, I pity them!' So let us accept the 3rd

Without seeking to know, and without further thought.[2]

Tell me simply that it's *agreed at last*. I say the 3rd, and not the 2nd, because, in spite of my high opinion of our Friend's stamina, it seems to me that the day after these receptions, her receptivity might well be a bit sleepy.[3] Ah! how unevangelical these Christians are! They always have their crops to bring in when one offers them the *necessarium*. You will tell me again that I'm *taking the beau rôle*. One cannot change one's ways! And one takes what one finds!

'Cahen' – I once knew an amiable lady of that name. But it's such a long time since she ceased to be amiable that I had come to the conclusion that she had *ceased to be*![4] Let us stick to the names we have mentioned. And above all, make sure that the thing retains the quite simple character of a bit of reading to beguile a slightly valetudinarian friend. That's it.

1. The Count's alliterative trio of verbs is impossible to reproduce in English.

2. 'Sans chercher à savoir, et sans considérer.' Victor Hugo, 'Ultima verba' (*Les Châtiments*).

3. Nonetheless, they reverted to Mme Lemaire's proposed date (see letter 138, n. 2).

4. Cf. letter 81, n. 2.

140   *To Robert Dreyfus*

[Monday evening, 29 May 1905]

My dear Robert,

You must feel that my eye trouble, my lung trouble, etc. cannot excuse my silence. And with reason. In a few words, since I'm shattered with exhaustion, here is the explanation. After endless vicissitudes M. de Montesquiou is coming to my house to read an extract from his new book to five or six people, and I shall of course have to be up that evening; so, thinking of how I could find a way of seeing you, I wrote to him asking whether, in addition to the few people designated by him, I could invite two or three of my friends including Reynaldo and the eminent historian of Gobineau.[1] But I've been met with a blanket veto, on the grounds that it would change the character of his reading etc. and as I've only just received his refusal (which applies not to you personally but to everyone), I haven't been able to write to you sooner. Nor have I seen my source again, whose name wasn't accepted either.

As for the person with whom I might have discussed the matter,[2] although she knows nothing about it, but is a friend of both Barrès and myself, Madame de Noailles, I haven't invited her to this reading either, and in a fortnight or so it will be a year since I last saw her.[3] I myself *could* write to Barrès, but I don't think it would be a good idea. The best thing to do, I think, when my most violent period of hay fever is over, will be to ask Mme de Noailles to invite me with Barrès and raise the subject with him incidentally. However, I feel that there's no substitute for a public rectification. For I remember a similar case several years ago. *La Libre Parole*[4] had said that a certain number of young Jews, among them M. Marcel Proust etc., reviled Barrès. To rectify this statement I should have had to say that I wasn't a Jew and I didn't want to do that. So I allowed it to be said as well that I had demonstrated against Barrès, which was untrue. Having met him, I told him that I'd felt it was useless to issue a denial. But I sensed that he didn't agree that it would have been useless. We'll talk again about this, but if you wanted something done immediately, I'm at your disposal to write to my source, confess my indiscretion and ask whether you or I may tell Barrès that I've repeated it all to you. If you'd prefer something else, I'm also at your disposal. *For everything*, indeed, quite simply.

Ever yours
Marcel Proust

1. Dreyfus himself (see letter 91).

2. A rumour, presumably heard from the source (unknown) just mentioned, that Barrès believed Dreyfus to have accused him of plagiarizing Gobineau.

3. Since the evening when Proust broke her Tanagra figurine. See letter 49, n. 16.

4. Anti-semitic newspaper.

141 *To Maurice Duplay*

[End of May 1905]

My dear Maurice,

I wanted to write to you again about your book[1] and about possible reviews. Here is why I haven't done so. M. de Montesquiou is coming to read me an extract from his new book and asked me to invite seven persons whom he named. I wrote back to him to ask if I could add a few others, you among them, and it's only this instant that I've received his reply in which he refuses everything *en bloc*, saying that it would alter the character of the reading, etc. etc. Actually I'm delighted, because had I invited you I think that on the day itself I would have begged you not to come. Already I don't dare see you in front of one person. But at a reading! I think it would have led to the most scandalous fit of the giggles imaginable. Only I was awaiting his reply in order to tell you everything at once. His refusal is a great relief to me, as I'm less afraid of the giggles without you there. But I shall get them nonetheless.

The book: I find it even more stunning than the first time. There are countless passages which reveal the great writer – the fellows lying on their beds as though they were tombs, the windows of gold and amethyst – hundreds of magnificent things. I don't care a damn about the thesis, which is fine and probably false, but that doesn't matter in the least. The theses of Rousseau, of Flaubert, of Balzac, of so many others, are doubtless false. And if absolute monarchy and clericalism are not the only remedy for France, does that make *Le Médecin de campagne*[2] a less good book? Or Barbey d'Aurevilly's novels[3] . . . Now the question is, did you have the right to say all that? I think you did, but I'm not a judge from that point of view. My feeling is in fact more favourable to you on this point. But I'm too tired to go into it. It's curious incidentally that we two, who are both on the whole kind people, should have seen the army, you as a prison, I as a paradise.[4] But what difference does that make? Because the sunset exists, it doesn't prevent the dawn being beautiful, another moment of truth; a painter can't paint everything at once, and my goodness you've painted quite enough as it is. In spite of the thesis, Lucien Daudet has probably made the Empress Eugénie buy the book. At least he said so as he was leaving me. – As for reviews, I shouldn't like you to think I'm unwilling to help because I swear to you it isn't true. But the thing is this: when I did *La Bible d'Amiens*[5] I thought my literary friends would go out of their way to help. The preface was dedicated to Léon Daudet, who writes in *Le Gaulois*, *L'Echo*, etc., and I hinted to him that an article would be welcome. *Never a word, not the slightest allusion*, no mention of my name in articles where he mentions everyone. Don't say so, but I still haven't been able to swallow it. I turned to Hermant[6] and made it known to him that it would give me pleasure. In each of his articles I could see the subject clamouring for a reference to *La Bible d'Amiens*. Not a word! What's more, he wrote at length on Ruskin in an article in the *Gil Blas* and didn't breathe a word either about Marcel Proust or about *La Bible d'Amiens*. What do you think of that? It isn't that I didn't get some excellent reviews. But they were *all* from people I didn't know, which came out of the blue, unforeseen and against all expectation. After all that, what can I possibly say to you? That I'll mention it to the two I know best, Hermant and Daudet? They'll do nothing. And it's more complicated because of the subject. Send your book to Brancovan without asking for a review, simply telling him that you met him with me, and he'll be touched. In the meantime I'll get someone to tell him it's excellent and it will seem quite natural; perhaps he'll get it reviewed, although he no longer has a real literary critic.[7] I shall tell all the artists I meet that it's superb, I'll try and get them to read it, it's the only thing that works. And I'll also try and get someone to mention it to *L'Human-ité* and *Les Essais*. If by any chance you have already had promises from either of those quarters, let me know because it wouldn't be worth while my tiring myself by writing to them. You've no idea how much you make me want to see your beautiful, fluvial, grape-laden countryside.

Your

Marcel

1. *La Trempe*, see letter 135.

2. Balzac novel.

3. Jules Barbey d'Aurevilly (1808–89), intensely Catholic, visionary novelist much admired by Proust.

4. Proust had greatly enjoyed his military service, when he was based at Orleans. Cf. vol. I.

5. See letter 20, n. 3 *et al.*

6. Cf. letter 25, n. 4.

7. Cf. letter 6, n. 4.

142   *To Madame de Noailles*

Monday evening [19 June 1905]

Madame,

Stop being so nice, I beseech you, for I cannot bear it any longer; the burden of happiness, gratitude, emotion, stupefaction is too overwhelming and I might die of it.[1] There is also the fear that the whole thing may be a joke, for nothing can penetrate the armour of my sadness, my conviction that all those pages are execrable, a sort of indigestible nougat which sticks between one's teeth. And yet as I know that you always hate everything I write, I wanted you to read a few pages of this (infinitely worse written than what I used to do before being ill) so that you could see that after all I did think a bit and wasn't quite so sottish as people say. Hence the joy it is for me, far beyond my hopes, that you should have read the whole thing. But forgive me for talking about myself. I'm in a state of shame and confusion beyond words, enhanced by the Beaunier piece[2] which I strongly suspect you dictated. To think that you've written to me twice, that you've tired yourself, that she without whom the sun would have no radiance nor the world any meaning lavishes this divine attention on me. 'Goddess, gaze at me no longer,' says Ulysses, 'for I feel my legs tremble beneath me.'[3]

Madame, although I am overjoyed each time I read your name and hear you praised, I don't know why I wasn't at all happy when I read Ballot's review.[4] That way of describing *La Domination* like any other novel filled me all the time with inexpressible unease, as though its individuality were gradually evaporating, as though Antoine had been told that Elisabeth was *of medium build*, had *pretty eyebrows*, at any rate things that could be applied to anyone. It seems to me that he is usually better inspired. But the truth is (don't repeat it, I beg of you) that when it was good his criticism was far superior to the man himself, whom I've never much prized from the point of view of taste. And in such cases, he keeps it up for a time but there's always a final nose-dive as Gabriel de L[a Rochefoucauld] would say. – As for the end of the article, I really thought it absurd. Or else I didn't understand a thing about *La Domination*. Never has a male character (and when I say never I'm thinking of all the greatest novels I know) seemed to me so essentially, so profoundly, the character of a *man* as Antoine, hard, individual and striking, like a bronze medallion by Pisano. I see the profile, the pre-ordained soul, which your whim was powerless to alter. His very softness, at the end, is the softness of that same metal and its marvellous glow. That there is in him an element of your genius, since everything that is admirable partakes of it obscurely, is axiomatic. But to speak of transvestism! He's a man from top to toe. If one took off his clothes – forgive me, Madame, but I mean that one would find a man, a complete man, and not remotely a woman's body. Am I wrong or is Ballot? I think it must be Ballot because I believe my taste is surer than his, although I'd be quite incapable of writing his articles. I remember once hearing him say things about Baudelaire (I may be mixing it up with conversations with Paléologue[5] but I don't think so) which made me think that all his fine sagacity since was an artificial veneer. If Antoine is a woman I feel I'm witnessing the demise of male humanity. The world is losing the most masculine man, the most authentic male I know. It's a great loss, a great sorrow. It's possible that I'm wrong. I'm still so dazzled by *La Domination*, it's a book so distinct from any other, such a marvellous planet won over for the contemplation of mankind, but still so recently, and so different from everything else we see on earth, that I can't judge very clearly. And yet I still think of the despotic profile engraved on the Pisano medallion and I know that it isn't a transvestite. Of course there's a lot of you in it, but it's a creation, not an incarnation or an abstraction. Quite the opposite. Or else I've misunderstood the book completely. Forgive me, Madame, for all these letters. I send you all my infinite gratitude and respectful admiration.

Marcel Proust

1. She had written to congratulate him warmly on 'Sur la lecture' (his preface to his translation of *Sesame and Lilies*) published on 15 June in *La Renaissance latine*.

2. In *Le Figaro*, Beaunier (see letter 138, n. 3) had welcomed 'M. Marcel Proust, the incomparable translator of Ruskin' whose preface is 'charming, moving and often marvellous'.

3. Homeric pastiche, rather than an actual quotation.

4. Review of Anna de Noailles's novel *La Domination*, published on 7 June, by Marcel Ballot (1860–1930), literary critic of *La Vie littéraire*. Its chief characters are Antoine, whom Ballot calls effeminate, and Elisabeth.

5. Maurice Paléologue (1859–1944), diplomat, French Ambassador in St Petersburg 1914–18. Described without being named in *Cities of the Plain* (*RTP*, II, 670); 'notoriously inadequate in Serbia' according to Norpois in *The Fugitive* (III, 646), though in fact he never served in Belgrade, but was *en poste* in Sofia 1907–12 (see letter 235).

143   *To Madame de Noailles*

[Tuesday evening, 20 June 1905]

Madame,

I am so unwell this evening, and I have just written such a long letter to the Princesse de Chimay, that I simply want to tell you that the *Arabian Nights* existence you have been giving me for the past two days has extinguished my sense of reality to such a degree that I was not in the least surprised when I was told that *you* had been to *my house*! It seemed to me to be another of those marvellous enchantments with which you've bewitched and deluded me for the past two days and which will make life seem very insipid later on. I fall asleep, I dream that the greatest poet of all time writes to tell me that I'm wonderful, that I have some talent, I wake up, I'm already impatient not to have received the letter from this sublime poet. But I don't have to wait for more than two seconds: I ring, and her letter is brought to me. And I have no need to wish to receive one from the most beautiful woman, because they are one and the same person through a miracle of incarnation which really ought to make you the object of a religion with a metaphysic and an incarnation. Moreover the letter was still not enough: I fall asleep again, I fall asleep with the thought of you engraved in my mind, I wake up again, and sure enough your shadow still haunted the house. And I had the most moving thing of all, your handwriting on this writing paper, a presence withdrawing, an urge to run after you. Since you cannot perhaps imagine exactly what you are to me, please understand that all that represents my *mythology* and makes me understand perfectly how Minerva, Jupiter or Venus appeared from time to time to mortals, took cognizance of their actions and offered them congratulations. I have nothing to envy Ulysses because my Athena is more beautiful, has greater genius and knows more than his, and she comes to me while I'm asleep. But I nevertheless read Homer and the *Arabian Nights* more readily, being no longer disturbed by the improbability which forms the crimson thread of my new life.

Madame, may you be blessed.
Your respectful
Marcel Proust

144 **To Georges Goyau**

45 rue de Courcelles
Thursday [22 June 1905]

Dear Sir,

I cannot wait until I have read, and in many parts re-read, the whole of this book[1] to tell you what the first volume which I have just finished already enables me to affirm, that it is more than an admirable work of history, that it almost represents (through a masterpiece) the inauguration of a new genre, as elevated as an expression of faith, as subtle as a psychological analysis, as real as a sociological inquiry, as living as a contemporary political and historical narrative. What a fine psychological chemist you are, and how well you succeed in following the simplest elements of Catholicism through to the most complex living compounds and aggregates.

Temperamentally no doubt I was particularly struck by the chapter in which you auscultate the thawing of the last Voltairean icicles (I say Voltairean in order to simplify), in which you detect the voices of the first swallows of Romanticism which are about to alight on the towers of hitherto despised cathedrals, in which throughout the whole of nature you discern each sign and each germ, evaluate exactly each symptom (and as a somewhat pessimistic consultant on the whole, when for instance you detract from the value of Goethe's admiration for Strasbourg cathedral on the grounds that he sees it as an individual work of art – and no doubt it is indeed the wrong way to feel, but when one compares it to Stendhal's journey through France where he revolts at seeing a barbaric church at Autun side by side with Roman ruins which he finds wonderful!). Cathedrals have in fact inspired you to write some charming things and I'm touched to think that Cologne cathedral played a part in the conversion of Dorothea Schlegel.[2] Those curious figures Stolberg,[3] Princess Galitzine,[4] Furstenberg,[5] the curio-hunting Boisserée brothers,[6] but above all Novalis[7] (your literal interpretation of the *Hymns to Christ*, pretty well corroborated by your few scattered quotations – the one about the Jesuits is very amusing – interests me a great deal) are infinitely engaging, and you're a great painter not only of the nuances of thought and the frescoes of history but also of the touching colours of individual life. In this regard the following chapter, the Munster Circle, is one of the most fascinating. The figure of Schlegel (I'm going back), at first so detestable, is ultimately very interesting. You are pretty hard on Goethe in emphasizing all the egotism and arrogance in his paganism. But when I say you're hard, need I add that it's combined with the greatest intellectual deference, the most absolute reverence for the judgment of one of the most powerful and subtle minds that ever existed, after whom there is nothing more to be judged – and nothing more to be written. You say that you are opening paths, and it's true if it means that your book will be fertile. But you also embrace and apprehend the whole horizon, which is to say that your book is perfect. Thus each of your ideas raises others, but it turns out that you have already anticipated them. What will remain to be done afterwards, to understand you better, to follow the new paths that you open, only to find that the man we come across who has forestalled us and who dominates them is still you?

I read your book like a novel, that is to say by beginning at the end, for although I've still only read the first volume I've already skimmed through the last chapter of the second. It's intellectually very alluring, an entire study devoted to one year, and what a year! I believe this book will not only provide a great deal of aesthetic pleasure, and will not only do a great deal of spiritual good (a great deal of spiritual good because in singing the glory of God in heaven it also brings on earth, practically and wisely, the possibility of peace to men of good will – this book being destined to be the breviary, I hope, of the Church in France, which is going to have to organize itself), but I believe it will also ensure that the already celebrated name of its author will be uttered with a tinge of even greater respect, of even more marked approval, of even more conscious and justified admiration. I think it will be his masterpiece, his most consummate achievement, his magnum opus. Considerations which are totally immaterial to the man who examines and describes and analyses all these rays that play upon this earth without losing sight of the fact that they emanate from the heavens, but which are nonetheless immensely dear to his friends, among whom would be so proud and happy to be numbered

Your most grateful and devoted
Marcel Proust

1. *L'Allemagne religieuse. Le Catholicisme (1800–1848)*. Cf. letter 69, n. 1.

2. Dorothea von Schlegel, *née* Mendelssohn, a Jewess who converted to Christianity, and married the Romantic writer and scholar Friedrich von Schlegel (1771–1829).

3. Friedrich von Stolberg (1750–1819), eccentric poet and Hellenist, one of the founders of the *Sturm und Drang* movement, converted to Catholicism in 1800.

4. Princess Galitzine, *née* Amélie de Schmettau (1748–1806), practised several religions before returning to Catholicism.

5. Franz von Furstenberg (1729–1820), Vicar General of Munster.

6. Sulpice Boisserée (1783–1854) and Melchior Boisserée (1786–1854), collectors of religious art (now in the Pinacothek, Munich).

7. See letter 93, n. 6.

145   *To Louisa de Mornand*

[Second half of June 1905]

My dear Louisa,

What a beautiful, what a delightful letter you wrote me! How I shall treasure it. If you knew how I re-read it, how I admire it, how it touches me! Momentarily withdrawn from the restless surge and froth of Paris, you looked into the depths of your heart, now calm and clear again, and discerned there pictures of the past. It is to that that I owe your letter, it is that that moved me so much. I mean nothing to you except as one who was involved in some sweet and painful moments of your life. I am like the man who held the horse or stood beside the carriage at some great historic event.[1] Nobody even knows his name. But in all the 'views' of the event, he inevitably appears, because chance, or destiny, placed him there. In the same way our memory often presents us with 'views' of the historic events of our own lives, not always very easy to discern, a little like those one strains to distinguish through the tip of a shell-encrusted pen-holder, a souvenir of the seaside. But in these views which memory presents to us of the happy or tragic days that still control our destinies, we inevitably glimpse the secondary character, the supernumerary who happened to be there, the Marcel Proust of whom the memory is thus tinged for us with the colour that bathes the whole picture. The supernumerary asks for nothing better and silently rejoices in these windfalls of friendship, in his share which in the words of the Gospel cannot be taken from him. As for the subject that preoccupies me, my little Louisa – and which hangs not so much on the question 'Shall I be able to see Louisa one of these days?' as on this other one: 'Shall I soon, after the holidays,[2] be able to see Louisa regularly and often?' – it is not at all isolated in my mind. It is only an aspect, more particularly preoccupying for me, it's true, of that other question which it's time to resolve at last: 'Shall I continue to the end of my days to lead a life that even invalids who are gravely ill don't lead, deprived of everything, of the light of day, of air, of all work, of all pleasure, in a word of all life? Or am I going to find a way of changing?' I can no longer postpone the answer, for it's not only my youth but my life that's going by . . . And in this respect I'm on the whole in favour of not delaying the solution which would enable me to see you often and properly by committing such an imprudence as to see you now in the state I'm in, when even *as a man* it embarrasses me that you should see me disguised in a long, unkempt beard. But will I have the courage not to say good-bye to you? It would be more sensible to send you in writing an immediate good-bye in the hope of frequent and regular future meetings. But as always I shall probably do the least sensible thing.

Tenderly yours
Marcel

1. Cf. letter 92 to Georges Goyau.
2. She was going to Trouville. See letter 137.

146 *To Marie Nordlinger*

Saturday [24 June 1905]

Dear friend,

I've been meaning to write to you for days and days, and all the more so because a practical necessity, already urgent at the time of your arrival in Paris,[1] makes each day's delay in writing to you a loss of time which aggravates the sadness of not being in touch with you, but my health has been such that everything, practical matters as well as sentiment, has been stifled by the terrible incapacity I was suffering most of the time. You arrived in Paris like the Messiah but you left like a demon and your passage was like a dream at once dispelled. What an idea not to have warned me of your imminent departure; I would have tried to see you instead of being faced with the accomplished fact of a detestable 'Manchester, Victoria Park, etc.'.[2] I also very much regretted not having met and got to know Mr Freer,[3] which was for me in every respect a very natural temptation, out of curiosity and fellow-feeling. My respect and admiration for him were infinitely increased, as you can imagine, when the other day, braving death (and alas almost meeting it) at the hour of my bedtime, utterly exhausted, I took a cab and went to look at the

Whistlers. It's the sort of thing one wouldn't do for a living person. But one does it for a dead one, perhaps from a mad unacknowledged conceit which tells one that one's eyes may be among those by which this nomadic beauty on its way back to Boston would have liked to be beheld. But it would have been a hundred times better to go to Boston.[4] It would have made me less ill. When I saw that all the finest Whistlers belonged to Mr Freer, I exclaimed to myself what Ruskin said of the unknown artist who sculpted and painted the delightful archivolt of St Mark's: 'I don't know who this man was. But I believe the man who designed and the man who delighted in that archivolt to have been wise, happy and holy!'[5] This I would say of Mr Freer except for 'happy', not being as convinced as Ruskin that taste brings happiness. But it can happen. You know that at the moment there is a terrible reaction against Whistler among the artistic élite in France. He is regarded as a man of exquisite taste who because of that was able to pass himself off as a great painter although he's nothing of the kind. Jacques Blanche[6] in the *Renaissance latine* (have you read my article in the same issue which Beaunier praised so exorbitantly in the *Figaro*?[7]) expressed the same opinion, though with more justice and even fervour. It's not at all mine. If the man who painted those Venices in turquoise, those Amsterdams in topaz, those Brittanies in opal,[8] if the portraitist of Miss Alexander, the painter of the room with the rose-strewn curtains and above all of the sails at night belonging to Messrs Vanderbilt and Freer (why does one see only the sail and not the boat?), is not a great painter, one can only think there never was one.

Dear friend, I'm already a little tired of writing and I still haven't said anything about practical matters. I've waited for months for your return from America and then your weeks in Manchester, to send you my manuscript, there being a great deal to ask you about *Lilies*.[9] And now I'm obliged to send it off. It can no longer appear now, it's too late, but I shall get it printed anyway so that if I'm in a nursing home this winter, it can appear without me, *vox silentiae*. I'm sending it to the publisher but I'd like your criticisms and if you send them to me I shall ask for the manuscript back for two days to make corrections. Would it be possible to send to you at 'Manchester, Gladville and all the rest' my copy of *Sesame*, in which I've put crosses and underlinings wherever I was in doubt? I won't swear that you'll resolve my doubts but it's probable. They never apply to more than one or two words at a time. Does this seem feasible to you? But you'd have to be careful not to lose my copy, which contains all my wisdom, and return it afterwards.

If you find this too complicated, would you mind if I turned to someone else for help with *Lilies*? You know that I have English-speaking friends, less knowledgeable than you, but who might perhaps be able to help a little. If not I shall send you the book. Let me know what you think.

The note you protested against[10] was not the one for the book but the only note I put in the review in which the translation appeared (*Les Arts de la vie*), reserving the other notes for the book. I don't regret not having been able to heed you (it had already appeared) because *Les Arts de la vie* is read by all the really intelligent artists in France. Besnard and Carrière and Rodin[11] often contribute to it and read it assiduously and I'm pleased that your name should be well-known there. In the book I shall be as moderate as you wish; I shall say that you're completely devoid of talent and even natural aptitude, if you like, and I shall exhort you to pursue another career. Is that what you'd like? If the charming Venetian aunt,[12] the passionate and meticulous friend of art and virtue and comfort, and full of benevolence for the signatory of this letter, is living at the moment in that mysterious Victoria Park which I can't picture to myself, please offer her my most respectful regards and believe, dear friend, in my sincere affection.

Marcel Proust

1. She was on a flying visit from the United States. See letter 40, n. 1.

2. Marie Nordlinger's family lived in Victoria Park, Manchester.

3. See letter 102, n. 1.

4. Charles Freer's collection was then in Detroit, not Boston (it was left to the Smithsonian, Washington).

5. Proust quotes freely from Ruskin's *The Stones of Venice*.

6. Jacques-Emile Blanche (1861–1942), painter and art critic. It was when he and Proust were staying at Les Frémonts (see letter 137, n. 7) near Cabourg that he did the sketch for his famous portrait of Proust aged twenty as a young dandy (1892); the friendship cooled over the Dreyfus case. See vol. I.

7. See letter 142, n. 2.

8. For the painter Elstir, in *RTP*, Balbec bay was 'the gulf of opal painted by Whistler in his *Harmonies in Blue and Silver*' (II, 23).

9. Ruskin's *Sesame and Lilies* which she had helped him translate.

10. A charming acknowledgement, expressing his gratitude (cf. letter 68, n. 4).

11. Albert Besnard (1849–1934), painter and engraver, Eugène Carrière (1849–1906), painter and lithographer, Auguste Rodin (1840–1917), sculptor.

12. See letter 52, n. 2.

147 *To Robert Dreyfus*

[3 or 4 July 1905]

Dear friend,

You are too nice to remind me of the good intentions which I couldn't fulfil à propos of the Barrès incident. I still hoped to see Mme de Noailles and, through her, Barrès himself. But on the 17th of July it will be a year since I was in a fit state to go to her house or receive her here. I'm still at your disposal if I can be of use to you. In any case Mme de Noailles wasn't 'my source'. That mysterious source would I'm sure agree to my disclosing his or her name to you. But I haven't been able to ask for permission since you forbade me to say I'd told you[1] . . .

You are very nice about 'On Reading'.[2] It's the article I wanted to send to you. Then when it appeared I was so disgusted with it I no longer dared. I felt that you who know how to say so much in half a line would be exasperated by sentences that run to a hundred. Ah, how I should like to be able to write like Mme Straus! But I must perforce weave these long silken threads as I spin them, and if I shortened them the result would be little fragments rather than whole sentences. So that I remain like a silkworm, and indeed live in the same temperature, or rather like an earthworm ('in love with a star', that's to say contemplating the un-attainable perfection of Mme Straus's concision). Speaking of Mme Straus, the desire to see her again will, I think, force me to change my hours a little. As soon as this reform, if it comes off, has been accomplished I shall write to you to try and arrange for us to meet at last.

Affectionately yours

Marcel Proust

1. See letter 140 for the explanation of this paragraph.
2. 'Sur la lecture', his preface to *Sésame et les lys*.

148 *To Louisa de Mornand*

Friday [14 July 1905]

My dear little Louisa,

Thank you with all my heart for your letter. I didn't see you again. But how could I possibly blame anyone but myself. All I can hope is that on the grounds of my health I shall be forgiven for the strange behaviour which is imposed on me by the force of circumstance. But far from having to forgive you, all I can do is to beg you to forgive me. It is more than natural that when one telephones a lady at ten o'clock to ask her to come round at midnight, she should not be there on the alert waiting for you. I have more grievous sins against friendship to reproach you with, my little Louisa. But I've given up this kind of attitude and correspondence with everyone. People are what they are, and no amount of resentment on our part has the power to change their hearts.

I'm glad to know you are in Trouville as it gives me the pleasure of picturing one of the people I'm fondest of in one of the countrysides I love best.[1] It condenses two beautiful pictures into one. I don't know exactly where your Villa Saint-Jean is. I suppose it's on the heights between Trouville and Hennequeville but I don't know whether it looks out to the sea or the valley. If it overlooks the sea, it must glimpse it through the foliage, which is so soft and lovely, and in the evening you must have wonderful views of Le Havre.[2] There is a combined scent of leafage, milk and sea salt on these paths that seems to me more delicious than the most refined 'mixtures'. If you look over the valley I envy you the moonlight which turns it into an opalescent lake. I remember one night coming back from Honfleur by those upland paths. At every step we stumbled into pools of moonlight and the damp mist in the valley seemed like a huge pond. I recommend to you a very pretty walk called Les Creuniers (I don't answer for the spelling).[3] From there you'll have a splendid view and a sense of peace and infinitude that makes you feel you are completely dissolving. From there all your cares and all your sorrows will seem to you as minute as the ridiculous little people you can see down below on the beach. One is truly right up in the sky. For a drive I recommend an even more beautiful place: Les Allées Marguerite.[4] But once you've arrived you must open the little wooden fence (otherwise you'll have seen nothing), drive in (if the present owner isn't in residence), and wander for hours in that enchanted forest with the rhododendrons around you and the sea at your feet. Beaumont, too, is a splendid walk, but on the other side.[5] The entry into Honfleur by the old Caen road, between the great elms, is very beautiful. However there are countless other walks and drives which you must already know far better than I who spent so little time there and so long ago. Trouville incidentally is still strewn with villas belonging to friends of mine, the Prévosts[6] who must be quite near you as well as the Clos des Mûriers.[7] There was also the villa of that poor Mme de Galliffet[8] but I don't know what's happened to it since she died, as I don't know her children.

My little Louisa, your letters are always exquisite. But the last ones are even more marvellous. For some time an extraordinary change has been taking place in you, a change in which, I think, the different life, the precocious maturity of your reflections, the theatre too and reading have played a great part. A propos of the theatre, Robert d'Humières[9] tells me he met you at the Vaudeville and greeted you but you didn't respond. He was distressed by this. I think however that he'll forget it in the contentment of marriage. For he is about to be married. To judge by my little Louisa's ravishing letters, and the great feeling for life which I have always recognized in her, I shouldn't be surprised if one of these days a book by her were announced and she added this literary fleuron to her crown of art and beauty.

I go from bad to worse. It's three weeks since I set foot outside the house, and I have a beard which doesn't even look dirty any more it's so long. If you go and see the dear little church of Criqueboeuf[10] nestling under its ivy, give it an affectionate greeting from me, and the same to an old pear-tree, battered but indefatigable like an aged servant, which holds up with all the strength of its gnarled but still green branches a little house in the neighbouring village from whose only window the pretty faces of little girls often smile, although perhaps they are no longer either little or pretty or even girls, for it was a long time ago.

All my tender thoughts,
Marcel

My respects to your mother, if she is with you.

1. Cf. letter 137.

2. Cf. the views from La Raspelière (i.e. Les Frémonts: see letter 137, n. 7), the villa near Balbec rented by the Verdurins (*RTP*, II, 837–8, 1030 sqq.).

3. Cf. the rocky cliff of the same name near Balbec where in *RTP* the narrator goes for a walk with Andrée (I, 962, 983, 986–7).

4. See letter 137, n. 2.

5. Cf., in *RTP*, the 'very special, very remote, very high' spot which impresses the narrator during his excursions in Mme de Villeparisis's carriage, and later with Albertine in a motor-car (I, 760–1, II, 1037–8).

6. The novelist and his wife (see letter 136, n. 1).

7. Le Clos des Mûriers was the villa at fashionable Trouville rented by M. and Mme Straus.

8. Widow of General the Marquis Gaston de Galliffet, personality of the Second Empire.

9. See letter 19, n. 7.

10. This became the ivy-covered church of Carqueville in *RTP* (I, 761, 768–70). In this letter Proust is relying on his memory of his last visit to Normandy in 1884 (cf. vol. I).

149 *To Marie Nordlinger*

[July or August 1905]

Dear friend,

I received your precious parcel,[1] for which I thank you with all my heart. It's wonderful! I'll write to you tomorrow (if I'm well: I should have written to you two days ago: tomorrow) to ask you what two or three explanations I didn't understand precisely mean. But I didn't want to wait any longer to tell you how deeply grateful I am. Your irritation with the *Moral Union*[2] amused me greatly and when I reached the final explosion: 'The *Moral Union* gets on my nerves' I had a good laugh. My notes were madly illegible and I see there are several of them that you misread, or rather read correctly since you read what they represented to the eye which bore no resemblance to what my absurd hand had intended to write. I cannot understand this prolongation of your exile if one can so term a sojourn in one's mother country. As for me, my imprisonment, my isolation (which has nothing splendid about it like that of the land of the Marys) go from bad to worse, without my even having, like the Pope, with whom I hold the record for voluntary incarceration, the Vatican gardens to stroll in, which however would doubtless be very bad for me.

This evening I had an unexpected visitor whom I received in spite of my fatigue since for once my breathing was not too bad. It was Reynaldo, back from a rather amazing assortment of beaches where he led the life of a vaudeville actor rather than a musician if one is to believe his accounts of how, thinking he was going to M.X's by the front door, he found himself entering M.Z's by the service stairs, which didn't matter as he also knows M.Z etc. etc. etc. He has come back extremely enamoured of the *Iliad* which by a coincidence which is not in the least strange I was reading at the same time. I say not in the least strange because it was reading Lemaître's *Contes*[3] and wanting to reply to it that made me take up that wonderful old book again and in Reynaldo's case it was the sight of beaches which he 'felt' were Homeric that prompted him to read it. He confessed to me how sad he was to discover that the *Iliad* was an anonymous, collective work and not the work of 'old Homer'. I consoled him by informing him that this theory of a collective work was no longer credited by scholars and moreover never had been by sensible people. He was still dubious but I gave him a copy of the *Revue de Paris* in which an excellent article by M. Bréal[4] would prove to him conclusively that Homer existed just as surely as Massenet and that the *Iliad* was composed like *Sesame and Lilies* and indeed written not recited. All this seemed to cheer him. But how, in fetching this liberating review from quite nearby, I caught a chill which will prostrate me for weeks is something which my eyes are too bad for me to describe to you and which in any case concerns literary history and indeed just plain history to a lesser degree than the paternity of the *Iliad*, the opinion of Bréal and the opinion of Reynaldo. What particularly struck Reynaldo in the *Iliad* was the heroes' politeness. Since they spend their time calling each other 'dog' and bashing each other over the head I don't agree with him. But it's true that even when insulting one another they say things like: 'Know then, O magnanimous Hector, that I intend to kill you like the dog that you are', or 'Vile soul, worthy of the bitches of Hell, irreproachable Helen'.

Dear friend, I wanted to blow as far as Gladville a little of the air of Courcelles and Alfred de Vigny.[5] But really I'm too exhausted for this kind of undertaking and too lacking in 'puff' myself this evening. Thank you again, and yours as ever.

Marcel Proust

1. Proust had sent her some queries on *Sesame* (see letter 146, n. 9).

2. Bulletin of the Union for Moral Action founded by Paul Desjardins (1859–1940), philosopher and critic.

3. A collection of short stories of which the first two parody Homer. Cf. letter 142, n. 3.

4. In which the distinguished French linguist Michel Bréal (1832–1915) rebuts the theories of the German philologist Friedrich-August Wolf (1759–1824) referred to above. Cf. *Within a Budding Grove*, where the narrator cites 'a popular expression of which, as of the most famous epics, we do not know the author, although, like them, and contrary to Wolf's theory, it most certainly had one' (*RTP*, I, 549).

5. Gladville was the name of Marie Nordlinger's family home in Manchester (see letter 146, n. 2), Courcelles is the Paris street where the Prousts had lived since 1900, Alfred de Vigny the street where the Hahns had lived since 1898.

150   *To Hélène de Caraman-Chimay*

[10 August 1905]

Princess,

Very quickly because I'm very unwell today.

1. On the question of your genius.[1]

You know that one can infer the presence of arsenic in the body indirectly through a reaction or by deduction – or directly by extracting it. Similarly with your genius. One can prove it indirectly through the fact of your sister's genius which is easier to extract because she eliminates it more copiously and more constantly, and by reasoning as follows. Madame de Noailles has genius. According to various people both eminent and modest (this means me) Madame de Chimay is superior to her sister. These people think it, I say it but I don't think it, I think that all things considered she is equal to her sister. Therefore she has genius. But this is such a primitive process that it leaves glaring possibilities of error. I prefer to extract your genius directly. Now although this accusation of genius seems to vex you, you are undeniably full of genius. A doctor who, in deciding on your regimen, did not take into account the presence of genius in you, would lay himself open to the gravest miscalculations.[2] And you can be absolutely sure that when you are unwell etc. it's solely due to insufficient elimination of genius. There's a lot one could say about that, but I can't go on. I possess some incomparable nuggets of your genius in the *Notes sur Florence*, and in *L'Exclue*, in your *Portraits*.[3] Even if you had never written a line or opened your mouth it would silently bestow its proud and lofty spiritual significance to the sober and powerful mural painting that is you, which enables one to guide one's actions by your eyes as by two black suns of righteousness (sun of righteousness, *soleil de droiture*, is in the Bible, English translation[4]). Naturally this idea must not be confused with the stupid idea of beauty as unconscious genius. I'm speaking here of the genius of thought. But I'm so tired!

As for the dedication, here goes. It's simply that I don't admire, or no longer admire, the other person with whom you demand similarity of treatment. And if I put 'her admirer', everyone will understand that it's a woman, that her mother is of all living creatures (apart from my family) the one who has done most for me, that I should already have dedicated everything to her, and because that embarrasses me, I hope to achieve the same result by transferring to the daughter, etc.[5] But lumping her together with the person I admire so much bothers me. In reality I'm fifty dedications behind: France dedicated a novella to me,[6] Montesquiou a poem,[7] Robert de Flers a short story,[8] etc. etc., and to all of them I dedicated things in *Les Plaisirs et les jours* as well as to M. Hervieu[9] etc. etc., and Madame Lemaire compelled me to take all these dedications out so that I haven't yet dedicated anything to anyone.[10] All those people must take precedence over anyone else and that is why for these two translations I've chosen the two most pressing, Reynaldo[11] and the lady in question. But I'd like to put in front the great writer I so admire (the Princesse de C-C) to whom I want to dedicate in this form which will preclude any thought of snobbishness in Madame Arman's mind:[12]

'To the author of *Notes sur Florence* and *L'Exclue*
Her respectful admirer'

As for *friend*, I'd never dare! That would seem to me the height of ill-breeding. I don't know you well enough, Princess. It would look as though I were boasting of too grand a friendship (it's true that I then seem to have less friendship for you than for Suzette Lemaire, but there's no way out of it – anyhow nothing would hire me to put friend or friendship[13]). I should be grateful if you'd burn this letter in which I say of people who have been so kind to me that I can't admire them as I should like to, or at least as much as I should like.

Your respectful admirer (and secret friend)
Marcel Proust

1. In an earlier letter (not included in this selection) Proust says he intends to dedicate the preface to *Sésame et les lys* to the Princess 'in admiration of her genius' despite her protestations that she would prefer to be named as a friend.

2. Cf. *RTP*, I, 614 where Bergotte says that Cottard cannot be a good doctor for intelligent people: 'He has made allowances for the difficulty of digesting sauces . . . but . . . no allowance for the effect of reading Shakespeare.'

3. *Notes sur Florence* apparently refers to an article by the Princess published in June in *La Renaissance latine* (cf. letter 6, n. 7). *L'Exclue*, which was eventually omitted from Proust's dedication (see note 13), was probably an unpublished novel.

4. *Malachi*, IV, 2: 'But unto you that fear my name shall the Sun of righteousness arise with healing in his wings; . . .' The usual French translation is *soleil de justice*; Proust prefers the English version.

5. Allusion to Mme Madeleine Lemaire and her daughter Suzette. Proust dedicated Part II of *Sésame* to the latter.

6. 'Madame de Luzy' in *L'Etui de nacre* (The Mother-of-pearl Case, 1892).

7. 'Sérée' in *Parcours du rêve au souvenir* (Journey from Dream to Memory, 1895).

8. 'La Comtesse de Pripoli', published in *Le Banquet*, No. 8, 1893.

9. Paul Hervieu (1857–1915), novelist and playwright.

10. Madeleine Lemaire, late with her illustrations for Proust's *Les Plaisirs et les jours* (1896: see vol. I), had held up publication.

11. Part I of *Sésame* was dedicated 'To M. Reynaldo Hahn, composer of *The Muses Mourning the Death of Ruskin*, this translation is dedicated as a token of my admiration and friendship. M.P.' See letter 18.

12. Mme Arman de Caillavet (real name Mme Albert Arman), *née* Léontine Lippmann (1844–1910), mother of playwright Gaston de Caillavet (see letter 276, n. 1) and notable society hostess. She was Anatole France's Egeria.

13. The final dedication read: 'To Madame la Princesse Alexandre de Caraman-Chimay, whose *Notes on Florence* would have delighted Ruskin, I respectfully dedicate these pages, collected here because they pleased her, as a token of my profound admiration for her.'

151 *To Madame de Noailles*

[Wednesday, 27 September 1905]

Thank you, Madame.

My poor Mama[1] admired no one as much as you; she was infinitely grateful to you for your kindness to me. She has died at fifty-six, looking no more than thirty since her illness made her so much thinner and especially since death restored to her the youthfulness of the days before her sorrows; she hadn't a single white hair. She takes away my life with her, as Papa had taken away hers. She wanted to survive him for our sakes, but wasn't able to . . . As she didn't give up her Jewish religion on marrying Papa, because she regarded it as a token of respect for her parents, there will be no church, simply at the house tomorrow Thursday at 12 o'clock sharp, and the cemetery; but don't tire yourself, don't come, or at any rate only to the house. But in that case at 12 o'clock because there won't be any prayers at the house so we'll leave at once. Today I have her still, dead but still receiving my caresses. And then I shall never have her again.

Your respectful friend
Marcel Proust

I wanted to catch the person you sent but she had already gone as I had said I wasn't to be disturbed for anyone.

1. Early in September Proust went with his mother to Evian where she became seriously ill with nephritis and had to be taken back to Paris by her doctor son Robert. She forbade Marcel to accompany her, but when her condition worsened, a telegram from Robert summoned him home. Mme Proust died, aged fifty-six, on Tuesday 26 September. Like the majority of Marcel's friends, the Noailles went to her funeral at Père Lachaise, the famous Paris cemetery where she was buried in the family grave. Fifteen years later, on 22 November 1922, Marcel was buried beside her. Cf. the description of the narrator's grandmother on her death-bed (*RTP*, II, 357).

152   To Robert de Montesquiou

[Soon after 28 September 1905]

Dear Sir,

I don't know how I shall ever be able to thank you for so many kindnesses. When I am, not less unhappy, for that I shall never be, but less acutely ill than I am now, as soon as I can talk and get out of bed, I shall come and see you. Your pity for my distress is a new and magnificent interpretation of Hugo's 'For the petal of the lily is opening outwards'.[1] And it is at such moments that you are 'more splendid than Solomon in all his glory'.[2] For 'the greatest of these is charity'.[3] My life has now forever lost its only purpose, its only sweetness, its only love, its only consolation. I have lost her whose unceasing vigilance brought me in peace and tenderness the only honey of my life, which I still taste at moments with horror in the deep silence which she knew how to enforce all day around my sleep and which, through the habit she imbued in the servants she had trained, survives, inert, now that her zeal has ended. I have been steeped in every sorrow, I have lost her, I have seen her suffer, I can well believe that she knew she was leaving me and yet could not give me instructions which it may well have been agonizing for her to hold back; I have the feeling that because of my poor health I was the bane and the torment of her life. The very excess of my need to see her again prevents me from visualizing anything at all when I think of her, except for the past two days two particularly painful visions of her illness. I can no longer sleep and if by chance I doze off my sleep itself, less forbearing of my pain than my awakened intelligence, overwhelms me with agonizing thoughts which, when I'm awake, my reason tries at least to temper, and to contradict when I can no longer bear them. Only one thing has been spared me. I haven't had the torment of dying before her and experiencing the horror that that would have been for her. But leaving me for eternity, feeling me to be so little capable of struggling with life, must have been a very great torture for her too. She must have understood the wisdom of parents who kill their little children before dying. As the nun who nursed her said, for her I was still four years old.

Forgive me, dear Sir; as Hesiod said, unhappy people are garrulous and complacent in talking of their troubles. But there is a sort of fraternity among sorrows. 'In that, the poor man is the brother of Jesus.' I shall never forget your gentleness, your kindness, your magnanimous pity.

Your profoundly grateful
Marcel Proust

1. 'Car la feuille du lys est tournée au dehors.' Victor Hugo, *La Fin de Satan*.
2. *Matthew*, VI, 29; *Luke*, XII, 27.
3. *Corinthians*, XIII, 13.

153 *To André Maurel*[1]

[Boulogne-sur-Seine,[2] January 1906?]

Dear friend,

The state of my health unfortunately prevents me from telling you at length (in fact I'm absolutely forbidden even to write) that your letter filled me with gratitude – and astonishment. *The Stones of Venice*[3] is a hymn to the beauty of St Mark's, which Ruskin regards as one of the most beautiful monuments in the world, precisely because of its colour for which, like you, he finds the justification in the Venetian sky. He is so much of the same mind as you that he contrasts St Mark's Square with the gloomy square from which an English cathedral rises and even the sombre crows that caw around the one and the silky pigeons that coo at the foot of the other – all this to show the necessity of St Mark's and its marvellous beauty. So there is a misunderstanding between you, and I thought I must, though ill and unable to write, fling you into one another's arms. – As for the painting, I recognize that the theory you ascribe to him is exactly as you say and yours for its part is perfectly defensible. But all that is just theory, and on the whole it didn't hamper Ruskin much since in practice the two painters he admired most and to whom he devoted entire works were Tintoretto and Turner, neither of whom is a primitive. He greatly admired Giotto, but so do you, and moreover he admired just as much Reynolds, Holbein, Carpaccio, Giorgione and Gainsborough. And then in the last analysis don't you think that the most beautiful thing about Giotto is not a matter of ideas but a *certain feelings for form* which is no longer the same with Veronese. Veronese is a very great painter but it's true that his mosaics make a very bad impression in St Mark's.

Your devoted
Marcel Proust

1. André Maurel (1863–1943), author of a series of books on Italy of which the first, on Tuscany and the Veneto, was about to appear.

2. Proust was in a clinic for nervous ailments at Billancourt near the Bois de Boulogne run by Dr Paul Sollier (1861–1938), where he was to spend six weeks, only to leave as ill as when he arrived.

3. Maurel had written about the translation of *Stones of Venice* by Mme Peigné (see letter 77, n. 4) which had been published in December 1905 and which Proust had been re-reading for his review in *Chronique des Arts*, supplement to the *Gazette des Beaux-Arts*.

154   *To Madame Catusse*

[Soon after 6 May 1906]

Dear Madame,

To vary my habitual state with more specific sufferings, I have just been completely ill[1] and this prevented me from thanking you for your charming card. It is very ironical about poor Ruskin who did not indeed write his masterpiece on Florence. I think that if one used only his *Mornings* as a guide one would see nothing of what there is to see and would wear oneself out climbing to the top of a ladder at seven o'clock in the morning to make out some entirely repainted Giotto which wasn't worth the effort. No matter; if I ever go to Florence it will be to follow in his footsteps. Venice is too much a graveyard of happiness for me to feel strong enough yet to go back there. I very much want to, but when I think of it as a definite plan, too many pangs are aroused which stand in the way of its immediate realization. If when you return I feel a little better – for at the moment my hours are deplorable – and if you are good enough to come and see me once more, I'll show you a beautifully illustrated edition of *Mornings in Florence* which I've just received and in which it would give me so much pleasure to see you recapture your impressions and talk to me about them.

I haven't seen Lucien Daudet since I saw you, not having been in a state to receive anyone. I don't know how he took the 'blow' of the elections[2] which I take very badly. I make this confession while your dear Charles,[3] whom I find so charming, isn't listening to us, since he is the zealous servant of this petty-minded regime.

Your respectful friend
Marcel Proust

1. He had had 'flu over Easter.

2. The May elections had resulted in a landslide victory for the combined socialist and radical parties.

3. Mme Catusse's son, Charles Catusse (1851–1953).

155 *To Lucien Daudet*

[Early June 1906]

Dear Lucien,

What an enchanting letter. 'Rigging'[1] is delightful and of a delicacy all your own. There is nothing that you don't transcend in talking of it. And God knows I don't mean my own poor productions which you are so much above. Your unbelievable dive away from me, identical to the one that followed the Dreyfus case and then the Casa Fuerte friendship,[2] delighted my psychological Machiavellism. But you've done right, it's much too complicated trying to see me, or rather it wasn't but has just become extremely so, as I've resumed my noxious routine and moreover I'm going to try at last to take a few steps outside in the evening soon, as I've weakened myself too much by so many months without air. Thank you with all my heart for everything you say about Mama. She is absent from this preface,[3] and I even replaced the words 'My mother', which was in any case fictitious and didn't apply to her, by 'my aunt'[4] so that there should be no mention of her in anything I write until I've finished something I've begun which is exclusively about her. It's terrifying how my grief has been transformed in the past few months; it's even more painful now.

My dearest Lucien, how I envy you everything, but especially your talent, but only your talent. It would be so comforting for me before I die to do something that would have pleased Mama. So when I hear you say that you feel discouraged and that you'd like to smash your frames[5] I say to myself how ungrateful you are towards life and towards God. Thank you again, dear Lucien; I won't go on any longer about your too nice but above all marvellous letter (I haven't allowed myself to be carried away by its kindness, as I know that with your intelligence it's easy to pay compliments that ring true), because I'm exhausted by all this parcelling which I wanted to do myself, the Mercure not having done it very well for *La Bible d'Amiens*,[6] so that for the last few days I've been doing a grocer's job all the time with balls of string, wrapping paper and volumes of *Tout Paris*.[7] Forgive me for having sent the book for Mme Allard[8] to your address, but I couldn't remember exactly the number (27?) in the rue de Bellechasse and was afraid of getting it wrong.

Tenderly yours

Marcel

---

1. *Gréer*, to rig a boat. Presumably Proust is alluding to some inventive usage by Lucien.

2. Marquis Illan Alvarez de Toledo de Casa Fuerte (1882–1962), an exceptionally handsome youth; said to be the reincarnation of Wilde's *Dorian Gray* (which was published in 1890), he was first introduced to Proust when still a schoolboy by Daudet (cf. vol. I).

3. Proust's preface to his translation of *Sesame and Lilies*.

4. In 'Sur la lecture' there are two allusions to 'my aunt', and one to a 'great-aunt' whose mother instilled in her 'the rules and principles of making certain dishes, playing Beethoven sonatas and being a gracious hostess'.

5. Daudet, a painter, was preparing for his exhibition on 17 June (Mme de Noailles wrote the catalogue).

6. Proust was sending out complimentary copies of *Sésame*. Cf. letter 23.

7. *Who's Who* in Paris.

8. Mme Jules Allard, *née* Léonide Navoit (1822?–1909), Daudet's maternal grand-mother.

156   *To Madame Straus*

[16 or 17 June 1906]

Madame,

If my response to Ruskin was 'of the same quality'[1] (which your friendship makes you believe but which I can't), your response to me is of an infinitely rarer quality as well as being delightful and simple. What you say about the wit, the gaiety of melancholy people is exquisite, and profound. You are a marvellous person and your talent goes hand in hand with your niceness. It makes it more precious. But it also gives it a supreme distinction, a rare psychological charm.

Madame (this has nothing to do with the thanks which I should have sent you anyway even if I had nothing to ask you), do you know what Dr Sollier charges for his visits? I owe him for two which he paid me, one on the day I went into the sanatorium, the other the day of my return home. They weren't included in my bill. And when I raised the matter Mme Sollier[2] said it was nothing to do with her and I could send whatever I liked. But since I've asked several times already, I wouldn't want you to ask Sollier. That would embarrass me. I'd simply like you to tell me, if you know, how much he charges for his visits in general. He himself is exquisite but I'm convinced that he has nothing to do with the bills from the sanatorium, for Mme Sollier's was fearsome (which of course I didn't tell her!). He came to see me the other day but I was asleep. However he left me an extremely witty note. Since then he has written to give me the names of his assistants for which I asked him. Unfortunately I found it impossible to read them! I've been told some very amusing stories but so improper that it's impossible for me to write them. This one however is perfectly proper but you possibly know it; it comes from Degas, reported by Forain (not to me!), and it's about Gustave Moreau: 'He has taught us that the gods wore watch-chains.' Actually I don't find it all that marvellous and I feel that it sounds too like Forain to come from Degas.

Reynaldo was extraordinarily touched that you should have taken the trouble to go to that gathering at Madame Lemaire's[3] where he had to pretend not to know what people were there for, and especially to appear delighted to receive some Mozart scores which he wonders how he'll be able to get rid of! Which he certainly won't. He repeated to me some delightful things you said to him about me and I owe you even more gratitude for not having said them only to me.

Your respectful and grateful admirer
Marcel Proust

1. She had written praising his preface to *Sésame*.
2. Mme Sollier was also a doctor. Mme Straus had been with Proust and Dr Sollier when he had decided to enter the sanatorium (see letter 153, n. 2).
3. A *soirée* at which Camille Saint-Saëns (1835–1921) played the piano and Hahn was reported to have had a great success.

157 *To Robert Dreyfus*

[Wednesday evening, 20 June 1906]

My dear Robert,

How delighted I am to hear of your great academic success![1] I know it's inadequate to your talent and your work. But if an award had to measure up to your *Gobineau*[2] it would need to be a unique award and one that was bestowed only very rarely. One mustn't be too demanding but rejoice when one sees talent honoured and rewarded in a field where it usually goes unrecognized. It's all the more pleasing to me because I've followed and observed very thoroughly the progress of your success and been perfectly aware of the great impression your *Gobineau* has made on a public normally resistant to influences from such a high level, until finally, in all that academic froth which you have the right to regard as vain but which I find delightful, I see a powerful wave of success emerge to create an enormous stir. I can't tell you how happy about it I was and unhappy about Beaunier's article on *Sesame*.[3] Knowing Calmette's charming intentions towards me – I've never been able to understand why, knowing me so little, he has been so kind to me, and with such admirable persistence considering that he only sees me once in two years – I had asked him, if he was going to mention me, at least not to ask Beaunier to do it since he had already done so last year. He didn't listen to me and it was Beaunier who did the 'snapshot'.[4] I was overcome by it but I had no further fears. For how could I suppose that it was only a beginning and that Calmette also wanted me to have an article. So much fuss for a translation! I'm afraid of Beaunier's taking a dislike to me and I like him so much that that would greatly upset me. I shall never again have a line of mine published (which alas is in any case only too likely) without first asking him for a written promise not to write about it. Moreover it's a terrifying thing and one that shows how little interest there is in literature to see the extent to which the most intelligent of our contemporaries, to say nothing of the others, are incapable of *reading*, even a *newspaper*. You know that I see nobody. But up till the day before yesterday (when I made an effort to leave the house for a quarter of an hour without dressing which made me so ill that I shall probably never try again), I've managed to receive one or two of my most faithful friends. Not one of them (and they're all subscribers to the *Figaro*) had read Beaunier's article! Reynaldo, whose eager and inquisitive mind you know, came into my room the following day and asked me: 'Have you read yesterday's *Figaro*?' I modestly answered 'Yes'. 'So you read Varenne's article?'[5] After talking about Varenne, seeing that he wasn't going to mention anything else, I said: 'What did you think of Beaunier's article?' 'I didn't see one.' Albufera, whose whole family subscribes to the *Figaro* and who takes it himself, wrote to me to say how strange it was as I knew Calmette that the *Figaro* hadn't mentioned *Sesame*. And when I said that on the contrary they'd talked about it too much, he said to me: '*You must be mistaken*, because my wife reads the *Figaro* from beginning to end every morning and there was absolutely nothing about you.' Finally, I referred to it in a letter to Lucien Daudet who replied: 'I haven't seen either the paragraph or the article you mention' (he reads the *Figaro* every morning), 'could you send them to me.' – Ah! I don't read much, but I don't read like that. I assure you (I *swear* to you) that it isn't because it's to do with me that this strikes me so forcibly; on the contrary, it's because it's to do with me that I shall cut short my examples there and omit the most striking one of all.

Forgive me for going on about such trivialities. I only wrote to you to tell you how much pleasure it gave me to see your name among the academic prizes. I was so delighted! And then if I weren't so tired I should like to thank you again for the letter you wrote to me the other day which touched me so much and which I re-read so often. What made me so happy about Beaunier's article (still giving the word *happy* its sad connotation with its counterpart of heartsick anguish which I mentioned to you the other day[6]) and the one he wrote last year, when the Preface appeared in a review, was that it seemed to have been written in collaboration with the friends I most value. Last year I received a wonderful letter from a woman whose great genius I think you recognize as much as I do, Madame de Noailles, saying some exquisite things about that Preface and adding 'We read it yesterday with Beaunier, etc.'[7] Next day came Beaunier's article in which I felt I could still breathe the 'fragrant memory'[8] of the flowery praise of the day before. Similarly this year you write to say 'I was reading it yesterday with Beaunier etc.' and the next day or the one after there is a great flourish of praise from Beaunier of which your light and graceful lines had been a no less auspicious harbinger. As a matter of fact it isn't the things people tell me that I like best about my Preface. But enough talk about me. I'd like you to know that my object was mainly to talk about you and my grateful friendship for you.

I pictured you to myself in the Bois this afternoon, at a tea-party where there was a woman you admire and who is indeed beautiful.[9] Naturally I couldn't go, but I'm touched that people still think about me. It tires me so to write that this will have to do for a long time.

Ever yours

Marcel Proust

1. The Académie Française had just announced the award of the Prix Montyon, worth 500 francs, to Dreyfus and others.

2. Cf. letter 131, n. 1.

3. In a long article on 14 June entitled *Sésame* Beaunier wrote of the fidelity, intelligence and skill of Proust's translation and commentary, making comparisons to the detriment of Ruskin. Proust reverts to the subject in letter 215.

4. Gaston Calmette (1858–1914), editor of *Le Figaro*, had published Beaunier's paragraph announcing the publication of *Sésame* and praising both the footnotes – 'a model of erudition' – and the preface: 'original, delightful, moving . . .'

5. A complex, colourful and amusing court report by the legal correspondent of *Le Figaro* who signed himself Henri Varennes.

6. He had written to Dreyfus: 'Since Mama's death [happy] is a word which no longer holds any meaning for me . . .'

7. See letter 142, n. 1, n. 2.

8. *L'odorant souvenir*: from a poem by Mme Desbordes-Valmore.

9. Mme Numa Jacquemaire, daughter of the statesman Georges Clemenceau. The tea-party was in honour of Mme de Noailles, whom Mme Jacquemaire had expressed the desire to meet.

158   *To Robert Dreyfus*

[Around 25 June 1906]

Dear Robert,
    You described him, I recognized him:[1]

> A humble countenance
> A modest look and yet a glittering eye[2]

But frankly I can't understand why you were reluctant to talk about the tea-party. For in that case where will it lead? Already the whole legislature which saw Chaumié[3] as Minister has been thereby vitiated in all its work, and 'as for him', as M. Léon Blum would say, he cannot think well of a law of separation which was voted at a time when a man who had called him a 'poetaster' was tolerated as a member of the government.[4] But if he can't even have tea with her any longer![5] *She* about whom he has been telling us for several years: 'She's got genius, that girl', with the illusion that he was detaching that genius from his own to give it to her. Only she kept it and it appears that that wasn't part of the game. But really, a tea-party, in the Bois! I'll explain to you what I think about that but not tonight, I'm too ill. Too ill to reply to anything in your letter, to which I've so much to reply. But I wanted to write to you simply to say this trivial thing: You say that your prize gave you pleasure only because of the thirty louis. Do you then need money at the moment? As you know, I now alas have my little fortune at my own disposal. It isn't very large since I can't keep the only thing in the world I care about, the flat in which I've lived with Mama these last few years. But thank heavens it would enable me to put whatever you wanted at your disposal, if you had the least need of money. And with what pleasure. Don't tire yourself answering me. Just a figure, if you want anything. Nothing if you want nothing. But I hope it will be the first alternative. I'm very sad about the condition of M. Sorel, who is going to die.[6] Alas, I shan't even be able to go to his funeral. Today I haven't been able to get out of bed.
    Do you admire Léon Blum,[7] two of whose books Reynaldo has

brought me (incidentally uncut) about which he's very enthusiastic. I don't.

Ever yours

Marcel Proust

1. Fernand Gregh (see letter 58, n. 1). Cf. following letter.

2. '. . . une humble contenance / Un modeste regard et pourtant l'oeil luisant.' La Fontaine, *Fables: The Cockerel, the Cat, and the Young Mouse.* The quotation refers to the cat.

3. Joseph Chaumié (1849–1919), Minister of Education in the Combes government 1902–5, then Minister of Justice in the Rouvier cabinet.

4. It seems impossible to clarify all the allusions in this letter, but Gregh appears to have been slighted by Chaumié and perhaps compared unfavourably as a poet to Mme de Noailles.

5. i.e. Mme de Noailles (see previous letter).

6. Albert Sorel died on 29 June.

7. Proust had never admired Blum, his exact contemporary (cf. vol. I). The books were collected theatre and literary criticism.

159 **To Robert Dreyfus**

[About 26 June 1906]

Dear Robert,

You didn't understand me, I didn't understand you, and it's exhausting to write to each other every day. Don't let's write any more: you have too much work to do and I too much suffering to endure. But what is one to make of life! What is one to make of the judgment of the Wise! Just imagine, while writing to you yesterday I was thinking (I really don't know why!) of Gregh.[1] And when it came to addressing the letter I automatically wrote on the envelope: M. Fernand Gregh. Fortunately I noticed it in time! Think of all that might have happened as a result; with the generalization of his rancour the whole of Ruskin, the *Entente Cordiale* itself, would have been condemned in all those conversations in which, having run into Barthou or Briand,[2] he gives them advice, especially the advice, in their own interest, if they want to make a mark on history, to take him on as a minister, as anything at all! I think as you do about him, a little more kindly perhaps: I admire his superb intelligence, his superb talent, a mastery of his profession (his own profession) which has long since achieved perfection, elements of real stature, even down to the impenetrable crust of naïvety which has preserved in him a delightful freshness of spirit. But for all that, you can never listen to him (and quite often read him) without a smile. When there are several of you it's impossible to avoid exchanging winks. One could say of him, à propos of almost any one of us, what Caumartin said to M. de Clermont-Tonnerre, whose absurd self-conceit made him a laughing stock, when welcoming him ironically (without his remotely suspecting it) to the Academy: 'When you are with the King, no one can help noticing the look of joy on his face, etc.' M. de Clermont-Tonnerre swallowed it all with a simper, and only the cattiness of the Archbishop of Paris several years later (all this is a bit vague in my memory) disabused him and let him know he'd been made a fool of.[3] I know a very different example (from that of Gregh) close to us of an even more excessive self-conceit which makes one laugh but in this case on purpose. Or at least, just as Baron[4] may perhaps have had a naturally comic voice but noticed it and used it deliberately, M. de Montesquiou (it's he) makes a work of art of his own absurdity and has stylized it marvellously. Alas, Gregh isn't up to stylizing his: he isn't aware of it! I don't know by what convention, as on the stage, he doesn't even hear the laughter around him, or so vaguely that he might ask, like Harpagon in Plautus hearing the audience laugh (that is to say people who are unreal to him): 'What are all those people laughing about?'[5] Personally he doesn't make me laugh, because I understand so well everything he says – and I find him so endearing. But I can understand why people find him priceless. I remember that at first I didn't know that Montesquiou was aware that he was ridiculous until, at his first lecture when everyone expected to see him dressed in pink and green and he appeared in black like a notary's clerk, he said to me: 'I wanted to arouse this feeling: the expectation of the ridiculous disappointed.'[6] It's a disappointment that Gregh doesn't often provide. And yet he is charming, exquisite and admirable. And it's perhaps only to you that I'd say he was absurd, because . . . you told me so!

Ever yours

Marcel

1. See the veiled allusions to Gregh in the previous letter.

2. Louis Barthou (1862–1934), since 14 March Minister of Public Works in Sarrien's cabinet. Aristide Briand, see letter 97, n. 1.

3. Proust quotes from memory Sainte-Beuve's account in his *Causeries du Lundi* of the reception into the Academy in 1694 of François de Clermont-Tonnerre, Count-Bishop of Noyon. The same scene is related in Saint-Simon's *Mémoires*.

4. Cf. letters 130, n. 11, and 328, n. 1.

5. The reference is to Plautus's *The Pot of Gold* which was the source of Molière's *L'Avare*.

6. Cf. letter 80, n. 3.

160 *To Lucien Daudet*

[Thursday morning, 5 July 1906]

My dearest Lucien,

I have two favours to ask of you which perhaps you can do for me.

1. Do you still see (or can you still see) your footman who came from the Swedish legation and if so would it bother you to ask him the exact name of the servant from the same legation about whom I think I've spoken to you. I need to talk to him and can only find out his address if I know his full name. If you agree, I'll send you the details (his approximate surname and Christian name), and won't bother you with them in advance in case you can't.

2. Could you find out from Léon[1] or from Flament who signs himself Pepper and Salt in the *Gaulois*, and, if it isn't always the same person, who wrote the article on *Sésame et les lys* signed Pepper and Salt two or three days ago.[2] This article, which in regard to me is indeed rather pepper and salt and neither one thing nor the other, is very remarkable about Ruskin. It's for that reason and that alone that I'm upset that it isn't more obliging about me because it's the first discordant note. But anyhow I'd still like to know who it is; I feel I ought to express my thanks, because the fact of having written an article on the subject is very amiable. Perhaps it was a friendly act on the part of Meyer,[3] the execution of which was entrusted to someone who only half likes me. Perhaps my memory's at fault, but I thought you told me it was Flament who did those pieces. But it doesn't seem to me at all his manner. Anyhow if you can find out without going to too much trouble I'd be glad. As it happens I'm just about to write to Léon to thank him for his *Primaires*[4] and I could have asked him. But that would mean asking him to reply. And for something like that I'm less embarrassed with you, my dearest Lucien, than with him (you won't take that amiss, will you? I mean that I'm closer to you though I'm embarrassed all the same). The surprising thing is that the article is called 'Un Professeur de Beauté' – which is the title of my article on Montesquiou . . .[5]

Lots of love

Marcel

1. His brother Léon Daudet.
2. On the front page of *Le Gaulois*, 4 July.
3. Arthur Meyer, editor of *Le Gaulois*. See letter 64, n. 8.
4. *Les Primaires, roman contemporain*, published on 3 July.
5. Proust's appreciation of Montesquiou's works had appeared in *Les Arts de la vie* in August 1905.

161 *To Lucien Daudet*

[Soon after 5 July 1906]

My dearest Lucien,

What a scoundrel you are not to have told me that the 'Pepper and Salt' article was by Léon. I know very well that you have other things to think about at the moment but since you were writing to me it's a bit much. And to think I've let all this time go by without thanking him; what must he think of me? I only discovered yesterday. Until the moment when I calculate that your brother has received my letter I shall feel very embarrassed. Because I'm so grateful to him. He has already been so kind to me that the thought of his being nice to me once again touches me to an extraordinary degree. Especially as one can sense that he only did it out of kindness and not inclination, that Ruskin and I get on his nerves and that he went at it like a dog that has been whipped or rather that whips itself. So that the little stings that seemed to me less than kind from a stranger are what touch me most from him, a sign of his gruff kindness. And such a splendid article for me, so long, so prominent, which otherwise I'd never have got! My dear little Lucien, you really ought to have told me!

Tenderly yours

Marcel Proust

162 *To Madame de Noailles*

[Monday, 16 July 1906]

Madame,

Post-scriptum to my letter[1] to tell you that I think Barrès was extremely courageous and noble the other day in the Chamber.[2] Tell him so, and that I'm not writing to him because I should have to add at once that Dreyfus is innocent nonetheless and that in spite of my great pity for General Mercier[3] he's an out-and-out scoundrel, and so many distinctions would be exhausting both to write and to read. All the same, when I think that I organized the first list in *L'Aurore* to demand a revision[4] and that so many politicians who were then wild anti-Dreyfusards now mortally insult on the floor of the House this old man of seventy-five who was courageous enough to appear, surrounded by a hostile pack, with nothing to say, knowing that he would have no argument to put forward except that the procedure (of the Court of Appeal!) had been *irregular*, *illegal*, and *in camera*! It would be unbelievably comic but for the fact that the newspaper says: General Mercier very pale, General Mercier even paler.[5] It's horrible to read, for even in the wickedest man there's a poor innocent horse toiling away, with a heart, a liver and arteries in which there's no malice, and which suffer. And the most splendid moment of triumph is spoiled because there's always someone who suffers.

Your respectful friend
Marcel Proust

1. An unpublished letter on the subject of the poet Francis Jammes (1868–1938).

2. On 13 July the government had apprised the Chamber of Deputies of the unconditional pardon, and reinstatement and promotion, of Captain Alfred Dreyfus (1850–1935), hero and victim of the long drawn-out Affair (cf. vol. I). In his speech to the Chamber, Barrès, a virulent anti-Dreyfusard, had deplored the gratuitous insults flung at Dreyfus's accusers.

3. Auguste Mercier (1833–1921), Minister of War at the outset of the Dreyfus case and one of Dreyfus's principal accusers in the second trial of 1899; author of the notorious and false 'Mercier bombshell'. Cf. vol. I.

4. A list of signatories to the Picquart Protest of November 1898 (cf. vol. I) objecting to the treatment of Colonel Picquart. Cf. following letter.

5. *Le Siècle* of 14 July gave a graphic account of the brutal heckling of General Mercier.

163   *To Madame Straus*

[Saturday evening, 21 July 1906]

Madame,

It was too kind of you to write me such exquisite letters and I thank you for them with all my heart. What you say about the Dreyfus case is naturally the funniest, most profound and best written thing that could be said on the subject. You have the infallibility of grace and wit. It's odd to think that life – which is so unlike fiction – for once resembles it. Alas, in these last ten years we've all had many a sorrow, many a disappointment, many a torment in our lives. And for none of us will the hour ever strike when our sorrows will be changed into exultations, our disappointments into unhoped-for fulfilments, and our torments into delectable triumphs. I shall get more and more ill, I shall miss those I've lost more and more, everything I aspired to in life will be more and more inaccessible to me. But for Dreyfus and Picquart[1] it is not so. Life for them has been 'providential', after the fashion of fairy tales and serial stories. This is because our miseries were founded on truths, physiological truths, human and emotional truths. Their misfortunes were founded on errors. Blessed are those who are victims of error, judicial or otherwise! They are the only human beings for whom there is redress and reparation. I don't know who, incidentally, is the stage manager of these final 'tableaux' in this particular reparation. But he is incomparable and even moving. And it's impossible to read this morning's closing scene: 'In the courtyard of the Ecole Militaire, with five hundred participants' without having tears in one's eyes.[2] When I think of the trouble I had smuggling *Les Plaisirs et les jours* into Mont Valérien where Picquart was imprisoned, it almost takes away my desire to send him *Sésame et les lys* now as being too easy. I haven't managed to find in the newspapers the story which is mentioned in *Les Débats* of an army corps commander who on assuming his functions made a speech about the Dreyfus case to the President of the Tribunal. I suppose it was General Galliéni[3] but I have no idea and I can't find it anywhere. I'd also like to know the names of the deputies who voted for the promotion of Dreyfus and Picquart. Where I know I would irritate you, but I hope that in a letter it won't, is on the subject of Mercier. I don't like seeing Barthou (whose Dreyfusism is only a few weeks old) currying favour by insulting with a violence that makes one wince that obscene blackguard who is nevertheless an old man of seventy-five who had had the courage to appear on the floor of the Senate in front of a howling assembly, knowing that he had absolutely nothing to say, not an argument to put forward except the priceless one that the Court of Appeal had tried the case *in camera* and that the procedure was irregular![4] But to express my thoughts on all that would need pages which I shall spare you. A man who must feel profoundly happy and who deserves it, the most enviable man I know for the good he desired and achieved, is Reinach.[5] I regret that his triumph in the newspapers and in the Chamber is so modest. He did a great deal more than Zola . . .[6]

Your respectful and grateful friend
Marcel Proust

1. Georges-Marie Picquart (1854–1914), Colonel in the Intelligence Service at the time of the Dreyfus case, later General and Minister for War. In 1898, having expressed doubts as to Dreyfus's guilt, he was arrested and held incommunicado in solitary confinement for a year. He was reinstated and promoted at the same time as Dreyfus. See previous letter.

2. On 21 July, at the Ecole Militaire in Paris, Major Dreyfus was invested with the cross of Chevalier de la Légion d'Honneur.

3. Joseph-Simon Galliéni (1849–1916).

4. See previous letter.

5. Cf. letter 94, n. 5.

6. Emile Zola (1840–1902), author of the famous open letter *J'accuse* addressed to the President of the Republic attacking the General Staff over the Dreyfus case.

164   **To Madame Straus**

[Versailles, Wednesday evening, 8? August 1906]

Madame,

I've been so ill for two days that I haven't been able to write to you; I simply want to tell you that, being unable to go away because of my uncle's condition,[1] I settled into the Réservoirs in Versailles, where I immediately fell ill. I don't think I shall stay here and as soon as I'm fit enough to leave Versailles I shall do so. But even if I were to go to Trouville, it wouldn't be immediately because I'm too tired. I hasten to tell you all this because in your kindness you might continue to look for houses for me and it would be pointless.[2] However if ever you hear by chance of something nice and inexpensive to let you could always tell me. But I don't think I'll go, at least for some time. Every time I make an unsuccessful effort I feel discouraged, but once I've recovered I forget and am ready to start again. I hope that the fatigue you complained of the other day, of which I'm afraid of having been partly (or even entirely) the cause, has disappeared. I have a vast and splendid apartment at the Réservoirs (which will certainly cost me more than Trouville!), complete with pictures, hangings and mirrors, but terribly gloomy, dark and cold. It's an apartment of the historic type, one of those places where the guide tells you that Charles IX died there and where you cast a furtive glance around as you hurry to get out again into the light and warmth and the comforting present. But when you not only can't get out but have to make the ultimate submission of sleeping there! It's enough to make one want to die. I don't know how they managed to orient it so that the sun never penetrates, at any hour. I asked if the chimneys smoked and was told they didn't, which is Jesuitically true, for indeed when you light a fire the chimney in which you light it doesn't smoke, but at that very moment all the others begin to smoke with such violence that the apartment is one huge cloud. Forgive me for all these boring details; in any case I don't want to tire myself as I have a raging fever.

Your respectful friend
Marcel Proust

It appears that my brother, thinking I was at Trouville, went there on Sunday. He found the Hôtel de Paris excellent, and not madly expensive. If Robert Dreyfus is with you please give him my most affectionate regards.

I have two pianos, with a ravishing bust on one, which make me hope that if Mme de Saint-Paul,[3] who is apparently in Versailles, discovers that I'm here too, she will come and play some exercises for me.

1. Georges Weil (1847–1906), Mme Proust's brother, died on 23 August (cf. letter 173, n. 1).

2. Proust had recently been corresponding with Mme Straus about the possibility of taking a house near Trouville.

3. The Marquise Diane de Saint-Paul, musical hostess. See letter 122, n. 2.

165 *To Reynaldo Hahn*

[Versailles, Thursday, 9 August 1906]

My dear Marquis de Bunibuls,

I have to apprise you of a thing that is the least incredible, the least important, the least trivial, a thing that has existed for four months but that you will not believe in ten, a thing that was true yesterday but that is perhaps no longer true today. (If it is.) Guess who has established himself as an antiquary, valuer and dealer like Molinier,[1] has purchased a vast warehouse which he has filled with his collections at Versailles, spends several hours a day selling and discussing the price of curios and works of art, makes a commerce of it, a science of it, applies his knowledge also to what is not his, appraises the period and the authenticity of everything, dares to assert that the *Deposition from the Cross* of Rubens might well be by Van Dyck, whom he pronounces Vamm Dike like Mme Oppenheim and like Henraux, and has, by the by, aged a hundred years. Who is it? I give you two guesses. 'Well, of course,' says Mme Lilli Lehmann,[2] 'it's M. de Nolhac.'[3] Nay, not he. 'Then it is Robert de Montesquiou.' Nor yet he. 'Gracious, how foolish we are, say you; it can only be Lobre,[4] if it be not Tenré.'[5] Nay, do you give it up, it's, it's . . . once more, do you guess? But no, you cannot, it's . . . I want to tell you but not to have you believe it, it's, it's Hector . . . 'Hector? Hector who?' It's Hector . . . Hector . . . Hector . . . Well then I shall tell you, it's Hector, simply Hector, Hector of the Réservoirs, Hector of la Potocka,[6] your Hector, my Hector, Hector the head-waiter, in a hundred words as in one, it's Hector. Expostulate, accuse me of falsehood, say I am a liar, I shall bear you no grudge for I myself did the same thing, etc. – Mark you, I don't have Mme de Sévigné here, and I'm telling you all this while pressing all too clumsily on the stirrup of a tottering memory, and on the other side on the stirrup of reconstructive imagination in order to set my pony's hooves[7] in the immortal hoofprints left by the Pegasus of the Marquise (who wasn't flying that day!).[8]

Reynaldo, (1) I fell ill on arriving at the Réservoirs on Monday, but so ill that I can scarcely write to you. I have a lugubrious ground-floor apartment, not yours,[9] bigger, near the theatre, full of pictures, mirrors, etc. I've already been pitilessly moved in spite of my fever, etc. etc. What should one give by way of tips to the chambermaid, to the telephone operator, to the hall porter with powdered hair as in the Théâtre Français? In the hotel itself I had a very nice one with a red nose who knows you and was a servant of the Duc de Fezensac but I've already settled with him. What a lot I could write to you about if I were less ill. But alas! Let me simply tell you that I've learnt on the most positive authority that Robert de Rothschild[10] hasn't one jot to reproach himself with as far as you are concerned, that they have a formal embargo on accepting anything at all, whether a single sou or a million, that he refused a month ago to take on his best friend's fortune of ten million, that you must get it into your head that there isn't a shadow of unkindness in it all, that we assessed the whole thing quite wrongly. I'll explain it to you when I see you, but I assure you that what I've told you has been rigorously checked and is absolutely positive. A propos of Robert, think how a letter like the one about hay-making – 'When one can do that one can toss hay', 'I ween that he is the most disagreeable man in the world and one who mislikes hay-making'[11] – resembles what old mother Gustava says.[12] Don't despise Sévigné on that account but ask yourself whether one might not say like Marmontel:[13] 'Gustava or the Sévigné of the masses.'

Fondest love from
Buncht

Burn this letter at once and in replying to me don't mention the name Hector.

This morning's *Figaro* says that M. Reynaldo Hahn is at Territet.[14]

When you reply to me tell me you've burnt this letter (and the Bréval[15]). And make sure that it's true. Antoine is in San Francisco, to forget.[16]

1. Emile Molinier (1857–1907), palaeographic archivist, Keeper at the Louvre.

2. In March Lilli Lehmann (1848–1929) had sung at the Paris Mozart festival directed by Reynaldo Hahn; in August, she was to sing Donna Anna in several performances of *Don Giovanni* conducted by Hahn at the Salzburg Festival.

3. Pierre de Nolhac (1859–1936), Keeper of the National Museum of Versailles.

4. Maurice Lobre (1862–1951), painter dubbed by Léon Daudet the 'French Vermeer'.

5. Henry Tenré (1864–1926), genre painter and illustrator.

6. Presumably a reference to Comtesse Potocka (see letter 64, n. 5).

7. Cf. letter 18, n. 5.

8. Proust has been parodying a letter from Mme de Sévigné to M. de Coulanges (15 December 1670). In *The Fugitive* the narrator's mother refers to the same letter, as well as the one about hay-making (*RTP*, III, 672). See note 11 below.

9. Hahn sometimes took a room at the Réservoirs in which to compose.

10. Baron Robert de Rothschild (1880–1946). Proust had recommended Hahn to bank with Rothschilds.

11. Cf. letter from Mme de Sévigné to M. de Coulanges (22 July 1671).

12. Baroness Gustave de Rothschild, *née* Cécile Anspach (1840–1912), a famous beauty.

13. Jean-François Marmontel (1722–99), writer of tragedies, moral tales and novels.

14. Spa on Lake Geneva.

15. Allusion to a burlesque song about Lucienne Bréval (1869–1935), dramatic soprano at the Opéra, which Proust enclosed with his letter.

16. Antoine Bibesco had had a long love affair with Lucienne Bréval. See also letter 167.

166   *To Madame Straus*

[Versailles, Tuesday, 9 October 1906]

Madame,

You must be at a loss to understand my silence and alas I cannot break it yet. Having been compelled to leave the rue de Courcelles,[1] for the past month and more I've had people house-hunting for me and my hesitations, my anxieties, my tentative leases broken just as they were about to be signed, have robbed me of sleep so that I scarcely have the strength to write to you. Finally I couldn't make up my mind to go and live in a house that Mama wouldn't have known and so for this year, as an interim arrangement, I've sub-rented an apartment in our house in the boulevard Haussmann[2] where Mama and I often came to dine, and where together we saw our old uncle die in the room that I shall occupy. Of course I shall be spared nothing – frightful dust, trees under my window, the noise of the boulevard, between the Printemps and Saint-Augustin![3] If I can't stand it I shall leave. And the apartment is too expensive for me to stay there forever. But this year, since it was sub-let to a tenant who paid for it without living in it, I've got it for relatively little.

What a lot of things I have to say to you, what affection to lavish on you, even some amusing stories to tell you. But I'm too tired now and simply want to say that I've never been so fond of you.

Marcel Proust

During the two and half months I've been in Versailles I've succeeded (*unberufen*[4]) in not seeing Mme de Saint-Paul.[5] A few letters and that's all. Your card (your cards) from Bayonne intoxicated me. Ah, there's a house I'd like to live in. And Jacques's cards from Brittany! I haven't had the strength to thank him.[6]

---

1. His lease expired on 30 September.

2. 102 boulevard Haussmann had belonged to his great-uncle Louis Weil (see letter 75, n. 6) who left it to his nephew Georges and his niece Mme Proust; Proust, who now owned half his mother's share in it, had taken the first floor.

3. The first a popular department store, the second a fashionable church, at opposite ends of the boulevard Haussmann.

4. The equivalent in German of 'touch wood'.

5. Cf. the final paragraph of letter 164.

6. Her son, Jacques Bizet.

167 *To Georges de Lauris*

[Versailles, soon after 20 October 1906]

My dear Georges,

I've been so unwell these last few days that I haven't been able to tell you how much your letter moved me in every way.[1] The possibility of an imminent operation disturbs and torments me a great deal, as you can imagine. On the other hand, what you say about your mother from the point of view of strength, eating, appearance and resistance also impressed me greatly and filled me with joy and hope. In the obscure drama of a hidden malady[2] the only clear, or at any rate reliable and authentic, interpretation we can go on is precisely the general condition; that is very important, indeed everything. I beg you to keep me informed.

My housing problems get more and more complicated. I shan't explain how because if I began I should need ten pages and I have to write to the architect, the manager, the second-floor tenant, my brother and my aunt as the owners, the concierge, the telephone company, the upholsterer, etc. I must leave you for these graceless tasks.

Affectionately
Marcel

I'm told that Mlle Bréval is inconsolable at Antoine's departure.[3] But to that phrase of Fénelon's (warmest regards to ours) I reply with La Fontaine's advice:

Lovers, happy lovers, will you not roam?[4]

Why do these people always leave one another if they're in love? But are they?

Have you any idea of M. Berenson's fortune (in the most vulgar sense of the word)? I'll tell you why I'm interested. I should also be glad to know what there is by M. Berenson in French or translated into French.[5]

---

1. Lauris's mother (whom Proust had never met) was suffering from a biliary calculus.

2. Proust will develop this idea of the 'obscure drama of a hidden malady' when describing the narrator's grandmother's illness in *The Guermantes Way* (*RTP*, II, 308–14, 324–5, 327, 333).

3. Cf. letter 165, n. 16.

4. From *The Two Pigeons*.

5. Bernard Berenson (1865–1959), leading authority on Italian art, had already written several important books, including *Florentine Painters of the Renaissance*, which had appeared in a French translation in 1896.

168 *To Bertrand de Fénelon*

[Versailles, November 1906]

For a long time, my dear Bertrand, I've wanted to thank you for your charming post-card from Göttingen,[1] which gave me great pleasure. And there's a great deal I could say to you about it – the charm with which you can talk about cities like this one, like Veere, like Delft, is so rare that I found myself thinking this: if I had more talent, more life, more time, if I were a better friend, I would do, with these really exquisite post-cards and whatever I myself might have been able to gather in the course of my dealings with their author, what Sainte-Beuve did for the outstanding personalities of his time who without him would be forgotten, since they never wrote anything but those letters or said those words of a superior quality which needed to be linked and interpreted and are nowhere recorded. For it seems unfortunately probable that, whether because of too much work, or indolence or disdain, you will never write any work of the imagination, any book of your own. So the only recourse left to you is a portraitist, a memorialist with the talent, the devotion, the affection and the application to pin down the physiognomy which might otherwise remain unknown. Since mine was too early shattered by grief, find another pen, a golden pen.

I'm putting all this very badly, my old friend, in the bustle of a move (though I'm staying another month in Versailles) and perhaps a lawsuit against some people whose flat in Paris I've sub-rented and who haven't finished the repairs in time,[2] but I wanted you to know that I think of you very often. As you know, the desire to travel and the impossibility of doing so are always with me. So each letter I receive from places I should like to see gives me a sense of poetry by setting my imagination in motion. But when the letter is from you, the impression is intensified, because it was with you that I experienced my principal travel impressions, so that a letter from you sends me not only on an imaginary journey but on a remembered journey too.[3] I hope you are happy, as they say, and I send you my friendliest remembrances.

Marcel

1. Fénelon was *en poste* in Berlin.
2. Proust was having problems with a Dr Gagey on the ground floor, Arthur Pernolet, an acquaintance, on the second, and the outgoing tenant of his own apartment.
3. See letter 3, n. 3.

169   To Georges de Lauris

[Versailles, towards the end of November 1906]

Georges,

Do you possess *A Lover*[1] (forgive the strangeness of this form of words) and *Jude the Obscure*.[2] If so, and if it isn't a nuisance, could you send them to me. If not, if you pass Mme Paul Emile's could you tell her to send them to me together with the bill.[3]

Antoine[4] didn't write to me today as I hoped he would with news of your mother.[5] I wait, I hope, I think of you tenderly.

Marcel

I should be particularly grateful if Mme Emile Paul could send me Gabriel Mourey's *Gainsborough* (published by Laurens) which I badly need.[6]

1. *Un Amant*: title of the first French translation (1892) of *Wuthering Heights* (later re-translated as *Les Hauts de hurlevent*).
2. Translated (1901) as *Jude l'obscur*.
3. Bookshop in the faubourg Saint-Honoré.
4. Antoine Bertholhomme, concierge at 102 boulevard Haussmann.
5. See letter 167, n. 1.
6. His review of the book for the *Chronique des arts et de la curiosité* was long overdue, the review copy having been mislaid in the move from the rue de Courcelles.

170   To Marie Nordlinger

[Versailles, Friday evening, 7 December 1906]

Dear, dear, dear, dear Mary!

First:

Did Reynaldo tell you that I had sent letter then *Sesame* to the strange place and to the address where, as I said to him, '*Detroit* is the name of the town, isn't it, *Avenue* of the Province, and *Lake Ontario* of the country?' But never an answer, and I realize that nothing arrived, since you ask me 'What about *Sesame*?' Here it is enclosed in response.

Dear friend, how close you are to my heart, and how little has absence separated you from me! I think of you constantly with such tenderness and an indestructible nostalgia for the past. In my ravaged life, in my shattered heart, you hold a cherished place.

Dear friend, permit me in a few words to explain to you a trivial matter. If you have so far received no royalties for *Sesame* it's because the review in which it first appeared, *Les Arts de la vie*, has gone bankrupt and paid nothing, and the publishers of the book, the Mercure, won't be settling with me until it's more nearly sold out. If between now and then you would like me to send you an advance, nothing could be easier; you know that, alas, I no longer have to answer to anyone for the use I make of my money.

You are in Manchester, I see. I, for the past four months, in Versailles. In Versailles, can I really say? As I took to my bed on arrival and haven't left it since, haven't been able to go even once either to the Château or to Trianon or anywhere else, and don't wake up until after nightfall, am I in Versailles rather than anywhere else – I've no idea. I ought to be in Paris but I've had apartment troubles, have started a lawsuit, and have rented from October an apartment which I can't get into. But anyway if you write to me either at the Hôtel des Réservoirs where I still am, or at 45 rue de Courcelles where I had the terrible wrench of being unable to stay because of the price of the apartment but from which my nice concierge will forward Mary's letter, or 102 boulevard Haussmann which is the address of the apartment rented but still uninhabitable, you will reach me. You will forgive me if I'm too tired to reply. But I shall try to. Are you working? I no longer am. I've closed forever the era of translations, which Mama encouraged. And as for translations of myself, I no longer have the heart. Did you see some beautiful things in America? What strange folly to have sent me back that little text-book![1] If I went out in the daytime I should love to see some of that Egyptian and Assyrian art which seems to me so beautiful. Does M. Bing sell Egyptian and Assyrian, and Gothic, things?[2] How are your family? How is your aunt,[3] to whom I beg you to remember me and who remains in my mind as one of the most curious *Stones of Venice*. Nothing could soften, nothing could budge the inflexibility of her principles. But how I liked her, and how she seemed to love you! And she represents for me the *Mornings in Venice*[4] which I never saw – she being so much the early riser who ignores the slug-a-bed. I have not ceased, dear friend, to think of you always, a great deal, constantly, and I shall never cease. I kiss your hands in infinite friendship.

Marcel Proust

1. A scholarly tract by the eminent Egyptologist Gaston Maspero (1846–1916).
2. Marcel Bing, son of the late Siegfried Bing (see letter 16, n. 1).
3. Mme Caroline Hindrichsen. See letter 52, n. 2.
4. Allusion to Ruskin's *Mornings in Florence*.

171   **To Georges Goyau**

Hôtel des Réservoirs
Versailles
[Sunday evening, 9? December 1906]

Dear Sir,

I don't mean to write to you every day![1] But knowing what a friend and admirer (whom you yourself admired no less) you lose in Monsieur Brunetière,[2] I wanted to write you these few lines of sad condolence. From the sorrow I feel at the disappearance of a man whom I knew only through his thought (but since his thought combined heart and action, never has one known a man so well by reading and listening to him), I realize that you must be deeply grieved. And I sympathize with you profoundly. And not knowing his family either, it is to the outstanding member of his true family, his spiritual family, that I send my heartfelt salutation. Mama admired him as I did, and would have been saddened as I am. And I feel that France never had greater need of this light which has now been extinguished. It will intensify the painful darkness of the present time. I suppose you will write a portrait of him, one of those portraits which simply transfer on to paper the feelings we have engraved in the depths of our hearts. And in addition to the power to move which such pure transcriptions of the heart always have, I am so well aware of the beauty of all those portraits which you draw that I am impatient to read this one. But it is in my innermost thoughts without reference to literature that I want to tell you how much I feel for you in your grief.

Please share with Madame Goyau my respectful and admiring respects.

Marcel Proust

1. He had written the day before on other topics.
2. Ferdinand Brunetière (1849–1906), literary critic, professor at the Sorbonne.

172   **To Madame Catusse**

[Versailles, Monday, 10 December 1906]

Dear Madame,

You are quite right: my apartment mustn't look like a furniture repository. And not only for aesthetic reasons but for the sake of my health. Every speck of dust suffocates me. Every piece of furniture gathers dust. And in a house where it's difficult to beat and clean, because of the hours when I sleep and my fear of noise, my sensitivity to the cold and fear of open windows, an apartment that was like a hospital would be the ideal. Since that's not available, I want at least to have as little furniture as possible, while still having a lot. So keep only the best, what is really of some quality. The rest can wait in a repository, until the time comes when I'm forced to live permanently in the country and will be able to have more room – or until some decisions can be taken later on in consultation with Robert.[1] Even my old desk can await its hour – mine. As for the desks to be kept, the one Robert wants is the one from the boulevard Malesherbes;[2] mine will be my uncle's[3] (but what use will a desk be to me if I never get down to work?). It will return to exactly the same place as it occupied during my uncle's lifetime, and perhaps in doing so, having once been accustomed to the aura of familiar things, it will recapture a little of the unaesthetic but touching life which it retained in my imagination, the only setting in which places remain the same.

Dear Madame, I don't want to tire myself writing, although it's an excellent thing that I've had so many good days when I've been able to write to you at length. It's true that I do absolutely nothing else and that it represents my whole life, intellectual, physical, moral, vegetative, for when I write to you I neither eat nor sleep! But I should never have been able to do it in Paris. So without tiring myself too much, I want to tell you this, which I've meant to tell you each time. Never did I think that you would do anything more than give five minutes' worth of explanation to the upholsterer. And I see that you yourself are the upholsterer and everything else, and a thousand times more, and I'm so unhappy about it – and at the same time so happy – that if I were to come into a fortune tomorrow and have a palace, I wouldn't ask your advice on the smallest detail, since I've bothered you enough in advance for my whole lifetime, and yours. But now that the process of which I had no inkling has been set in motion I can only let you carry on,[4] admire you without protesting, thank you wholeheartedly without saying so. I give an occasional start when I read: 'I've checked with the Place Clichy,[5] I've examined the lavatory, etc.', but I repress it.

How true Vogüé's article on the separation[6] basically was when it said that it was only a word, that Church and State would in no way be separated for all that, and that it wasn't viable, the law would have to be altered, there would have to be compromise, there should have been an understanding with Rome, etc. I had thought all that was untrue. But how true it was! Prophetic! What power even a mediocre Pope has, since all Briand's admirable intelligence, his sublime 'good will', are thwarted by it even though supported by material strength.[7] Indeed, what a difference between today and former times, when neither a Philippe-Auguste nor a Louis XIV nor a Napoleon would ever have accepted a quarter of what Briand puts up with from the Pope; when, too, the clergy would never have had the highmindedness or at least the disinterestedness which has made them renounce all their property in obedience to the Pope. Power amounts to very little, however, since now that the Pope no longer has an army or territory he is more powerful (even in France, where he is least powerful) than he ever was in the days of his material power. I hope nevertheless that this crisis will come to an end. And I hope that Briand's law, so fine, so wise, so truly and broadly equitable, will become the statute – so desirable even for itself – of the Catholic Church.

Marcel Proust

1. His brother; when consulted he invariably replied, 'It's whatever Marcel wants.'

2. 9 boulevard Malesherbes was the Proust family home for the first twenty-two years of his life. Cf. vol. I.

3. The late Louis Weil.

4. She had been inundated with letters since October, often repetitive or contradictory, requesting her to decide upon and oversee every detail of the furnishing and decorating of the apartment in the boulevard Haussmann.

5. Address of the repository where his furniture was stored.

6. An article in *Le Figaro* entitled 'L'Encyclique', by Eugène de Vogüé (1848–1910). Cf. letter 97.

7. Aristide Briand was now Minister of Education in the Clemenceau government, but shared (with Guyot-Dessaigne) the responsibility for Religious Affairs. His Law of Separation, though far less rigorous than most Catholics had feared, was denounced by the narrow-minded and reactionary Pope Pius X. Cf. letter 97.

173   *To Madame Catusse*

[Versailles, Wednesday evening, 12 December 1906]

Dear Madame,

I've been so ill for the past two days that it's only this morning that I was able to look at my letters and read yours. Then I fell asleep, slept until this evening, got up at about eleven o'clock, and am now writing to you. My letter will go out tomorrow at dawn, but in the meantime, what must you be thinking of me! You must think me thoroughly ungrateful. You must be saying to yourself: he was in a hurry to write to me when he needed me, but now that I've done him a favour he doesn't even take the trouble to thank me. Dear Madame, the idea that you may be thinking this up to the moment when you receive this letter grieves me. I'm so full of gratitude for your kindness.

Everything you told me seemed to me – through the haze of my vanishing distemper – to be perfect. I wasn't as attached to the red wallpaper as you thought. It was that Empire paper in particular which seemed to me nice although red. But it couldn't go there. And I'm not in the least hostile to red, on the contrary! I'm glad to see that the panelling can be used – I thought the ante-room was too small – and it will give me great pleasure to come across it again there because Mama took so much pleasure in it, was so fond of her ante-room. And even apart from that feeling, it will please me in any case because it was charming whatever my sister-in-law may think! Alas, what you say about the apartment in the boulevard Haussmann I know only too well. It's at least fifteen years since I saw it, but I remember it as the ugliest thing I've ever seen, the triumph of the bourgeois bad taste of a period still too close to be inoffensive! It isn't even old-fashioned in the charming sense of the word. Old-fashioned! It's too ugly ever to be old-fashioned. But I've explained to you the tender and melancholy attraction which drew me back to it, in spite of my even greater horror of the neighbourhood, the dust, the Gare Saint-Lazare,[1] and so many other things. The friends who house-hunted for me with such angelic devotion since I couldn't do it myself, and who knew my instructions, my tastes and my recommendations – no trees, no noise, no dust, preferably on high ground etc. etc. – haven't yet got over seeing me choose the 'splendid apartment' of a less rich and much later Nucingen.[2] But I've told you the reason, which perhaps I didn't tell anyone else. It cancels out all the others and reason knows nothing of it.[3] If I can ever decide to leave it it will at least have been a transition between the place where Mama rests, which isn't the cemetery but the apartment in the rue de Courcelles, and an apartment which she will never have seen, entirely unfamiliar. And besides, all that is mixed up with other things, too long to explain by letter. What you say about the beautiful golden autumn at Versailles pains me. For would you believe that except for the first few days when I saw the last rays of the sun from my bed, I have never woken up before nightfall and I know nothing of the charms of the season or the hour. I've spent four months in Versailles as though in a telephone kiosk without being the least aware of my surroundings. And in the old days I used constantly to travel from

Paris to Versailles so much do I love these incomparable purlieus which our nostalgia has reconstructed into something even more beautiful than they ever were in their pristine splendour and which have gained so much in beauty between Louis XIV and Barrès! I haven't your letter to hand but I seem to remember your saying the little drawing-room couldn't be made into a completely tapestried room. That's what I should have liked, to remind me of the ante-room in the rue de Courcelles or the tapestried study in the boulevard Malesherbes. So it will have to be for the dining-room.

Do you see Reynaldo? He telephones me constantly to suggest coming here, but I'm really too unwell, and I hope he doesn't take this for indifference. God knows that it's the opposite I feel for him! I don't know when I shall be able to move back to Paris; the incredible remissness of the manager of the building, and the lack of goodwill of the tenant to whom I've sub-let, have caused me to prolong indefinitely my sojourn in Versailles in conditions which are at once extremely costly and extremely uncomfortable. As soon as I arrive – even if I feel better in the apartment later on – I shall be ill for a few days, as after any change. I shall try and see you either at your house or mine. I shall re-read your letter tonight and if I find points on which I should have replied to you I shall do so at once. Now I simply wanted to thank you and to expend in your company the little strength that remains to me after these attacks. You're giving me the blue room as bedroom. It will be very painful for me. But if the bedroom has to be furnished I can see that there's no alternative.

Madame, I should like to stay with you much longer to thank you for the trouble you've taken for me but really I'm worn out. I thank you with all my heart and beg you to accept my respectful gratitude.

Marcel Proust

As regards the pictures, the only ones I want to see at all prominently displayed are the quaint little shepherdess with the freakish, blue-blooded look of a Spanish infanta, the portrait of Mama, and my portrait by Blanche.[4] However, the exact copies of the Snyders[5] will go very well in the dining-room. I know that the Govaert Flinck (*Tobias and the Angel*)[6] is a valuable picture and on the whole a very good bit of painting, though a little dark, by one of Rembrandt's best pupils. But I intend to leave it for Robert (and indeed anything else he likes), as well as the fine portrait of Papa by Lecomte du Nouy[7] which Jacques Blanche so admired, but I think Robert will be very pleased to have it. I shall also send him, if he's willing to house them (or I'll keep them for him in a corner), *Esther and Haman*,[8] the scene from Roman History and the Metsu.[9] However if he doesn't take them at once I shan't put them in a corner so that my sister-in-law won't decide they're worthless. But to my mind any picture that one hasn't *coveted*, bought with pain and love, is horrid in a private house. William Morris once said: 'Have nothing in your house except things that you find useful or consider beautiful.' A wardrobe or a table, however ugly, however useless, still evoke an idea of utility. But a picture that isn't pleasing is a horror. And I can say that of all those that will be there.

1. In August he had rushed to his uncle Georges Weil's death-bed, taking the train to Paris dressed only in a fur coat thrown over his nightshirt; on the way back he collapsed at the Gare Saint-Lazare and was helped by a railwayman whom he later tried hard to find (with extraordinary consequences: cf. letter 183).

2. The vulgar *nouveau-riche* Alsatian banker in Balzac.

3. Cf. Pascal, *Pensées*: 'Le coeur a ses raisons que la raison ne connaît point; on le sait en mille choses.'

4. Cf. letter 146, n. 6.

5. Franz Snyders (1579–1657), Flemish still-life and animal painter.

6. Govaert-Flinck (1615–60), Dutch painter.

7. Oil painted in 1885 by Lecomte du Nouy (1842–1923).

8. By Frantz Franck (1581–1642), Flemish painter of Biblical subjects. The picture was exhibited in London in 1935, under the title *Esther and Ahasuerus*.

9. Gabriel Metsu (1629?–67), Dutch painter of interiors.

174   *To Louisa de Mornand*

1 January [1907]

My dear little Louisa,

You are divine! How shall I ever be able to express to you my infinite gratitude and affection! That wonderful diadem enchased with as many precious stones as in the treasure house at Conques or Charlemagne's gospel-book is also a vigilant alarm-bell, delightfully practical for an invalid![1] Thus, my little Louisa, everything you do, so well symbolized by this admirable and ingenious present, is at once beauty and kindness, poetry and intelligence. You are marvellous, far too kind, and I'm overcome with confusion. I see everywhere, in all the magazines,[2] your dear incomparable face, your gentle, witty eyes, glorified.

May 1907 bring glory to your talent and sweetness to your heart. That is the wish of your grateful friend

Marcel Proust

1. A New Year's present (cf. letter 95).
2. Most recently a fashion photograph in *La Vie heureuse* (December 1906).

175   *To Reynaldo Hahn*

Monday [7 January 1907]

My dear little Reynaldo,

I'm sad not to be in a fit state to say this to you rather than write it. If you write to Montesquiou tell him that the truth is beyond his imagining, totally improbable for anyone who doesn't know about my life. The truth is that, having arrived in Versailles on 6 August, I wasn't able to go out *once* during those five months.[1] I didn't once go to the Château, not once to Trianon (but anyhow you know all that), not once to Les Gonards cemetery.[2] If I had had only one good day I should have gone to Les Gonards rather than to the Château or Trianon, especially as M. de Montesquiou wasn't at Versailles and couldn't go there, and I should have had the gratifying feeling of deputizing for him, coming to visit poor Yturri on his behalf as he so often came to visit me on behalf of M. de Montesquiou. And then I knew, from you and others, that it was a uniquely beautiful and moving tomb. And since I think of little else but tombs I should have liked to see what Montesquiou had done there and how his taste had succeeded in investing his sorrow with even greater nobility. When he is back in Paris or Versailles, I shall do my utmost to try to see him one evening, but apart from the fact that it's impossible with everyone, with him the difficulty becomes even greater, as he is the person with whom I'm most embarrassed, in the bad sense of the word. And even if for once he falls in with my odd hours, the possibility of an untimely attack will prevent me from daring to give him a rendezvous which I'd rather die than break, whereas others would understand. You can tell him that I was delighted to receive *Les Hortensias bleus*[3] which I never liked so much. The early pieces seem to me even more exquisite than before. As for 'Ancilla', a fragment of which I recently applied to you, it's an admirable thing, a magnificent pendant to 'the great-hearted servant'.[4] It seems to me (but I'm not sure) that the poem to Yturri has been touched up and perhaps not improved.[5] It remains perhaps the best thing he has ever written but I don't remember the crown being green the first time and I don't know that it's better thus. No point in telling him this, first of all because he wouldn't give a damn and secondly because it's a very vague doubt and I'm not at all sure I'm right.

Have you been interrogated by *Lettres* on the subject of Shakespeare and Tolstoy?[6] I'm too unwell to reply; I can't tell you how much even a letter like this exhausts me. Several people (notably Mme G. de Caillavet) have written to me to say that your *Noël*[7] was adorable. I should love to have heard it, Bunchnibuls, and am sad not to have been able to. Tell M. de Montesquiou that I wasn't even able to go to my poor uncle's funeral.[8]

Tenderly yours
Marcel

You can tell M. de Montesquiou that I wasn't once well enough to see Miss Deacon[9] who was living in the same hotel.

1. On 27 December Proust had left the Hôtel des Réservoirs and moved into his new apartment on the boulevard Haussmann.

2. The Versailles cemetery where Montesquiou's secretary and companion Gabriel Yturri was buried; he had died of diabetes in July 1905. Montesquiou was buried beside him in December 1921, in the splendid mausoleum he had built for him, beneath a statue of the Angel of Silence.

3. *The Blue Hydrangeas*, published December 1906, the first volume of the definitive edition of Montesquiou's verse. Nicknamed 'Hortensiou' by Léon Daudet, he was famous for his passion for hydrangeas.

4. 'Ancilla', a poem from the above, quoted by Proust in a letter to Hahn on 13 December, about an old servant whom Montesquiou's mother had commended to her son in her will; Baudelaire refers to the grave of 'La servante au grand coeur' in *Tableaux parisiens* (*Les Fleurs du Mal*).

5. Allusion to the sonnet 'In Memoriam' which appears at the beginning of the above collection (see n. 3) dedicated to Gabriel Yturri.

6. *Les Lettres* had reprinted Tolstoy's dismissive view of Shakespeare (quoted in Georges Bourdon's *En écoutant Tolstoi*: 1904) and asked certain writers and artists to comment.

7. A recital at Mme Lemaire's of a Christmas mystery play with piano accompaniment by Hahn.

8. Cf. letter 173, n. 1.

9. Gladys Deacon (1881–1977), brilliant American heiress who was to marry the ninth Duke of Marlborough in 1920. She is described in letter 228.

176   *Reynaldo Hahn to Robert de Montesquiou*

[About 8 or 9 January 1907]

Dear Sir,

I communicated your letter to Marcel. I enclose his reply. I haven't seen him for several days. It is alas too true that *not once* did he go out in Versailles.

Excuse this scrap of paper; I'm out of writing-paper and don't want to delay sending you the letter from the poor, innocent 'accused'.

Thank you for the pretty card from Potsdam on which your signature prevented the soaring fountain from falling back to earth.

Please accept, dear Sir, my affectionate and devoted regards.

                                                    Reynaldo Hahn

177   *To Auguste Marguillier*[1]

                                        102 boulevard Haussmann
                                                    [? January 1907]

Dear Sir,

Herewith the little notice on M. Mourey's *Gainsborough* which I've owed you for a very long time! I should be grateful if you could send me proofs and not put any other signature but M.P.[2] I've never put anything more at the bottom of these brief and hurried notices, but some mysterious hand always reinstates Marcel Proust, which rather annoys me. Especially just now when I'm so ill (indeed may I ask you not to say I'm in Paris and not to disclose my new address – people can write to me at 45 rue de Courcelles, please forward; 102 boulevard Haussmann is for you alone).

Please don't think it's the little dig at M. Groult[3] that makes me take refuge in anonymity. On the contrary he'll identify me at once and if he asked you who it was I'd be delighted for you to tell him. If I thought it was disagreeable I'd sign my name in full. The anecdote isn't by the way a disguised invitation either. If M. Groult asked me round every day to see his Turners (he has done so in fact), I couldn't take advantage of it as I no longer leave my bed. So it's quite disinterested. I preach for other people and for art, and I don't think a private individual has the right to possess and hide so much universal beauty.

                                        Your grateful and devoted
                                                    Marcel Proust

Have you any of the translated works of M. Berenson at the *Gazette*?[4] And any works on Sienese painting.

I don't know how you'll find your way in the mad pagination of this letter.

1. Deputy editor of the *Gazette des Beaux Arts*.

2. Proust's review, overdue by seven months (cf. letter 169, n. 6), appeared, signed M.P., in *La Chronique des arts et de la curiosité* (the supplement to the *Gazette*) of 9 March 1907.

3. Camille Groult (1832–1908), rich industrialist and art collector. In the article in question Proust refers to his collection as 'the Louvre of English painting – our Louvre, alas, contains so little of it'. The anecdote Proust goes on to mention concerned works by English painters which Groult acquired in Ruskin's memory and was, if anything, flattering.

4. Cf. letter 167, n. 5.

178   *To Jean Bonnerot*[1]

Wednesday [9? January 1907]

Dear Sir,

I am very ill, and have great difficulty in reading or writing. Let me however thank you most warmly for sending me your book; I was most touched. I shall read your verses with great pleasure, with feelings of affectionate involvement. My eye fell on a poem called 'Herbier' which I liked very much. I also noticed a sonnet, evidently a little inspired by Ronsard, in which it seemed to me that all the images very neatly harked back to a primordial rose, and through insistent allusions (*blossoming* poetry, *leafless* stanzas), reminiscent of a painted fabric, repeated it indefinitely. I don't know whether it's your first book, whether you are still very young. But I wish you many others, and much fame and happiness.

Yours sincerely

Marcel Proust

1. Jean Bonnerot (1882–1964), poet, Sorbonne librarian, later editor of Sainte-Beuve's letters. His first book of verse was published in 1905.

179   *To Léo Larguier*[1]

[January 1907]

Dear friend,

You will perhaps be surprised that, although I am too unwell to answer letters, I should write to you spontaneously about a work *on which*[2] (as they say nowadays as though it must add beauty to a sentence to employ an expression which one hasn't created and which means that everything one writes will date from 1906 or 1905 or 1904) you have not asked my opinion.

But one owes more respect to Truth, with a capital T, than to one's correspondents, and one must take more trouble on its behalf. And, dear friend, if you do not help to uphold the truth, who will? You are intelligent enough to understand everything, too lazy, or too impotent – I don't know which – to create anything. A creator or a mediocrity would have the right to be mistaken. You are neither. You have one of the most judicious minds I know, judicious in the appraisal of what is most difficult of all, that of truly delicate works. This is to say that you are a man of rare intellect, and there are so few of whom that can be said that it creates duties.

How sad I was, therefore, to find that you had debased the modest spiritual magistracy with which your taste, your discrimination and your high-mindedness invest you by lending your approbation to the most despicable creature, the most devoid of intelligence, of style, of grammar, of sensibility, of originality (dare one say after that, of talent?), that exists, the first publicist who makes me truly understand the meaning of the word unspeakable, for I have some repugnance in speaking his name: M. Léautaud.[3] Personally I do not know him; I know absolutely nothing about him. But I read a book of his called *Amours* and if you don't find it the most ghastly, the most imbecile thing that ever was, one of us must have gone mad.

Unfortunately you live among friends who so often use the word imbecile in speaking of very intelligent people whom they in fact think very intelligent, and the word despicable to designate works that are perfectly honourable and remarkable, that I realize that in declaring this book despicable and imbecile I'm saying nothing. I'm saying what you have heard said hundreds of times about books by Bazin[4] or Bourget or others, men of the highest talent among those who are not truly the elect.

But M. Léautaud's *Amours*! I won't speak of the book's moral baseness, because I should be incapable of doing so. I don't know any words that could express the pain I felt on seeing a human being feign sentiments beside which those of the cruellest murderer would be estimable. But from the point of view of *talent*? One could frankly say that by comparison there isn't at present a serial story in a single one of the newspapers of Paris that isn't a work of genius, or hasn't something more to offer. As for the people whom we don't particularly admire such as Messrs Prévost, Margueritte, Provins, E. Arène, Maindron, Commandant Z, Serge Basset, Interim, and so on, their writings glitter with knowledge, wit, grace, humanity, emotion, inventiveness, compared with these fetid pages.[5]

Dear friend, I hope we shall meet again one day. When that day comes, before shaking each other's hands, we shall go through an expiatory ceremony. I shall read aloud to you, if I have the strength, a few pages from *Amours*. After each sentence, after each word, you will give voice to your disgust, as one still used to say when I first started to read. And afterwards we shall embrace, rejoicing that we are men and not the unspeakable creature who was capable of writing such things.

Please don't give too much publicity in this form (spoken of course) to this letter, for M. Léautaud is one of the few people with whom I should be very frightened to fight a duel. I should feel as though I were up against the angel of darkness. And I'm not pure enough to be confident of winning.

Good-bye, dear friend. Don't let your privilege of being a good judge and arbiter of the arts go for nothing, and believe in my devoted affection.

Marcel Proust[6]

1. Léo Larguier (1878–1950), poet, critic and essayist.

2. Proust writes *sur quoi* (evidently a fashionable new usage) instead of the usual *sur lequel* or *sur laquelle*.

3. Paul Léautaud (1872–1956), whose novel *Amours* had infuriated Proust by treating family feelings with cynicism and of which Larguier had written that 'M. Léautaud . . . puts his courage at the service of his talent; he says what no one dares to say . . .'

4. Well-known middlebrow novelists René Bazin (1853–1932) and Paul Bourget (1852–1935).

5. Proust lists successful middlebrow writers: Marcel Prévost (1862–1941), Paul Margueritte (1860–1918), his brother Victor (1867–1942), Michel Provins (1861–1928), Emmanuel Arène (1856–1908), Maurice Maindron (1857–1911); and three journalists by their pseudonyms: Commandant Z. (military affairs editor of *La Libre Parole*: see letter 140, n. 4), Serge Basset (theatre editor of *Le Figaro*), Interim (sports editor of *Le Figaro*).

6. Proust never sent this letter, found among his papers after his death.

180   *To Robert de Flers*

[Monday, 14 January 1907]

My dearest Robert,

'Joy! Joy! Tears of joy! . . .'[1] These words of Pascal's à propos of another Cross describe very well my immense satisfaction about the one that has just been so rightly awarded to you.[2] I should like to embrace you and also your grandmother[3] but I'm in a state that is a thousand times worse than it's ever been, truly horrible.

Tenderly yours
Marcel Proust

If your name is written across this sheet it's because being able to write only with great difficulty I intended to send you a telegram but I preferred to express my joy to you at a little greater length. If you are with your parents and your wife tell them how happy I am; I shall write to your grandmother.

1. Cf. Pascal's *Mémorial*: 'Joie, pleurs de joie, joie.' Proust borrowed the same words to welcome Antoine Bibesco back to Paris (vol. I: letter 226).
2. He had been nominated Chevalier of the Légion d'Honneur.
3. Madame Eugène de Rozière, now a widow. Cf. letter 234.

181   *To René Peter*[1]

[Monday evening, 14 January 1907]

My dear René,

I'm writing to you with several objects.

1) To say to you: O René! you said that I'd never see you again in Paris. And ever since you came for a brief moment more than a fortnight ago, you've made countless excuses for having to leave while saying that you would be back next day. You've never come back, never come back!

2) is linked to 1). It's to tell you that nothing could give you the slightest idea of what this past fortnight has been like. For the first time in my life I've been laid low (four times already) by attacks which last thirty-six, forty, fifty hours! And during this time . . . death! You will say to me: 'So I was right not to come since I couldn't have seen you.' But René, there have been gaps in between and Reynaldo, who naturally

couldn't see me during the attacks, nevertheless came back the next day, and if I was still too unwell, the day after. And with the telephone you can even save yourself unnecessary journeys. What a deserter! Don't think I'm saying this reproachfully, I quite realize that my hours are inconvenient and my unpredictability irritating. But what would Albu say, since, being married, he can no longer come in the evening and I'd rather die here and now than see people in the daytime these days when I don't dine till ten o'clock (if I dine! for I sometimes go for three days without eating).

3) But dear René the real reason I'm writing is not that, it's this. On re-reading just now the note you sent me a fortnight ago, I saw some lines scribbled across the page that I hadn't noticed at first. And these lines which I've just read sadden me a great deal: you mention the grave state of your aunt Belin.[2] Please God she has recovered in the meantime. But for the moment I have to admit that I'm desperately worried, thinking of poor Monsieur Eugène, so if by telephone or letter (I don't say telegram because there's a danger that they might bring it up the main staircase and wake me) you could tell me how she is, that would make me singularly happy. Is that possible? I'm a bit better this evening, almost well, but I'm afraid it may only last a few hours, and I need to be well for a few days. The violence of these attacks drives me mad. My domestic worries have taken a turn such that it's comic, a case of 'Thanks to the Gods, my misfortune exceeds my hope'[3] which is beyond everything. On top of it all Félicie[4] made a frightful scene this evening which I shall relate to you if you have retained the compassionate spirit of Versailles. But 'shall I see you, and which you?'[5]

But I would have you know, O René, that places sometimes restore to us the spirit with which they imbued us. And as my doctor considers that if my present condition continues it would be sensible for me to give up Paris for good, it's possible that Versailles might become my habitual residence. And since it is yours for six months of the year I should have six months to see you and six months to be forgotten by you, which would be excellent for my emotional health. How is our old friend Constantin?[6] If you see him give him lots of friendly messages from me and believe me, dear René, yours (yours what? yours sincerely?).

Your

Marcel

1. René Peter (1872–1947), playwright, who lived in Versailles. While Proust was there in 1906 he and Peter had amused themselves by writing a pantomime and had also planned to collaborate on a more ambitious play.

2. Mother of Peter's cousin Eugène, mentioned in the next sentence.
3. Cf. Racine, *Andromaque*: 'Grâce aux dieux mon malheur passe mon espérance.'
4. Proust's cook.
5. Cf. Verlaine, 'Amour': 'Te reverrai-je? Et quel?'
6. A young man-about-town called Constantin Ullmann.

182   *To Louisa de Mornand*

[January 1907]

My dear little Louisa,

A few hours after sending you my telegram I've learned, between two attacks, or rather in the middle of a terrible one, that thanks to you the young woman has been placed!![1] My little Louisa, my admiration and gratitude for what you've done are beyond bounds. What an adorable and powerful person you are! I'm stunned (not at your power but that you should have used it for that). I don't know how to thank you, I'm overwhelmed. This action of yours is going to create a stir in the theatre world and the world in which Mlle Macherez numbered several protectresses. At this moment Sarah Bernhardt, Mme Le Bargy, Mme Georges Menier,[2] must be saying to themselves: 'What power, what graciousness, what kindness that exquisite Mademoiselle de Mornand has!'

As for me, what can I say to myself, except at each new proof of your kindness to feel that I love you even more.

Marcel Proust

1. Maurice Duplay, now director of the Théâtre de Cluny, was having an affair with a Mademoiselle Macherez who had a humble job at Paquin, the couturier where Louisa de Mornand happened to dress, and wanted to become a mannequin.

2. Mme Le Bargy was another well-known actress and Mme Georges Menier wife of the 'chocolate king'. *Née* Simone Legrand, daughter of Mme Gaston Legrand (see letter 20, n. 7), Mme Menier was soon to supplant Mlle Macherez with Maurice Duplay.

183   *To Madame Catusse*

[Saturday evening, 26? January 1907]

Dear Madame,

I hope that, with the work below nearing completion,[1] I shall be able at last to rest a little and to see you. For I've had no sleep either night or day, even when I happened not to be suffocating, because the hammering prevented me. As regards my antique dealer (who incidentally knows absolutely nothing about antiques), if he happens to come round to show you some mirrors, I'd be grateful if you could tell him bluntly what you think of their quality, their authenticity, and the exaggeration (for I imagine there's bound to be some) of their prices. I think his honesty will consist in transmitting your opinions to me, softening them a little.

The *Débats* is no less witty than *Le Temps* about the appointment to Berne,[2] which is indeed highly Ancien Régime. I remember the Lady very well, incidentally, from the far-off days when she and I used to move in fashionable society. She was charming, though she inspired a general mistrust which was perhaps completely unjustified. One never knows the truth of reputations, especially in such credulous circles, but she was pretty and had a great deal of charm. I imagine she's enormously changed. By the way, it's a curious manifestation of love to send the loved one far away, to intoxicate [*griser*] the Grisons, the other Grisons.[3] And indeed she must be pretty grizzled [*grisonnante*] herself. The unfortunate thing is that when she became acquainted with her powerful friend who at that time wasn't powerful, a number of diplomats who could not foresee either the liaison or the Ministry expressed themselves somewhat undiplomatically about him. And since he hasn't forgotten the insults of the Duc d'Orléans there are at the moment some remarkable diplomats who are marking time and might even do time at the first opportunity. One should never talk to any woman about any man and vice versa. The next day one learns of a marriage or the opposite and it's too late.

Can you imagine that I received ten days ago the most sensitive, the saddest, the most touching letter from that unfortunate Van Blarenberghe which would make him more to be pitied than Oedipus. What a shocking story![4]

I still have no news of Lucien.[5]

Dear Madame, I don't want to overtire myself, but it's good to chat with you for a while. It's true that it's perhaps less agreeable to chat with me!

Your grateful friend
Marcel Proust

1. Cf. letter 168, n. 2.

2. Where, by long tradition, France alone was represented by an ambassador; the Comte d'Aunay was the latest in a rapid turnover that according to the indignant Swiss made Berne 'an anteroom for ambassadors'. The 'Lady' referred to in the next sentence was Comte d'Aunay's second wife and Clemencea ; mistress before he became Prime Minister.

3. *Les Grisons* means both the Canton of Grisons and 'grey-beards'. Proust's running pun is impossible to reproduce in English.

4. Proust had corresponded with Henri van Blarenberghe (1867–1907), son of the late Chairman of Eastern Railways whom he had met socially some years earlier, about the railwayman at the Gare Saint-Lazare (see letter 173, n. 1). On 25 January *Le Figaro* reported a Grand Guignol crime: servants at the magnificent Blarenberghe house had found his mother covered in blood; with her dying breath she accused her son who had meanwhile stabbed and shot himself, and was interrogated by police as he lay dying on his bed, his left eye hanging on the pillow. The matricide inspired Proust to write a remarkable article for *Le Figaro*. Cf. letter 185.

5. Lucien Daudet.

184 *To Francis de Croisset*

[Towards the end of January 1907]

Dear friend,

What good news! I know that the girl you are marrying is delicious and I congratulate you wholeheartedly.[1] But she too is to be congratulated, for she is not only marrying a man of great and delightful talent but a person full of charm and wit, one of the rare people whose company one can enjoy. Again, a million affectionate compliments.

Your devoted
Marcel Proust

I'm truly very, very pleased at your happiness.

P.S. Dear friend, I'd be grateful if you could write and tell me what you would like as a wedding present. I've no idea and you would be doing me a great service.

1. His engagement had been announced to Mlle Juliette Dietz-Monin, member of an old Alsatian family.

185   *To Gaston Calmette*[1]

[Friday, 1 February 1907]

Dear Sir,

My deep sense of your kindness and my gratitude received an even more direct and more powerful, almost crushing reinforcement when I saw just now your charming *Figaro* encumbered by the compact mass of my unwieldy article,[2] and all the other articles, all the news, all the light flotilla of telegrams from every point of the compass held up by the enormous convoy to which your infinite kindness had accorded this special precedence of which I so unscrupulously took advantage.

One thing distresses me, however, because it increases even more the disproportion between the unworthiness of the article and your delightful benevolence. The only thing I had indicated to M. Cardane as being essential was omitted, though I said that he could cut anything he liked rather than these last few lines.[3] I had indeed in my hurry sent off the article in the morning without an ending. I added one on the proofs, a paragraph in which I gathered my reins, my scattered steeds, at once hurtling and floundering, straying. The article ended thus:

'Let us remember that for the ancients there was no altar more sacred, surrounded with more profound superstition and veneration, betokening more grandeur and glory for the land that possessed them and had dearly disputed them, than the tomb of Oedipus, at Colonus, and the tomb of Orestes at Sparta, that same Orestes whom the Furies had pursued to the feet of Apollo himself and Athene, saying: "We drive from the altar the parricidal son."'

Thus the word parricide, having opened the article, closed it. The article was given a sort of unity thereby. I dare not ask for an insertion tomorrow to the effect that a printing accident scuppered the final lines. Who will remember it all tomorrow? But to the extent that it may have made the article even more unworthy of the kindness which you so divinely bestow on its author, I am very unhappy, for nothing could be more distressing to me than to make you repent of your benevolence towards your grateful and devoted

Marcel Proust

1. Gaston Calmette (1858–1914). Editor and director of *Le Figaro* for many years, he published innumerable articles by Proust, who expressed his gratitude by dedicating *Du côté de chez Swann* to him in 1913. After running a virulent campaign against the Finance Minister Joseph Caillaux, he was shot dead in his office by Mme Caillaux on 16 March 1914.

2. 'Sentiments filiaux d'un parricide', inspired by Henri van Blarenberghe's murder of his mother, occupied more than four columns of the front page of *Le Figaro* of 1 February. Cf. letter 183, n. 4.

3. M. Cardane, the sub-editor, reportedly said to Proust's messenger: 'Does Monsieur Proust imagine that anyone will trouble to read his article besides himself and the few people who know him?'

186   *From Gaston Calmette*

LE FIGARO
26 rue Drouot
[Friday, 1 February 1907]

Your article was very fine, my dear contributor and excellent friend. Don't worry about those few lines: they frightened Cardane who thought they showed insufficient disapprobation for the unfortunate parricide's deed. Cardane was undoubtedly wrong: but there is not a reader who will not thank you and re-read your article with an enchanted heart.

Yours ever
Gaston Calmette

187   *To Gaston Calmette*

102 boulevard Haussmann
Friday [evening 1 February 1907]

Dear Sir,

Forgive me! It's my last letter! First of all it's too kind of you to have replied to me and I shan't dare write again. Secondly, if what I thought was the clumsiness of a make-up man, a compositor (the omission of my ending), was the deliberate act of a severe moralist (M. Cardane) I have nothing to say. Or rather I have: I have this to say to M. Cardane (but I don't know whether I'm supposed to know about his indignation – please don't bother to write and tell me, we'll talk about it when I see you), that one of his colleagues on the *Journal des Débats* of old, St-Marc Girardin, who was not known for his immorality, wrote in his *Cours de littérature dramatique* some very edifying pages on the Greeks' belief that the city which safeguarded the ashes of Oedipus and Orestes would always be victorious. He saw this as the effect of the high philosophy of the Greeks which required that the crime of these parricides, even if involuntary, should be punished in their lifetime, but that in order to re-establish a higher justice, since they had been involuntarily guilty, their memory should be honoured, consecrated. In my case the severe M. Cardane (who is, by the way, so kind and charming) must be aware of all the wars which Athens and Sparta waged in order to lay hands on the bodies of Oedipus and Orestes of whom the oracles had predicted that they alone could ensure the greatness of their cities. I don't want to bore you by quoting a passage from Herodotus on one of these oracles, although it's extremely interesting.

All the same, to take me for an apologist for parricide is a bit much! Forgive this self-defence which is not meant too seriously as I quite take the point of your letter and realize that M. Cardane didn't mean to censure me. But the tragedy of *Oedipus at Colonus* which revolves exclusively around the military glory which the possession of Oedipus's remains would bring to the Athenians has made these questions so popular, so topical, that I'm sure that if my article had arrived at the *Figaro* a little earlier, at an hour when one has a bit more time to remember the Greek tragedians, M. Cardane would have judged my ending in a diametrically opposite way. Dear Sir, please don't bother to write to me; forget me in order to forgive me, and believe me

Your *infinitely grateful and devoted*

Marcel Proust

188 *To Lucien Daudet*

[Early February 1907]

My dearest Lucien,

You have inherited from your father that genius for *motivated* kindness which makes of the reasons one gives to others for respecting and admiring themselves such a convincing masterpiece that they cannot resist, and do admire themselves. But alas, nothing can replace, whether as regards talent, or awareness, or health, or happiness, the direct and immediate testimony one gives to oneself. When someone is told 'You have everything to make you happy', if he is dissatisfied with life he would prefer a little unconscious happiness. Similarly, when I read you I think I have some talent, but when I read myself and especially when I write – for I never read myself (it's true I never write either) – I really *feel* I have none. This upsets me all the more because when by chance I happen to glance through *Les Plaisirs et les jours* (you know, the book which Mme Lemaire illustrated), I consider that I had some then. As for this article,[1] to finish with the subject as I'm ashamed to be talking so much about myself, I wrote it straight off without a draft since I had to deliver it that very day. And no doubt that can excuse its many faults but it can't excuse its quality. Although hasty it is also laborious. It's true that since my translation of *Sesame* I haven't written a line, apart from letters and accounts. And I had such a pain in my fourth finger from having written for such a long time that I had to leave the ending. However all that is too long and boring to explain so I'll come to what I wanted to say which is far more important. It's this: I swear to you that when I receive a letter from you about something I've written, I have the feeling, I can even say the certainty, that it's superior to what I've done. You may reply first of all that if what I do is bad, what you write may be superior without being very remarkable. But really we don't need to lie to one another and I can tell you that I'm not so modest, and if I consider that I haven't any talent, that for a variety of reasons I haven't been able to make the most of my gifts, that my style has rotted without ripening, on the other hand I'm aware that there are many more real ideas, genuine insights, in what I write than in almost all the articles that are published. Marcel Prévost, Rod, Margueritte, etc. will soon belong to the Academy.[2] And I know perfectly well that if my articles weren't by me and I read them beside theirs, mine would interest me much more. Don't think I'm Montesquiou if I tell you this; what I mean is that in regarding you as far superior to me I regard you as even more superior to a lot of highly respected people. You can also reply that my compliments can't convince you any better than yours can me. It isn't quite the same, because *your* talent, the field where what people say can't really cheer you if you're not satisfied with it, is painting. And I believe that in fact you ought to be not only satisfied but delighted with it. Moreover, *you* have never stopped painting, whereas I have never written anything more. It seems to me that you've made far better use of your time. I don't want to tire myself tonight but I'd like nevertheless to say something more useful to you, because I think it might make you happier and that would make me happy too. It's this: you are wrong to think of yourself always *inside time*. The part of ourselves that matters, when it matters, is outside time. It's all right for a Roqueplan[3] or an Alphonse Karr[4] to say to himself: I'm already such and such an age, I've passed such and such an age. Think of yourself simply as an instrument capable of whatever experiments in beauty or truth you wish to perform, and your gloom will evaporate.

Besides, La Fontaine began working at the age of forty at the earliest, and in your art look at Hals who did his really fine things only after he was eighty, and Corot his best things only after he was sixty. And to leave the field of art, don't think you'll be less loved when you are older. It's quite the opposite. What I say about Hals and Corot is to make you stop thinking of wasted time, not to make you waste more of it. But if you will try and think about all that I feel it might give you joy and a great eagerness for life and work, and I assure you that I've written it with deep feeling and if you could recognize its truth and efficacy my affection for you would be enormously gratified.

Your
Marcel

1. 'Sentiments filiaux d'un parricide'. Cf. letter 183 *et al.*
2. Popular middlebrow novelists: cf. letter 179.
3. Camille Roqueplan (1803–55), minor Romantic painter.
4. Minor Romantic writer (see letter 34, n. 2).

189 *To Robert Dreyfus*

[Sunday, 3 February 1907]

My dear Robert,

How nice you are, how fond I am of you, and how solid and justified our friendship is (as so few friendships are) since it has a truly rational, and indestructible, basis in a richly gratified self-esteem. But I'm slandering myself, for I assure you that if I'm fond of you it isn't because you appreciate my writings. I should be just as fond of you if you hated them, but since you like them I esteem you the more. If only I could hold myself in higher esteem. Alas, it's impossible and I wonder whether you aren't telling a charitable lie to an invalid when you say my article is good. At all events (and unfortunately this doesn't explain its deadliness, for it hasn't even the charm of improvisation, being laborious though hurried, and ice-cold though high-flown), to explain to you the repetitions of words and the innumerable slips (mainly the compositor's): Calmette asked me for the article on Wednesday morning in a letter which, because of an attack, I didn't read until ten o'clock in the evening. I rested until two in the morning without thinking about the article. At three o'clock I got up, started it at once, and wrote it without doing a draft straight on to the *Figaro*'s copy paper, until eight in the morning. As it wasn't finished and my hand (unaccustomed to writing) was too painful for me to continue because of cramp in the fourth finger, I went to bed leaving word that I was to be woken up during the day to finish it. But as the frightful building work that's going on below began at half past eight I felt so wretched that I gave up the idea of finishing the article and sent it off unfinished without having re-read it, to appear as it was. At eleven o'clock in the evening they brought me the proofs which had to be returned by midnight. I was about to correct them, but then I had an idea for a really rather good ending. As there wasn't time to do everything, I decided to forgo the corrections and wrote my ending at the bottom. At midnight the proofs were taken back to the *Figaro* (I neither go out nor get up – I get up without dressing once a week) and I made it clear that they could cut whatever they liked but not a word of the ending must be changed. They published all of it – or almost all – but omitted the ending, not a single word of which was there. And I dare not give the reason because I'm not supposed to know it, but here it is: Cardane thought my ending was immoral and constituted a eulogy of parricide. I assure you.

This point of history having been cleared up, I may tell you that I know what I'm exposing myself to by quoting a line of Rivoire's.[1] Indeed I'm a bit unhappy about it. But at that moment the line came to my mind so forcefully that it would have been dishonest of me not to quote it – as it would have been dishonest to quote at that moment a line I hadn't thought of, even if it was by Gregh. Besides, I don't know any, as my memory has gone. Gregh recited to one of my friends a very pretty poem he has just written called 'Letters'. The first line is 'A black E white I red O blue: vowels'.[2] He forestalled criticism by saying roguishly: 'I see that it reminds you of Rimbaud, but it's deliberate.' However I believe he's going to alter it a bit. We shall see.

Dear friend, I'm keeping my new address secret. Don't give it to anyone. My telephone number is 292 05. At half past ten in the evening if you've nothing better to do you can telephone me, and if I'm not in the middle of an attack I shall see you. And if you don't like telephoning and call round, you can ask the concierge. As soon as my terrible house-moving attacks have calmed down a little I'd prefer to see you earlier. At the moment that's the best time, on the days when I'm not having a major attack. On those days, no one comes into my room, not even the servants, because on those days I can't even drink.

Tenderly yours and thanks
Marcel

I don't know what you hear about Calmette from Beaunier and the *Figaro* people. I don't know myself how I should find him on closer acquaintance. But for amiability (which to this degree is genuine kindness and charm) I don't know anyone comparable. I can give you an idea if I tell you that this newspaper editor has written to me three times in three days, for no other reason than to give me pleasure and in this vein: 'Your article is admirable; it moved me more than I can say; it will be an ornament to the paper; there isn't a reader who won't re-read it and thank you for it with a sense of enchantment etc. etc. etc.' When one hasn't slept for a fortnight and when one's half crazy, I assure you that letters like that – even if he's saying to himself at the same moment 'What a bore that article is' – do one good, and one blesses this editor who is so different from Arthur Meyer.

1. In a passage of his article on parricide, Proust writes of days with snow in the air, citing what he calls 'André Rivoire's fine poem' (referring to the line 'And things have the air of awaiting the snow' from *Le Songe de l'amour*: 1900). André Rivoire (1872–1930) followed Gregh as editor of *La Revue de Paris*.

2. In fact the first line of Fernand Gregh's poem 'Lettres dansantes' (deliberately modelled on that of Rimbaud and first published in 1909) reads: 'A noir, E blanc, I rouge, U vert', and the first line of Rimbaud's famous sonnet 'Voyelles': 'A noir, E blanc, I rouge, U vert, O bleu: voyelles.'

190 *To Robert de Billy*

[February 1907]

My dearest Robert,

Once more I call upon your knowledge and your kindness. I want you to be good enough to send round and have left with my concierge this afternoon (or send me a *pneumatique*[1]) the following information. Imagine a framework of sentences roughly like this:

'When the late Marquis de Casa Fuerte wished to give his son[2] a christening present, he could find no rarer or sweeter jewel in the old Spain of the XIth century – not even in the leather of Cordoba or Arab bowls with their rose-pink or yellowish reflections – than the forename Illan which had not been borne since the capture of Toledo (1085 I think) and which seemed to have been carefully preserved ever since in the alveole of scalloped marble of some faintly Muslim cathedral by the light of candles that had not been extinguished for almost a millennium.'

Now, are there in Spain cathedrals (or churches) which date from the XIth century (or thereabouts)? If not, from when? Is the word alveole appropriate? Can Moorish bowls be rose-pink or yellowish? Are they glass or pottery? (I'd prefer glass, glass as in Venetian or Gallé[3] glass.) Give me some other words like bowl or glass etc. Can Cordoba leather have this sort of sheen? And by way of precious jewels of old Spain (XIth century or earlier) what would be the objects in Cordoba leather or beautiful fabrics that one might specify? I don't mention armour because it doesn't seem delicate enough unless it had a very delicate sheen but then I'd have to have very precise names. (If I knew the names of some weapons that would also be nice though less useful.) And tapestries: I suppose there weren't any before the XVth century. However at a pinch I could take objects later than the XIth and simply say old Spain, but I'd prefer Gothic or Romanesque (does the word Romanesque apply?). If you can think of some ecclesiastical jewel – but I'd prefer my rose-pink bowl. Do you know what Gallé's bowls were in, glass weren't they? Do you know (this has no connection with the things I've mentioned nor with Spain) if objects have ever been engraved (or any other process) in more or less precious stones not on the surface but at a certain depth, as Gallé did with his glasswork?

A million apologies and regards

Marcel

I must have the information before nine o'clock this evening, at the very latest. If you had a book with engravings of Spanish bowls etc. that would be wonderful.

1. An express letter on lightweight paper circulated around Paris through a network of tubes by means of compressed air.

2. Illan de Casa Fuerte, handsome young friend of Proust and Montesquiou. See letter 155, n. 2.

3. See letter 1, n. 2.

191   To Georges de Lauris

[Saturday, 16 February 1907]

My dearest Georges,

I feel as if I were losing Mama for the second time.[1] And how much more dreadful it is for you than it was for me after hoping for so long and still to have seen her suffer so. But all this vanishes before the fact of not having her any more, your mother whom I loved so much but did not know, whose face, radiant with heroic courage and sweetness and detachment from everything that was not you and your father, remained unknown to me but was so constantly present to me for the past year. And I think continually of your father and am haunted by his pain, but never having met him I cannot picture it. And this is less cruel because you, Georges, I see you, I feel you, I live you, and it's the most hideous torture for me. If only I could be of some use to you, be with you, do something appropriate to your grief and mine. But my helplessness rivets me yet more cruelly to it. Jean Blanc[2] will ask if there is anything I can do for you, other than think and weep with you unceasingly.

Marcel

1. His mother, the Marquise de Lauris, died aged fifty-three on 15 February. Cf. letter 167.
2. Proust's manservant.

192   *To Georges de Lauris*

[Monday evening, 18 February 1907]

My dearest Georges,

I didn't reply to your letter because I intended to come this morning, but I felt I couldn't go through with it, I'll explain to you later. I'm sending you this brief note this evening to ask how your father survived this day, and how you yourself survived it, my poor dear Georges.[1] There is one thing I can tell you now: you will know a sweetness that you cannot yet conceive. When you had your mother you thought a great deal about the days when you would no longer have her. Now you will think a great deal about the days when you did have her. When you have become accustomed to the terrible experience of being forever thrown back on the past, then you will feel her gently returning to life, coming back to take her place again, her whole place, beside you. At the moment it isn't yet possible. Be passive, wait until the incomprehensible force that has shattered you (of which doctors alas know little more than others) lifts you up again a little, I say a little because something will always remain broken in you. Tell yourself that too, for it is a comfort to know that one will never love any less, that one will never get over the loss, that one will remember more and more. I need not tell you, my dearest Georges, that I weep copiously as I write you this, bad tears that are more for me than for you, whereas until this evening I had been thinking of you alone. I hope you will be able to help your father to bear it, to be wholly with him.

My life has been turned upside down. Yesterday I had lunch at half past eleven, because I was preparing to go out today. If you come one day and I'm not so ill as I am today, seven o'clock would perhaps be best so as not to leave your father alone for the evening. A lot of people have written to me, knowing of my grief, even people you don't know, young Duplay for instance. I think Reynaldo and d'Albu must have been sad not to have been able to shake your hand. I can tell you at least that d'Albu, having heard of the terrible blow in a note from me, telephoned to inquire when the funeral was and said 'As long as it isn't Monday because I have to go to Compiègne', and Reynaldo, whom I saw, had a rehearsal he couldn't miss, but both of them are thinking of you a great deal, my brother too, and even my sister-in-law who telephoned me most touchingly. Everybody is shattered, but no one can feel the same sorrow as I feel, because no one has hoped and felt with you so much.

Tenderly yours
Marcel

1. Lauris's bereavement clearly revived in Proust agonies of unresolved guilt over his own mother's life and death; later that year he was to borrow and ponder over Mme de Lauris's photograph (see letter 243).

193 *To Joseph Primoli*

102 boulevard Haussmann
[Early 1907?]

Dear Sir,

Thank you most warmly for your kindness in sending me that post-card. It's extremely nice of you to think of my health. There is no possible cure for me. But I mustn't complain too much, since it has earned me such a charming attention and brought me an unexpected inkling of the charm of Rome and also of the charm I used to experience on seeing you from time to time. Such are the agreeable thoughts that your charming souvenir has aroused in me. And it is the privilege of solitary, tranquil, somnolent souls like mine to find, when an impulse is awakened in them, that it spreads and propagates itself indefinitely. And so the thought of all the pleasant hours that I shall never spend in Rome getting to know so many beautiful things under your guidance and in your company will be with me for a long time and will give me great pleasure. So I thank you again and send you my respectful regards.

Marcel Proust

194   *To Madame Straus*

[Mid-March 1907]

Madame,

How wonderful that Monsieur Straus should happen to know M. Katz,[1] that you were kind enough to speak to M. Straus and that he will be kind enough to speak to M. Katz. So will you allow me to give you a few details. First of all your telling me that if it were you, you would only allow them to bang after midnight was very nice and made me laugh a lot, but I see that you don't know that I've changed my hours and am now awake in the daytime. I shan't change them again unless the building work continues in this way, because they begin right behind my bed at seven o'clock in the morning and in the afternoon there's no more noise. So that I've done quite the wrong thing in choosing to be awake in the afternoon at precisely the only period when I could get some sleep. Since writing to you I've learnt that Mme Katz is in quite a hurry to move in because in the house she's leaving she has work going on above her head (which is less intolerable for her than hers is for me since she doesn't have the terrible attacks which I've been suffering for two months and which sleep alone can bring to an end so that the work she's having done here is infinitely more acceptable to her because it's not her that it affects and one accepts with alacrity the suffering of others if it can cut short a minor inconvenience to oneself). But she must be made aware of this: that her workmen (both those who are working for her and those of the proprietor) arrive at seven in the morning, insist on manifesting at once their matutinal high spirits by hammering ferociously and scraping their saws behind my bed, then idle for half a hour, then start hammering ferociously again so that I can't get back to sleep, then as soon as mid-day comes move further off and bang away in the distance, and by two o'clock there's no more noise. So if Mme Katz by agreement with her proprietor M. Couvreux, and without waiting for that as far as her own workmen are concerned, could arrange for them to come from Thursday onwards only at mid-day or one o'clock, or better still two o'clock (but twelve would already be something), and work until evening, she would gain time instead of losing it and even a bit of money, for you can tell her that I'll give her all the compensation she asks. I persuaded another tenant to have his decorating done between eight o'clock and midnight and it was all finished very quickly. I'm not asking as much of her. If she's prepared to have the work done between noon and eight o'clock in the evening, her alterations which are almost finished (alas! for if I'd known I would have asked her sooner but I'm at the end of my tether, and my doctor advises me to go away because my condition is too serious to go on putting up with all this), will be completed more rapidly. But for instance (excuse me, Madame!) they are about to install a basin and a lavatory seat in her W.C. which is next to my bedroom wall. Even if there's only half an hour's banging and they do it at seven o'clock in the morning, it will do me as much harm as if it took hours, whereas she could get it done (or ask M. Couvreux to get it done) after twelve or one o'clock. The same for her upholsterers who are going to be nailing carpets and hanging curtains in a few days' time. If she would even agree, when she has moved in and will certainly have things to be nailed down at the beginning, to get it done after twelve instead of the early morning. Finally (I'm deliberately asking a lot, knowing that I won't get everything I want) if one day I found that I'm really too ill I'd be immensely grateful if she would consent to suspend all work for that day. And if there are things that she is genuinely forced to have done in the morning (there aren't, but since she doesn't care she'll easily find excuses) I beg her to warn me in advance so that I don't take one and a half grammes of trional at six o'clock if the workers are going to start at seven. M. Straus mustn't tell M. Katz that I'm in the habit of sleeping in the daytime as though it were a caprice, but that I have terrible attacks of asthma at night which have got worse in the past month so that I'm in an acute state which improves whenever I can get a bit of rest. The trouble is (but one can't explain everything to M. Katz) that my doctor says that for reasons too difficult to explain to you it's absolutely essential that I start getting some fresh air again, and since my first outings will bring on severe attacks I don't dare try even one without an assurance of relative silence the following day. And I very much suspect that even if Mme Katz moves in very shortly she will have several months' worth of furniture moving and nailing up or nailing down a variety of things which she thinks beautiful or luxurious but which will drive me to the grave!

I came back from Versailles on the 27th of December and the hammering hasn't ceased ever since (it wasn't yet Mme Katz's) so that I haven't yet even once been able to risk that essential outing after which

I need an assurance of silence next day. I really think that in order that I should have as little noise as possible in the morning Mme Katz will have to resign herself to not bringing in any workmen at all. For however much one tries to persuade them to go and work on the other side of the house and not to make too much noise, however much one tips them and the concierge, their first ritual is always to wake up the neighbour and encourage him to share their high spirits, 'going at it hammer and tongs' with an almost religious intensity. Moreover at that hour in the morning there is no architect or proprietor to supervise them and moderate their enthusiasm. Unfortunately d'Albufera who undertook to approach Mme Katz (whom he didn't know) sent her an idiotic telegram saying that I was 'the son of *the famous Dr Proust*' which must have left her pretty indifferent, and had it delivered at one o'clock in the morning which probably didn't leave her indifferent at all (luckily it wasn't taken up to her until the morning). But there doesn't seem to have been any response from that quarter. As there's a M. Sauphar (?) in the house of whom the owner is rather afraid, I wonder if it isn't he who prevents the workmen from hammering in the afternoon and makes them stick to the morning. Madame, this subject is of very little interest to you and I apologize profusely for going on about it with such selfish insistence. But you are the only person who really understood. I'm so exhausted that I can't say anything more to you now. But I shall write to you soon on a more disinterested note.

Your respectful and grateful friend – you can imagine how grateful – to the point of delirium! Just think how joyful I shall be after the physical torture.

Marcel Proust

I think it's important to specify the *upholsterers* too. Otherwise she'll say that it's all nearly finished etc. Naturally I'll pay the costs of the electricity, or if her electricity isn't yet installed, the candles needed for working by artificial light (although it's daylight until six o'clock and they now knock off at three) and all the compensation she wants. But M. Katz must speak to his mother at once and get her to cancel *her* workmen without waiting to see the proprietor. Otherwise she'll temporize, they'll shift the responsibility from one to the other and in the meantime she'll keep up the hammering and I shall be obliged to leave (which I had decided to do because of the noise). And there's another gentleman who's moving in to the fourth floor of the same house, from which I can hear everything as though it were *in my bedroom*.

1. A judge, the son of Proust's new neighbour, Mme Katz. For Mme Straus's response, see letter 206.

195   *From Madame de Noailles*

109 avenue Henri-Martin
Wednesday morning [20 March 1907]

My dear friend,

I read the three columns[1] first with my fingers covering 'my passage' in order not to see it, not to get excited or bursting with vanity, and so it was without any preliminary gratitude or sentiment that I plunged into that divine fantasy, whose ironic, oblique, mysterious and profound lyricism drew me on from pleasure to pleasure, from sweetness to delight. It's the most tender account, and also the clearest, sharpest, most piercing, that could possibly be given of familiar life at the edge of dreams. The house of the friend who speaks – the rustic house surging into the dismal instrument with all its rustlings and wing-beats and gushing springs – strikes me as vividly, brings me as much air, as when Hélène[2] arrives at my house on foot in the morning, and the whole atmosphere of the avenue Henri-Martin – wind, scents, oxygen – still clings to the tight curls of her astrakhan jacket. The henceforth divine damsels of the telephone[3] – flower-maidens and Eumenides – bring back to me my vanished fairies; and intoxicated with dance and song I forget to protest against *Les Fleurs du Mal* and *Les Feuilles d'automne* whose sublimely sarcastic whirlwind will expose me to witches' fustigations after the manner of German ballads.[4] I should already, despite the early morning hour, have implored the faceless Goddesses to grant me your number, had I not been smitten since last evening with the most ferocious cough and thus capable of transmitting flu and laryngitis through the telephone, just like the lady framed in the doorway who cries 'May I come in?' and enters, smiling and benevolent, like a Father Christmas come to fill stockings with epidemics. Dear Marcel, what an enchantment for my friendship that you should write such beautiful, sinuous stories, interrupted, resumed, stopping to speak, to breathe, to laugh, to cry, stories that are the stuff of life, of actuality, of the heart, of happenstance, of dreams.

I thank you for all that, profoundly.

Anna de Noailles

I forgot to say that the whole of the end of the article is sublime in its poetry, translucency and fantasy.

1. Proust's essay 'Journées de lecture' on the pleasures of reading, in *Le Figaro* of 20 March, inspired by the recently published memoirs of the Comtesse de Boigne (1781–1866), which were peppered with the names of the great-grandparents of many of his aristocratic friends and acquaintances. Themes from this important essay were to reappear in *RTP*, in which the memoirs of the fictional Mme de Beausergent – next to Mme de Sévigné's letters the narrator's grandmother's favourite reading – are really those of Mme de Boigne. (Cf. *RTP*, II, 701.)

2. Her sister the Princesse Alexandre de Caraman-Chimay.

3. The passage on the telephone in the article in question was to be reproduced almost word for word in *The Guermantes Way* (*RTP*, II, 134–7).

4. In his article he had announced her forthcoming book of verse *Les Eblouissements*, saying it was 'truly the equal, it seems to me, of *Feuilles d'automne* and *Les Fleurs du Mal*'.

196   *To Madame de Noailles*

[Wednesday, 20 March 1907]

Madame,

I learnt from your sublime letter that my article had appeared, for I was still asleep and hadn't yet seen the *Figaro*. How distressing! I hadn't yet received new proofs, so that I tired you out to no purpose for two days asking you what you would like me to put and wasn't able either to add Ronsard's *Amours* or to substitute *La Légende des siècles* for *Les Feuilles d'automne*. As in Baudelaire, I raised my fists towards God who takes pity on me.[1] But it was even more frightful still; for as soon as I received your telegram, having asked for the *Figaro* and pounced on the passage that concerned you in order to indulge in my sorrow at finding only *Les Fleurs du Mal* there, I looked in the wrong place, couldn't find it at all at first, thought they had put in absolutely nothing about you, and felt I should die. But a moment later I found it, and thanks to that preliminary mortal anxiety I was pleased by what would otherwise have pained me, and *Les Fleurs du Mal* mixed with the *Feuilles d'automne* seemed to me sufficient. But if you knew the cuts they've made (I don't even mean the ones I agreed to but those made without my knowledge) – it makes the whole thing incomprehensible. For instance I said at the beginning that a less frank lady, after smothering us with kisses and caresses for two hours, would take out her watch and say, etc. But Cardane's prudishness must have taken fright at this because without consulting me he has cut the visit, the kisses etc.; it begins at 'takes out her watch' and no longer makes any sense.[2]

Madame, to think that you could write such an admirable page about such a bad article! You once – à propos of a letter from you to Gregh – explained to me so well the secret of your incomparable flatteries that I'm not taken in by them. But however incredulous, one is overwhelmed by the beauty of what you say; that trick of designating the different passages of the article by rewriting them yourself so much more beautifully, stamping them with a touch of genius where I wasn't even able to imbue them with talent: that image of the arrival of the Princesse de Chimay[3] bringing you the scents of the season in the fur of her astrakhan coat, what a divine thing, doubly moving for me as the Princesse de Chimay means so much to me, and I even experienced on the telephone a sensation I hardly dare tell you about when, thinking it was she who was speaking when it was you, I thought I was mistaken and felt as it were an interpolation into her person of another.

Madame, my hand hurts so much that I can't write to you and I don't know what I'm saying, but let me just thank you for your beatific letter after which I'm ashamed that an article so inferior to it should have appeared while your letter will be known only to me, and forgive me for the absurdly inadequate sentence about you which I intended to expand, merely tossing off this beginning after having failed on that first day before sending back the first proofs to extract from you on the telephone the secret of the books to which you would like yours to be compared. But such as it is, it's still capable of annoying other poets and that's already something.

Your admirer overwhelmed by the gifts of your phrases of this morning

Marcel Proust

1. Cf. the last line of the first stanza of 'Bénédiction': 'Crispe ses poings vers Dieu, qui la prend en pitié', from *Spleen et Idéal* (*Les Fleurs du Mal*).

2. Apart from this and other minor alterations, a long passage which Proust regarded as the article's *raison d'être* was omitted – apparently with his agreement, for possible future use. See letter 213 to Robert de Montesquiou.

3. Cf. letter 195, n. 2.

197 *To Georges de Lauris*

[Shortly after 20 March 1907]

My dearest Georges,

I cannot tell you what immense, and melancholy, pleasure (is there any other kind?) your letter gave me. Your sweet errors, even as regards Jammes[1] and myself, are very dear to me, for it seems to me that you must have a greater friendship for me than I sometimes think for your infallible critical sense to go so sweetly and so utterly astray where I am concerned. It used to be said that the proof of Leconte de Lisle's love for Mme Beer[2] (don't be jealous: he was seventy-five years old) was that he, such an accurate notator of every detail, gave her large eyes in his verses. The friendship that ascribes to me as much talent as to Jammes must be a great friendship indeed. But what moved me especially, and more deeply perhaps than I can make you realize, is that you do not keep to yourself the illusions you cherish with regard to me, you impart them by tender suggestion to those who love you. The thought of your father listening to you reading my article – for I'm sure you are telling me the truth – was infinitely sweet to me, because it showed me far better than anything you might have told me that the sweet spiritual life you shared with your mother was not finished for ever, that your father too was full of maternal tenderness for you and could also say to you as she did when you read her something: 'I shall say they are beautiful when your eyes have said so.'[3] If your over-generous error about me is precious to me because it proves your friendship, that into which you have readily plunged your father is even more precious to me because it's a proof of the tenderness, so willingly blindfolded, with which he surrounds you. I sense there a sweetness which even in your present terrible distress counts for something indeed, a sweetness which must recall to you something of the infinite disinterestedness, devoted unto death, the complete self-forgetfulness, that there was in your mother's tenderness. And it moves me perhaps less to think that it's a comfort for you than to think that it would have been a comfort for her, that often when she considered the possibility of leaving you both, she could have dreamt of nothing sweeter, more calculated to allay her anxieties, than that the two of you should achieve such a tender and perfect union that even goes as far as such idle fancies as 'admiring' (!) what I write. Only our parents can give us that tenderness. Afterwards, when we have them no longer, we never experience it again, from anyone. Except in the memory of the hours spent with them, which alone helps us to live, and above all will help us to die.

I love you tenderly, my dearest Georges

Marcel

1. He had written to Proust about 'Journées de lecture', and compared him to the poet Francis Jammes.

2. Mme Guillaume Beer (1874–1949), wife of the great-nephew of the composer Meyerbeer.

3. Cf. A. de Vigny, *La Maison du berger*: 'Je dirai qu'ils sont beaux quand tes yeux l'auront dit.'

198   *From Robert de Montesquiou*

[Saturday, 23 March 1907]

Dear Marcel,

I shall try to see you for a moment tomorrow, Sunday, in the afternoon, between 4 and 5. Perhaps we could agree on a day this Holy Week, to go and pay a visit together to our Friend, if your health permits.[1] I should also like to communicate to you some extracts from the book I am devoting to him.[2] All this, of course, as a token, longstanding and renewed, of friendship past, present and future.

I have been meaning for a long time to compliment you on your fine Blarenberghe article.[3]

Yesterday's is in quite a different vein, digressive but very agreeable.[4]

Till tomorrow, I hope

R. Montesquiou

But don't *trouble* yourself in any way about all that; just a word of reply *at the desk of your hotel*.[5] And if it's your hour of rest, simply say so.

1. i.e. to visit Yturri's grave. Cf. letter 175, n. 2.

2. *Le Chancelier des fleurs*, as yet unpublished, dedicated to Yturri's memory.

3. 'Sentiments filiaux d'un parricide' (see letter 185).

4. 'Journées de lecture' (see letter 195).

5. See following letter.

199 *From Robert de Montesquiou*

Monday evening [25 March 1907]

Dear Marcel,

As I was about to send you the enclosed letter,[1] written on Saturday, I learnt that you were no longer living in Versailles. If, therefore, I am addressing it to you today, it is now solely with the object of conveying to you my regards.

We shall take up again another time the project for a funerary pilgrimage; but, for the moment, I propose as a substitute a reading of the extract of which I speak.[2] Would you like me to come to see you, with that object, one day this week, at an hour which you will specify?

I shall send round to your house for your reply tomorrow, Tuesday evening, or Wednesday morning.

Please accept, in the meantime, my steadfast regards.

Cte R. de M.

1. i.e. the previous letter.
2. *Le Chancelier des fleurs* (see previous letter, n. 2): the reading never took place.

200 *To Robert de Montesquiou*

[Thursday evening, 26 March 1907]

Dear Sir,

You must have sensed, I hope, from the haste, incorrectness and brevity of my telephone message, the great joy your letter gave me and my desire to realize that project immediately.[1] Before sending the message, I had telephoned you for a long time without getting a reply from your house, and this prompted me to tell the Neuilly operator that since I couldn't get through I would dictate a message to him. But I wanted to explain all that to you in a longer letter. Only I had a very severe attack yesterday and I'm writing you just a few lines to thank you and to convey to you, among my many thoughts, those which seem to importune me most often, complaining of not having been communicated to you.

I was most touched by the fact that, having written an article on Beardsley[2]such that if I had written the first few lines on the *foresight* (is that the word you used? in any case it seems to me exactly appropriate) of the architect and the gardener,[3] I would tranquilly rest on my laurels without ever writing anything else, for what else could one write that was prettier? – I was touched, I say, that the author of that should think of congratulating me on *my* articles! And I'm touched by many other things, but am too tired today to say more than the essential.

It isn't really a bad thing that you weren't free for that reading,[4] if you can agree to lend me the extract, for I only take in what I read myself, alone. For me, reading means solitude (as you will know if you have glanced through my preface to *Sesame and Lilies*[5] – that preface about which you promised to speak to me, but never let me know that you had even read). Would you consent to having the extract sent round to me?[6]

I am about to start going out a little. God knows what endless attacks each outing will provoke, but still, as soon as I can venture any distance, I shall come to Neuilly. It will be better that way, for I shall choose a moment when I'm not feeling unwell and not risk the cruel possibility that you might come at a moment when I was unable to see you. Often however up to now, and still perhaps sometimes, except when I'm having an attack, you would find me at seven o'clock and at nine; I won't say eight as I shall have neither kitchen nor dining-room.

In sending you – with the utmost gratitude for the offer and the promise – these *Hortensias*,[7] I am, believe me, fully aware of all that the book represents in terms of thought, feeling, beauty and wit. I read it a great deal, know it very well, and believe I am capable of understanding and loving it.

These fine days, which I sense rather than see, give me a great desire for colour, for nature, for things the eye can appreciate. If you had some poetic photograph, some sensuous object which you might lend me with which to divert myself and muse on, it would be a great boon to me in my melancholy existence.

I send you the assurance of my keenest, most admiring, most grateful attachment.

Your

Marcel Proust

I thank you most profoundly for the kindly concern for my health to which your letter bore witness. I feel these things very deeply. And I have reason to. Just imagine, I get up (without dressing) only one day in seven!

Your

Marcel Proust

1. i.e. a reading from the Count's unpublished book on Yturri.

2. Montesquiou's article entitled 'Aubrey Beardsley' had appeared in *Le Figaro* of 21 February.

3. Montesquiou had written: 'It is, if not one of the glories, then one of the graces of Life, and one of the best, to see realized those things we have foreseen. Such a hope is to be found in the joy of the architect, in the pride of the gardener . . .'

4. Proust is being disingenuous: the Count had offered to come round 'one day this week'.

5. In 'Sur la lecture' there are two passages on solitary reading; in the second Proust writes of '. . . this fecund miracle of communication in the heart of solitude'.

6. Montesquiou replied: 'Thank you for the book' (i.e. *Sésame et les lys*) – nine months after having received it! – and 'Don't count *on me* to send anything.' (He brought it round nevertheless: see postscript to letter 202.)

7. Proust was sending him a copy of *Les Hortensias bleus* to be inscribed. See letter 175, n.3, and letter 204.

201   *From Robert de Montesquiou*

Neuilly
[Wednesday morning, 27 March 1907]

Dear Marcel,

If you wish *the thing to take place*, it must be on the one hand, *without delay*, as I am going away, and on the other, not *without warning me*, so that I do not come in vain.[1]

Would *tomorrow, Thursday evening, at 9 o'clock*, suit you, for you must, I believe, always prefer *the evening*?

If you would prefer this very day, likewise in the evening, let me know early enough for me to arrange everything; but I should prefer tomorrow.

Your friend
Robert de M.

N.B. I shall send round this afternoon for your reply.

1. Montesquiou had not yet received Proust's letter posted the previous day.

202   *From Robert de Montesquiou*

[Wednesday, 27 March 1907]

Dear Marcel,

Your letter charmed, touched and saddened me. What it tells me about your health grieves me. God forbid that I should tell you that you are not taking the proper treatment and thus range myself with those people who exclaim triumphantly: 'I told you so!' whenever anything untoward happens to us. – However, I'm inclined to believe that you are, I won't say an imaginary invalid, but a *spellbound valetudinarian*, in the fairy-tale sense, and that this spell can, will yield to a philtre or a bough, or a word.

Remember Madame Lemaire a few years ago, *semianimis*, half dead, and resuscitated by a glass of Forges water![1]

'The day will come when medicine will simply be *the science of migrations*.' May this splendid saying of Michelet's[2] come true for you, and may our planet contain a small corner, a piece of earth where your Antaeus' heel may recover a strength that will make you bounce back up to the stars.

Amen. This is my paschal wish.

Robert de M.

Your habits of life do not give me grounds for hoping that you may one day be able to come to Versailles.[3]

I shall be at the Pavillon at Neuilly in May and June.

In the meantime, I shall send round for the Book[4] which I want and hope to make more precious to you.

P.S. Certainly, I shall come and see you, bringing the extract.[5] We shall have to find a day and a time.

Meetings are not necessary between people who appreciate one another, but they are agreeable. It is merely a question of not abusing them, and indeed of seldom availing oneself of them.

With those to whom one is indifferent, the coolnesses that result from meetings are not to be feared – on the contrary!

1. Mineral water from Picardy.

2. Cf. *La Mer*, IV, 2 by the historian Jules Michelet (1798–1874): 'La médecine, de plus en plus, sera une émigration' which Montesquiou quotes incorrectly.

3. To visit Yturri's grave (cf. letter 198).

4. See letter 200, n. 7.

5. See letter 200, n. 6.

203 *To Emmanuel Bibesco*

[Thursday evening, 28? March 1907]

Dear friend,

I know that Parisians are generally unaware of the misdeeds of cutthroats or the cases of measles which from a distance give strangers the idea that Paris is uninhabitable. So that even supposing it to be true that there have been attempted strikes in Romania[1] (which in any case the newspapers alternately affirm and deny, and which must be less important than the bakers' strike[2] we read about but pay no attention to), I hope that you have been less aware of them than those who read the press. And besides, I've noticed that the infected areas have nothing in common with Strehaia, Corcova and Nehedjinski.[3] Nevertheless, dear friend, a word from you telling me that this news has not disturbed your Easter holidays would give me great pleasure. For as you know, you are constantly on the horizon of my thoughts, the perpetual object of their most affectionate scrutiny. And I should hate to think that your country might become revolutionary for in that case I should want to see you leave a troubled land and come to live evermore in ours, which, it's true, is even more revolutionary. Let us all go and live in Utah. It's true that bridge is forbidden there, a fact which you might not be able to come to terms with. I have often asked for 5 14 00.[4] But *5 14 zero zero doesn't reply* has always been the sole response.

With fondest regards, your
Marcel Proust

1. Where Prince Emmanuel Bibesco (see letter 49, n. 10) was on holiday at his family estate. A peasant insurrection had been reported in the French press.

2. There was a bakers' strike in France that lasted from 22 March to 3 May.

3. Sic. Corcova, near Stehaia, in the province of Mehedintzi, was one of the Bibesco estates.

4. Emmanuel Bibesco's Paris telephone number.

204 *To Robert de Montesquiou*

[Shortly after 27 March 1907]

Dear Sir,

How wonderfully kind of you to have written that for me.[1] As in Hugo's 'occasional verse', one recognizes the lion's claws – and his tears.

It seems to me that 'We twain are on the verge of heaven'[2] is not more beautiful. What pride, what gentleness, what poise. I feel that in spite of yourself you will have to include these delightful verses in one of your books, and that thus my name, which I should have been so pleased to see in these *Hortensias*, will perforce appear in that particular collection. Unless you are unkind enough to put 'to a friend of no interest' without a name![3]

I was most touched by the intelligent, noble and delicate things you had to say about my illness. Alas, the miracle that saved Mme Lemaire[4] cannot happen with me since it can only work as long as the essential organs haven't been affected. I dare say my illness itself might be suddenly carried away, and I believe this could be brought about by a migration[5] directed by an intuitive or experienced voice. But it would in a sense be too late. For the ravages of the disease would continue to grow, and even in the best of circumstances, could not be repaired. If in a fit of purely nervous convulsions a child breaks an arm, the arm remains broken, and the purely nervous character of the convulsions cannot change that. However, I definitely plan as soon as it's possible to return to the open air, and perhaps even to Versailles (where I cannot send you this letter as I don't know your address).

Your respectful and grateful friend
Marcel Proust

1. As promised Montesquiou had inscribed his *Hortensias bleus* for Proust. See letter 175, n. 3, also n. 3 below.

2. Cf. Victor Hugo, 'Les Sept Cordes', *Toute la lyre*: 'Et moi, je sens le gouffre étoilé dans mon âme; / Nous sommes tous les deux voisins du ciel, madame, . . .'

3. The verses dedicated to Proust appeared in a collection published in 1910 together with the definitive edition of *Les Perles rouges*.

4. See letter 202.

5. See letter 202, n. 2.

205   *To Georges de Lauris*

[Between mid-March and mid-April 1907]

My dear Georges,

I shall write to you at greater length; at the moment I'm dead tired, but I can't tell you with what tender feelings I think of your kindness to me. It isn't because you wrote to my neighbours[1] that I'm telling you this, my dearest Georges. That was charming of you and invaluable for me, but its real merit lies in the whole exquisite atmosphere of kindness and charm in which it is steeped like everything else. Georges, in the letter in which you told me about it, in the words which asked me to go out with you but only if it wouldn't aggravate my fatigue, there was a solicitude which I can only, with tears in my eyes, call maternal. My dear Georges, I don't know whether I shall ever understand your friendship for I'm not exactly sure that you have any for me. But I know that you have for me the kindness that you might show towards a child that had been entrusted to you. I am your virtue, I am the occasion for you to display a charm such as less altruistic actions do not encourage us to exercise. All this impresses and delights me to a degree that I'm describing very badly. I had various things to tell you but I feel too tired. I shall try and write to you tomorrow.

Your

Marcel

1. Proust had doubtless asked Lauris to write to M. and Mme Katz, as he had done in the case of other friends, notably Mme Straus. Cf. letter 194.

206 *To Madame Straus*

[Monday, 1? April 1907]

Madame,

Every day I have better news of Jacques (which I telephone to Robert) and my pleasure is doubled by the thought of yours on receiving the same news.[1] I wanted to write to you yesterday but I had an attack which lasted twenty-four hours during which it would have been physically impossible for me to put pen to paper. I wanted to write to you to say how wonderfully kind and delightfully comic it was of you to invite M. Katz to lunch.[2] It's one of those actions full of wit and kindness that are typical of you. His cow of a mother, alas, hasn't stopped building . . . I don't know what! A dozen workers a day hammering away with such frenzy for so many months must have erected something as majestic as the Pyramid of Cheops which passers-by must be astonished to see between the Printemps and Saint-Augustin.[3] I don't see it but I hear it. And when the hammer-blows intensify an attack which is already too acute and I feel that this lady is not only costing me a whole year in which I do little but suffer but curtailing my life by several years as a result of all the attacks and the drugs, I think of the cry uttered by the worker on that same pyramid in Sully Prudhomme. You must remember it:

> Sudden he cried out like a stricken tree,
> His cry rose, seeking Justice and the Gods,
> And beneath the vast pile for three thousand years,
> Cheops immutable in his splendour sleeps.[4]

Sometimes I *dream* that the work is finished and no more builders will come, but on waking up (or rather being woken by them) I discover (again as in Sully Prudhomme)

> The labourer told me in my dream:
> Brave workmen whistling on their ladders.[5]

(I'm not at all sure whether these are the exact lines, but I mean the poem where he dreams that there are no more workmen but is *glad* to find some when he wakes.)

However, the paint had to be left to dry and the judge's mother was forced to interrupt the work for a few days, though even during that period she changed two or three times the seat in her water-closet (too narrow I suppose) which I have the honour to be backed on to (and always between seven and eight in the morning).

The brief respite did me a world of good and I took a few steps outside in front of the house and on the balcony. Once she is installed I shall take on a new lease of life – and yet again perhaps no, for such an opulent person is sure to have countless pictures to be hung and she will certainly choose the morning for that. And then it will be the season of flowers and I shall no longer be able to go out! Still, I managed to take a few steps in the fresh air and in spite of the attacks it cost me, it gave me great pleasure. I found the sun a very pretty and a very strange object.

Madame, I'm writing to you as to a lady who has nothing to do in Dax[6] and thus has time to read boring letters. But I don't want to take advantage of you, nor of my own strength and my still very sore hand; I simply wanted to thank you, to tell you how much I think of Jacques, how much I admire you, how fond of you and how very grateful to you I am.

Marcel Proust

I think after all that the work has quietened down and it would be better not to say anything more to her. We shall judge of her goodwill when we see if she orders her upholsterers for the afternoon, now that it's light until seven o'clock.

1. Her son Jacques Bizet and, presumably, Robert Proust.

2. The judge, son of Proust's noisy neighbour. Cf. letter 194, the long letter to Mme Straus on the same subject.

3. Cf. letter 166, n. 3.

4. Proust paraphrases 'Cri perdu': 'Il cria tout à coup comme un arbre cassé. / Le cri monta, cherchant les Dieux et la Justice, / Et depuis trois mille ans sous l'énorme bâtisse / Dans sa gloire Chéops inaltérable dort.' Cf. letter 3 where Proust cites this same sonnet.

5. Proust quotes from different verses of the sonnet 'Un Songe': 'Le laboureur m'a dit en songe: "Fait ton pain ..." ... / De hardis compagnons sifflaient sur leur échelle ...'

6. She was taking the cure at the Hôtel des Thermes, Dax, in the Landes.

207 *To Reynaldo Hahn*

[Thursday evening, 11 April 1907]

My dear Nicens,[1]

You were so mopchant that I couldn't talk to you and say thank you thank you for bringing me to hear *Béatrice d'Este*[2] of which I shall simply tell you that I prefer the (I don't know which number) piece (you'll guess which) to the Beethoven sonata in which there's that lively pastoral song that I used to love more than anything in the world,[3] that I prefer the (number further on) piece to the *Siegfried Idyll, Forest Murmurs* etc., and the first and the last to the overture to the *Mastersingers* although here there's no connection. Anyway I'm mad about it and I'd like to know if Mama heard it. I think you led the first movement and all the last part admirably – brilliantly (better than Risler[4] plays the overture to the *Mastersingers*). Elsewhere you indulge in too many tricks, too many mannerisms, too many grimaces, and that way of bouncing up and down on your bottom which I don't find at all pretty.[5] I understood today for the first time what pretty orchestration means and I've never seen such power mixed with purity. Although you didn't have a baton, I could sense between your fingers the magic wand which flew to the far corner of the orchestra just in time to wake a sleeping triangle. I was impressed that you should have succeeded in forcing so many society people to stop and listen to a fountain weeping in silence and solitude.[6] The water rises in a systolic surge and then falls drop by drop. I didn't underschtand

Your gestures and cries with their hint of the shameless
As an inscrutable muncht may well be as blameless[7]

and I was very annoyed that you showed portrait of Polignach, by
Polignach.[8]

Who was turning pages? Where was Bardach?[9] D'Indy killingly
funny.[10] Mlle Leclerc[11] says *eunn* for *une*. The candles were loose and
several times, especially when you struck chords by dropping your hands
from a height of six feet and when you lunged at the orchestra as though
with a sword, they nearly fell and set fire to the paper roses on the
footlights (a fraction more and everything would have gone up).[12]
Legonidec now looks like Doudeauville, or M. de Biencourt.[13] He has
caught and overtaken La L. whose drunken laugh seemed to me very
painful. However her voice is less mannish than that of La Murat[14] who
moreover followed up or rather accompanied every phrase of the music
with a stimulating commentary: 'Ah! Les Roses d'Ispahan, the whole of
the East! Bravo, it's exquisite! One can smell the peppermint of the
seraglio! Bravo, bravo, it's exquisite! It's lighting up. How exquisite,
bravo etc.'

I've never seen anything so beautiful, kind, intelligent, sensitive,
meditative and sweet as M. Lister's face.[15] He's a four-square, golden
Régnier, delightful. How all the people I used to know have aged! Only
La Polignac[16] has at last attained a youth which she combines with the
gentleness of maturity. And also a few rudimentary and ferocious old
divinities in their summary delineation haven't managed to change.
Mme Odon de Montesquiou, Mme Fernand de Montebello, Saint-
André etc.[17] remain immutable in the barbaric hideousness of their
Lombard effigies.[18] They are portraits of monsters from the time when
people didn't know how to draw.

Hello.

M.P.

1. Genstil, for *gentil*: cf. letter 64, n. 1. Mopchant in the opening sentence is a variant
of moschant (*méchant*): see letter 64, n. 3.

2. See letter 133, n. 2.

3. Apparently the *Sonata in E minor*, No. 27, Op. 90.

4. The pianist Edouard Risler (1873–1929), old friend of Proust, later professor at
the Conservatoire.

5. Hahn conducted from the piano.

6. Hahn's choral work based on Horace, *La Fontaine de Bandusie*, had been
performed.

7. Cf. Molière, *Le Misanthrope*, Act III, scene V. 'Vos mines et vos cris aux ombres

d'indécence / Que d'un mot ambigu peut avoir l'innocence.' (Proust substitutes 'qui' for *que* and 'muncht' for *mot*.)

8. The performance was given at the Polignacs'. The portrait in question may have been one of the lighthearted sketches which Proust often drew for Hahn.

9. Henri Bardac, Oxford graduate, friend of Hahn, who later became one of Proust's greatest literary fans.

10. The composer Vincent d'Indy (1851–1931) appears not to have been in the audience, though one of his works was performed.

11. Jeanne Leclerc, Opéra-Comique singer who sang the solo part in Hahn's *La Fontaine de Bandusie* (see note 6) and some Fauré songs including 'Les Roses d'Ispahan' which Proust goes on to mention.

12. Cf. the narrowly averted accident at the performance of the Vinteuil Sonata at Mme de Saint-Euverte's in *Swann's Way*, when the young Mme de Cambremer dashes out to retrieve the candlestick (*RTP*, I, 366–7).

13. Comte Guy Le Gonidec, Sosthène Duc de Doudeauville (respectively member and president of the Jockey Club) and the Marquis de Biencourt.

14. Princesse Murat, *née* Cécile Ney d'Elchingen (1867–1960).

15. The Hon. Reginald Lister (1865–1912), elegant diplomat (nicknamed 'La Tante Cordiale', a pun on *l'entente cordiale*), Counsellor at the British Embassy, whom Proust compares to the novelist Henri de Régnier.

16. That evening's hostess, Princesse Edmond de Polignac (see letter 10, n. 1). Her soirée was Proust's first appearance in society for nearly two years.

17. Comtesse Odon de Montesquiou-Fezensac, *née* Princesse Marie Bibesco, aunt of Antoine and Emmanuel, and Comtesse Fernand de Montebello, *née* Elisabeth de Mieulle; Saint-André is unidentifiable.

208   *To Marie Nordlinger*

[Friday evening, 12 or 13 April 1907]

Alas, dear friend,

I live in such continuous pain and worry of every kind that although the thought of dear Mary is constantly present in my heart, I haven't the strength to transcribe on to paper the countless speeches full of friendship which I address to her unceasingly, and so she is without news of me. Dear friend, I don't even know, so full this winter has the cup of my afflictions been (that of my sorrows can no longer be increased), if I've thanked you for the calendar which is never out of my sight.[1] I pluck off the poetic wisdom of the days while thanking you each time, as well as Ruskin, for the thought which each leaf brings me.

Dear friend, I'm sending you two articles I've written,[2] but O horror, in case I ever want to publish my articles in a book, I must ask you to return them (no matter when, there's no hurry!) to 102 boulevard Haussmann. I'm so tired that I haven't the strength to write any more.

Will you allow me (from so far away!) to embrace you?

Your respectful friend

Marcel Proust

I notice I've written on two different sheets. Will you find your way? I'll number the pages. But I'm afraid you'll get them wrong. I'll tear them to simplify matters.

1. He had thanked her in his last letter, dated 2 February.
2. 'Sentiments filiaux d'un parricide' and 'Journées de lecture'.

209 *From Robert de Montesquiou*

Versailles
19 April 1907

Dear Marcel,

You will discover, before the end of this month, *one* of the *ways* in which I am pleased to associate your name with mine in the march of time.[1]

But there will be others; and your mild, legitimate and flattering request has struck home to me, moved me. It is true that you ought to be *One of the Saints of the Calendar of the Hydrangeas*.[2] Some names have fallen from it, among them *Eyragues*. Only one new canonization: *Bataille*, above a place that was his by right.[3] But your patronage was no less imperative, and we shall decide upon it. Soon, I hope.

I'm staying here until the end of the month. Some friends have taken the opportunity of coming here to see *the Mausoleum*[4] and to accompany me on that pilgrimage of memory. I wish you were among them; but I no longer dare hope . . .

But is it not best to say 'Till soon', when one can say 'Till always'!

Robert de M.

(Address, still, Neuilly.)

1. He had arranged for Proust's review of his works, 'Un Professeur de beauté' (cf. letter 160), to be published as a postscript in his book *Altesses sérénissimes* (cf. letter 129, n. 1).

2. Cf. letter 175, n. 3.

3. He had dropped a dedication to the Marquise d'Eyragues (a favourite act of spite: a dedication to *my cousin Claude* in the first edition became by the second *To an ungrateful cousin*). The poet and playwright Henri Bataille (1872–1922) was deserving because he had devoted several articles to Montesquiou.

4. Yturri's tomb (see letter 175, n. 2 *et al.*)

210   *To Robert de Montesquiou*

[Early May 1907]

Dear Sir,

How grateful I am to you for thinking of me! I for my part think of you a great deal and during my nights of fever and insomnia I weave variations on your theme, which I hope one day will admit of a satisfying transcription.[1] For I know you better than many others do! I hope the day on which you've invited me will be one of those rare ones when I was better. I say was because there are now virtually none. I have a great desire to see you.

Your friend and admirer
Marcel Proust

I should very much like to know of which woman of our day you said that she was like the Duchesse de Langeais.[2]

1. See following letter, n. 1.
2. Heroine of Balzac's novella of that name.

211   *To Robert de Montesquiou*

[Wednesday, 8 May 1907]

Dear Sir,

Amazingly, or rather quite naturally, at the very moment *Altesses sérénissimes* arrived I was writing to thank you for your delightful letter, and to tell you that, having thought recently of some essential features of your work which I had omitted from the article in *Les Arts de la vie*,[1] I was hoping for a moment when journalistic circumstances were less un-favourable than they are now, in order to try to express them if I could. If I could have imagined for a moment that you would do me the great honour of inscribing my name to perpetuate it in your Pyramid, I would have asked you to let me revise and complete that essay. But still I was profoundly touched by your thought and I thank you for it from the bottom of my heart. I would have done so yesterday were it not that when one is with you one prefers to listen to you than to reply to you, and I revelled in 'Encycliques mondaines',[2] re-read the splendid 'Essai sur Moreau'[3] – of which I knew only a part – I began to read and admire you all over again. However I shall write to you again, when I've read more; today I simply wanted to express my gratitude and admiration.

Your respectful friend
Marcel Proust

1. 'Un Professeur de beauté' had first appeared in *Les Arts de la vie* in August 1905. Cf. letter 209, n. 1.

2. Chapter of *Altesses sérénissimes*.

3. A chapter of his latest work which Proust had read in the catalogue of the Gustave Moreau exhibition in May 1906.

212 From Robert de Montesquiou

[Thursday, 9? May 1907]

Dear Marcel,

I saw Calmette yesterday and gave him my book, and he seemed eager to have it written about. If the idea suited you, it could be arranged, and besides, whatever you did would be well done. But above all, do not feel in the least constrained, obligations becoming painful (and mutually so) when they appear in this form; and if I thought of it at all, it is because of what you said.[1]

I shall await your reply before proceeding with further negotiations on the subject.

Sincerely yours in mind and heart
R.M.

1. In the previous letter.

213   *To Robert de Montesquiou*

[Saturday, 11? May 1907]

Dear Sir,

Especially as it's to tell you that the suggestion seems to me impossible, forgive me for not answering you as soon as I received your letter. But I've been more or less killed by an article I've just finished,[1] and I rested for a day without reading or writing any letters, for I no longer knew what I was doing. You distress me by talking of *constraint.* How could there be any when there's such keen admiration and such profound pleasure in expressing it. If I feel I cannot do it, it's for two categories of reasons. The first is that I don't, I can't talk about books in the *Figaro*. Apart from the article I've just done, which M. Calmette commissioned by way of an exception. And even then I don't know whether it will appear since they find that I'm always ten times too long, and however much I try to compress, to remove from myself, a bit here and a bit there, Shylock's pound of flesh in order to weigh less, I can't seem to arrive at the required length. Already I have articles outstanding at the *Figaro* which don't appear and go on waiting for room, notably the end of the piece about which you said such nice things.[2] I think that one will appear, as I announced it at the end of the other, but I don't know when.[3] As it happens, in the article which I've just finished and which will go in before the others, if it goes in, the name of Gustave Moreau[4] gave me a perfect opportunity to refer to *Altesses sérénissimes*. But I couldn't find a way, without its giving the impression, especially added to the quantity of proper names already in the article, of a pure visiting card. It didn't work. When I speak of the Lion, I like to give him '*partem leonis*' if not in quantity, at least in quality. So if I'm to do the article, it would have to be commissioned by M. Calmette, and I don't think it would suit the *Figaro* for me to do that sort of article, for which they have regular incumbents or deputies or perhaps the occasional fortuitous offering from a Vogüé etc.[5] Added to that, their irritation with my longwindedness, and the difficulty it's causing them at the moment with the article they want but which they don't seem to be able to find room for, though I've taken out as much as I can, and am too exhausted to compress it further, all that would make the thing more improbable.

My second category of reasons stems from the extraordinary fatigue which the present article has caused me, as a result of which I feel incapable of writing anything passable (this one is in any case execrable) before having had a complete rest. I feel that preoccupation with an article on a subject of this sort, together with the fear that the *Figaro* wouldn't take it, would be beyond my strength. There are other reasons I could give you, arising from the subject itself (not the *book*, but *you*), which would make it particularly tiring. But precisely at this very moment as I write to you I'm so tired that I hardly know what I'm writing. In a few hours, when I've tried to get some sleep, I might perhaps write you quite a passable letter, but I don't want to wait, being already ashamed at not having thanked you immediately. And at the moment the words slip away from under my hand and my mind. And yet I only wanted to express my gratitude to you. To think of me, when you can have articles by such famous and talented people! What misunderstanding or what disregard for your own interests, but also what benevolence, what unjustified and touching predilection!

Your grateful admirer who thanks you and leaves you, too tired to write a more affectionate letter.

Marcel Proust

1. A review of Mme de Noailles's *Les Eblouissements* (cf. letter 195, n. 4).

2. See letter 195, n. 3.

3. At the end of 'Journées de lecture' Proust wrote: 'Alas! here I am on the third column of this newspaper and I haven't even begun my article. It was to be called "Snobbism and Posterity" . . . It will have to wait for the next time.' The article never appeared, but it is printed as an appendix in the Pléiade edition of his *Essais et Articles* (Paris 1971). Cf. letter 196, n. 2.

4. Cf. letter 211, n. 3.

5. Vicomte Eugène de Vogüé, distinguished Russian scholar and novelist (see letter 172, n. 6).

214   *From Robert de Montesquiou*

Neuilly
12 May 1907

Dearest Marcel,

Your letter touches me, and also amuses me, when I succeed in forgetting that you are unwell, which would kill my pleasure. I am convinced (I don't know why . . .) that the article whose subject you conceal from me must be (like every other article, indeed, and as is fitting) devoted to the new book by Mme de N., which, incidentally, must be charming.[1]

If, then, I am not mistaken, I shall make sure of the fact by *reading you*, and that will give me a *double pleasure*. Nor do I know why this idea that I've taken into my head *delights me so much*. Doubtless it is because I am in a sympathetic mood. Encourage me.

As regards myself, I know of only *five or six persons* whose expressed opinion on my works seems to me desirable for them, and for the moment. And if you are not among them, that is because, *as you may suppose*, you have already done it once, with breadth and acuity, and *that is your true reason* for abstaining. If, therefore, I wrote to you in another sense, it is because you yourself seemed to be contemplating a post-scriptum of the kind to which Sainte-Beuve reverted for Mme Valmore, and which has often been inflicted on me, witness Helleu.[2] It is not impossible that you may do likewise, with regard to my books; but it is not *indispensable*, and above all, does not *depend on you*. When they manifest themselves, indeed, as I know well, *these obligations are imperious*.

Let us say no more about it, then, my dear friend, except to congratulate ourselves for being what we are, and perfectly immune from the contingencies of ink-pot, calamus and papyrus.

R.M.

1. He was aggrieved not only at Proust's refusal to write about him but at not having received a complimentary copy of Mme de Noailles's *Les Eblouissements*.

2. Sainte-Beuve summed up Desbordes-Valmore's verse, Montesquiou the works of the painter Paul Helleu (see letter 19, n. 6) who had participated in the decoration of his Pavillon des Muses.

215   *To Robert de Montesquiou*

[Wednesday, 15 May 1907]

Dear Sir,

Forgive me for not having replied to you at once, but I've had such a terrible attack during the past two days that nothing could have given me the strength to write.

You tell me that my letter amused you. Yours pained me because you say in it, very deliberately, two very disagreeable things among other nice ones for which I'm most grateful to you.

As for the subject of the article, you guessed right. But I don't know whether I've misunderstood the sentence in which you speak of it: you seem to believe that I concealed it from you on purpose (I thought on the contrary that I had told you). Now in the first place I cannot see what reason I could have for wishing you not to know. And then such a ruse would be really too stupid since the article was intended for the *Figaro* which you read every day, and Mme de Noailles's name and mine, however glorious the one is already, however obscure the other is forever destined to remain (at least with my forename), both being too familiar to you not to arrest your attention, even without prompting you to read the article. I know only too well that people have such a peculiar way of reading newspapers that one can never be sure of anything in that respect. Beaunier wrote article after article about me – publicity announcements, etc. – and my friend d'Albufera, a subscriber to the paper, saw none of them, he assured me.[1] Finally Beaunier wrote one, this time on the front page, which was headed *Sésame*, the title of the very book which d'Albufera had just received. Since he compared me to Montaigne and various other people of quality, I wasn't averse to finding out the effect it had produced. I spoke about it to d'Albufera, who maintained that I was mistaken, that there had been no article about me in that day's *Figaro*, that moreover his wife hadn't seen it, etc. And then recently a lady said to one of my friends who had told her that I'd published two or three articles in the *Figaro*: 'You're mistaken, it must have been another paper because I read the *Figaro* every morning from beginning to end, and you can imagine that if there had been an article by M. Proust whom I know, the name would have struck me.' So perhaps it's not a bad idea, when one wants to conceal from someone the fact that one has written an article, not to mention it, since they'll never see it or read it. But in this particular case I could not but be happy that you should know I was writing an article about Mme de Noailles; she's an admirable subject of whom you were the discoverer as of everything else, and this sharing with Jupiter, in whose steps I follow, alas, after fifty others, is not in the least dishonourable. Except that the article is idiotic. It was already that when I wrote to you the other day. But no sooner had my letter been posted than the article was returned to me by the *Figaro* who said it would take up the whole paper and I'd have to cut it by two thirds. So that the few stumps that remain no longer have any human shape and I could be made to eat these remains of my children slaughtered by myself without recognizing them any more than Pelops did his.[2]

I apologize to you above all for speaking with such naïve exorbitance about something that concerns me alone and is of so little interest. If I were less exhausted on reaching the end of this letter I should have asked you to be kind enough, if you ever had occasion to write to me, to provide me with some information about the Chalet Shickler,[3] or rather to tell me if there exist elsewhere, less far, and above all less high, other dwellings of the kind, which (for another civilization at least) give a similar impression through their preservation or restoration. But if I can get to Neuilly one day, I shall be very interested to listen to you on the subject – and on every other subject. That whole chapter on St Moritz is delightful. I've read very little yet, being extremely unwell, but I find it all infinitely enjoyable.

Your faithful and sometimes dejected admirer
Marcel Proust

1. Cf. letter 157.

2. Put to death by his father Tantalus, the Phrygian king, Pelops was served up as a feast for the gods.

3. A chalet in the Engadine which Montesquiou often visited; in a chapter of *Altesses sérénissimes* he deplored a new railway which had ruined its peaceful site.

216   *To Francis de Croisset*

[Tuesday, 28 May 1907]

My dear Francis,

I was sad to read the news of the breaking off of your engagement.[1] Not that I don't often think these first sketches of future happiness fairly enviable and necessary and that it's good to start afresh before reaching the final decision, so that perhaps your friends, among whom I believe I may count myself, will be thankful for it. But how could one not be sensible of all the melancholy feelings which the abandonment of a romantic dream can bring to a sensitive heart even if it is after having recognized that it would have been mad to persevere with it. And so I hope it will not have any repercussion on your health – about which I should like some time to speak to you seriously, and I think *very usefully*.

I shall keep, shall I not, the pretty box[2] which is merely a premature harbinger of a happiness which I hope will not be long in coming. And likewise you will I hope accept my souvenir when you receive it in a few days; I shall simply change its character, making it entirely personal and not matrimonial, so that it won't prevent me from recapitulating when it comes to a definitive engagement.

Feelingly yours
Marcel Proust

1. See letter 184. Croisset was eventually to marry a daughter of Mme de Chevigné.

2. Croisset seems to have sent a token to mark his engagement and Proust to have ordered a wedding present.

217   *To Robert de Montesquiou*

[Tuesday evening, 25 May 1907]

Dear Sir,

I was *very, very happy* to see you and hear you again.[1] And in the meantime before the horrible attack which will begin as soon as the effect of the caffeine wears off,[2] I have been continuing to read *Altesses sérénissimes*, more in a state of grace after having partaken of the sacrament in your tangible kinds.[3] And I cast a furtive, proud and grateful glance at the inscription at the end like the 'Arsène Houssaye' of whom you spoke.[4] Thank you again for all your kindnesses towards me. Please don't believe that it's 'age', as you said, that gives you the pink and wrinkled face of a moss rose. You know that it has its beauty.

Your

Marcel Proust

1. On 28 May there had been a reading of Montesquiou's poems at Mme Lemaire's.

2. Caffeine was the stimulant he took at night, before work; it counteracted the veronal he had taken in the morning.

3. Proust refers, of course, to Holy Communion; ecclesiastical metaphors are frequent in his correspondence with Montesquiou.

4. Arsène Housset (Houssaye) (1815–96), writer, director of the Comédie Française, presumably mentioned by Montesquiou at the Lemaire soirée.

218   *From Robert de Montesquiou*[1]

31 May [1907]

. . . of the *Two Sisters*.[2] Oh, the fine progress of Science! It will come to the point, you need have no doubt, of *lying through the cinematograph!*

Now, in the just comparison you make between my complexion and that of a *rose* (that's what I call talking!), if I do not much like the word *wrinkled* which you use, on the other hand the qualificative 'frothy' which, knowing your authors, you have borrowed certainly, and knowingly, from Mme Daudet, puts everything back into perspective. Indeed, if ever a rose deserved to be compared to me, it is not so much a *moss* rose [*moussue*] as a rose which froths [*mousse*], an astonishing hybrid of Poestum and Epernay,[3] revealed by our amiable lady friend in her *Notes on London*.[4]

Your 'moss rose'
R.M.

(I hope you have received *Les Chauves-souris*.[5])

1. The first page of this letter is missing.

2. A reference to a painting by Théodore Chassériau (1819–56) discussed in *Altesses sérénissimes*.

3. Presumably poetry and champagne. At fifty, according to his biographer Philippe Jullian, Montesquiou's face was deeply lined with furrows, and painted.

4. In *Notes sur Londres* (1896) Mme Alphonse Daudet describes Dover Street window-boxes with 'frothy roses tinged with purple against a tender green'. The play on words in this letter is difficult to convey in English. The French for moss rose is *rose moussue* or *rose mousseuse*. *Mousseux* can mean both 'mossy' and 'frothy, sparkling'.

5. *The Bats*, a revised version of an early collection of published poems.

219 *To Robert de Montesquiou*

Monday [3 June 1907]

Dear Sir,

I am too tired to thank you as I should like for your delightful letter. I simply want to tell you that I still haven't received *Les Chauves-souris*. As regards the moss rose, I owe my knowledge – as of many other things – to you alone. In the now distant era when the first fine evenings of spring brought us together at the rue Monceau,[1] as we took off our overcoats beneath the arborescent lilacs, I remember seeing you with a ravishing flower in your button-hole, at a time when it wasn't fashionable to wear one, but became so because of the elegance with which you sported it. You told me then that it was a moss rose.[2] Round and flat, it did not stand out too much from the button-hole, wasn't too protuberant. And is this not true of your face, in which nothing protrudes beyond the spherical and astral section.

What you say about my ill-chosen friends[3] cannot in my case apply to visitors. Reynaldo is the only one. But how could I demand of you (*'Maxima debetur magistris reverentia'*[4]) what I demand of him, who even though he comes at the most propitious hours, sometimes has to come back three or four times in succession at an hour's interval until an unexpected fumigation is over, and finally, when I see him, often has to speak alone, my answers coming to him on scraps of paper. Even my brother is too busy to put up with these habits and I haven't seen him for months. Perhaps when I mentioned d'Albufera you thought I saw him but I haven't seen him at all this year. Once or twice I've seen Illan *'in whom you are well pleased'*[5] and who has a great admiration for you, and Lauris and E. Bibesco. Madame de Noailles to whom I wrote to say that you hadn't received her book and who cannot believe it tells me that she will be sending you another copy.[6]

Your respectful friend
Marcel Proust

1. At Mme Lemaire's.

2. Cf. the narrator's first glimpse of the Baron de Charlus at Balbec in *Within a Budding Grove*: 'He darted a final glance at me that was at once bold, prudent, rapid and profound ... and ... turned towards a playbill in the reading of which he became absorbed, while he hummed a tune and fingered the moss rose in his button-hole' (*RTP*, I, 807).

3. Presumably at an earlier meeting.

4. Proust reverses Juvenal (*Satires*, XIV, 47): 'Maxima debetur puero reverentia' ('The greatest respect is owed to the child').

5. Cf. *Matthew*, XVII, 5: 'This is my beloved Son, in whom I am well pleased ...' à propos of Montesquiou's protégé Illan de Casa Fuerte.

6. Cf. letter 214, n. 1.

220   *To Robert de Montesquiou*

[6 or 7? June 1907]

Dear Sir,

I've received *Les Chauves-souris*; but, together with my copy, that of Mme Ed. André.[1] I haven't dared send it on to her, for fear that there might be some couplet about her intended for me on the title page which it would be inopportune to transmit to her. But I haven't dared look and see either, feeling that I could not take the liberty of opening a parcel which bore not my name but hers, even though attached to mine. In the uncertainty resulting from so many conflicting duties I have left Mme André's copy with my concierge where it will wait until either you instruct me to send it on to her, which couldn't be simpler, or send round for it. I envy this book which will perhaps find its way into the room in which there is a replica of Mantegna's fresco from the Eremitani, one of the paintings I like best in the whole world, which I glimpsed one day in Padua, and whose Parisian sister I should very much like to see.[2]

Your respectful friend
Marcel Proust

1. Mme Edouard André, *née* Nélie Jacquemart (1840–1912), genre painter who had a grand house on the boulevard Haussmann, today the Musée Jacquemart-André.

2. While in Venice (see letter 13, n. 1) Proust saw the Mantegna frescoes in the Eremitani church in Padua. His favourite among them is evoked in *Swann's Way* (*RTP*, I, 353). The copies are still in the Jacquemart-André museum.

221 *To Robert de Montesquiou*

[8 or 9 June 1907]

Dear Sir,

Forgive me: I couldn't have talked about Mme André, as I don't know her.[1] I have so far looked only at the first page of *Les Chauves-souris*, with the splendid image[2] beneath Vigny's sublime line:

You alone appeared to me that which we ever seek,[3]

but I've cast a furtive glance at my favourite places, afraid of not finding everything. Naturally, after the vision of the other evening,[4] the 'Reszké brothers'[5] were the first to appear, vibrantly! But I thought that my admiration for the book went without saying, and, if it were better said, that I ought to re-read it first, and that my admiration for Mantegna cannot be greater than the admiration you no doubt feel for him, and my regret at having a whole series of his works quite close to me without my knowing them is a perfectly natural feeling.

Your respectful friend
Marcel Proust

1. Evidently the Count had complained to Proust that he had talked about Mme André instead of praising his book.

2. The line from Vigny appears, in the definitive edition of the book, above a poem of four quatrains preceded by the dedication 'To the memory of Flavie de Balserano, Marquise de Casa-Fuerte'.

3. A. de Vigny, 'Eloa ou la soeur des anges': 'Toi seule me parus ce qu'on cherche toujours.'

4. Cf. letter 219, n. 1.

5. Allusion to 'Manière', from *Les Chauves-souris*: 'Ne me demandez pas de peindre des batailles / Ou de vibrer ainsi que les frères Reszké!' The Polish brothers Jean de Reszké (1850–1925), tenor, and Edouard de Reszké (1853–1917), bass, were world-famous singers.

222   *From Robert de Montesquiou*

10 June [1907]

Dear Marcel,

There's no misunderstanding. I spoke of Mme André, because you spoke of Mme André. That's better than if you had spoken of Mme Moore.[1] The former has, at least, *a substance*, all the more appreciable for the fact that she is unaware of it.

The *Book* has remained the same; and yet it is very different. You will realize this on re-reading it. A poem like 'Lunebourg' (p. 249), for example, may help you to register, by comparison with the previous text, the work, and the kind of work, which I have had to undertake. And also 'Seule à seule', which you will like, page 365.[2]

And since you are not a *Stullifer*, you will not pretend not to have read 'Les Miroirs malins'.[3]

Your friendly Mirror
R.M.

1. Mrs William Taylor Moore, a rich American hostess of grotesque appearance whom Montesquiou called 'a bed-bug afflicted with elephantiasis' and despised for her social climbing.

2. 'Lunebourg' was shortened from 31 to 23 quatrains, 'Seule à seule' from 14 to 5.

3. 'The Malicious Mirrors' was the title of Montesquiou's *Figaro* review on 10 June of the Salon des Humoristes (exhibition of cartoons and caricatures). 'Stullifère' is his coinage for d'Albufera (who failed to notice Proust's reviews: cf. letter 215).

223 *To Robert de Montesquiou*

[Monday evening, 10 June 1907]

Dear Sir,

(I wrote to you the day before yesterday, but the letters pile up around my bed, it's impossible to find them and one has to start them again.[1]) I shall not fail to do the work of comparison which you recommend and I'm sure it will give me an even greater and loftier idea of you. What you say about Mme André and Mme Moore, already the target for deadly arrows in *Professionelles beautés* – à propos of Sem – and in *Altesses sérénissimes* à propos of other Highnesses (or was it rather in the *Gil Blas* article) at whose photographs, on a chimneypiece not designed for them, the lady casts sidelong glances, delighted me.[2] Naturally I had read 'Malicious Mirrors' at the crack of dawn, just before I normally go to sleep – a matutinal feast, an orgy of wit to which you frequently invite me – and if I dare pay such a compliment to you, I found it *even better written!* than the other articles. That whole introductory comparison, and more particularly the sentence about certain characteristics of Sem and his mallow-flower tints, is delightful.

'Stullifer' is capitally 'turned'. But Albufèra is not, I assure you, a Stullifer. Living outside all literary or artistic culture, he is nonetheless a highly fructiferous seedling in his way, which is not in the least common, since he distils a pure essence of kindness, very delicate too, a cordial and an elixir of friendly wisdom. But he reads nothing, and I used every pretext in the world for two whole years to get him to read a few lines of *La Bible d'Amiens*[3] without success. The article on Mme de Noailles, cut to pieces, execrable, is going to appear at last, I believe, but in the Supplement. The thought that I should be the cause of her not getting the article which otherwise she could easily have had distresses me infinitely.[4]

Your respectful friend
Marcel Proust

1. This letter should be read in the context of the previous one.

2. In the first book mentioned Montesquiou devotes a chapter to the famous caricaturist Georges Goursat (1863–1934), known as 'Sem'; the next reference belongs to an article printed in another book where, contrarily, he *praises* Mrs Moore; it is in the *Gil Blas* article that he sneers at her weakness for signed photographs of royalty.

3. Proust's first Ruskin translation. See letters 215 and 269.

4. See letter 213. For Mme de Noailles's reaction, see letters 225 and 226.

224   *To Emmanuel Bibesco*

Tuesday–Wednesday night
[11–12 June 1907]

My dear Emmanuel,

I have some important things to tell you. Unfortunately as I went out this evening (indeed that's the reason why I have these important things to tell you), tomorrow Wednesday (today by the time you receive this note) I shall have a bad attack and be unable to see anyone. Will you however, just in case, telephone me around half past nine in the evening. If the telephone is with the concierge tell him to come up and tell me you are on the line if I'm not sleeping or fumigating. But in any case if your evening is booked it can perfectly well be put off for a few days. I need hardly tell you that if I say important things I mean things that concern you. I know alas that those that concern me are no longer of importance to anyone since I lost my parents.

Fondest regards
Marcel Proust

225   *From Madame de Noailles*

Saturday [15 June 1907]

Dear friend,

Among all the follies contained in this book of my youth, I have just seen the divine article shining through.[1]

I have read and re-read it with infinite emotion and gratitude. I realize that you are one of the all too rare people for whom I write.

I admire you tenderly and I think of you.

Anna

1. Proust's review of *Les Eblouissements* which had caused him so much angst appeared in *Le Figaro* of 15 June.

226 *From Madame de Noailles*

Ritz Hotel
Piccadilly
London W.
Tuesday [18 June 1907]

My dear friend,

I have written you a number of letters which I haven't sent, because I mixed up too many things with the thanks which alone must occupy a heart so touched by that wonderful article. It has earned me more congratulations than the book itself. I keep receiving little messages of participation in my happiness. Every day since Saturday evening I have re-read the beautiful history of myself that this long, intoxicating, inexhaustible article represents, and I walk (too hurriedly not to start again a hundred times) through the gardens you have described, which are so rich, so varied, so tangible that mine seem only a geranium blob, stupidly glaring and flat, in the sunlight.[1]

I'm saying very badly and feebly what this article means to me, for I am too moved by this divine incense which I owe less to your taste and your judgment than to a friendship for me which, the article once written, has diminished.

These columns of tender phrases tower so serenely above our obscure and tense discussions on the telephone. What grandeur there is in the beauty of words, which stay still, whereas hearts, below them, weave and unweave an endless tapestry.

Thank you, my dear friend, profoundly
Anna de Noailles

1. In his review (see letter 227) he had written: 'In a book I should like to write and which would be called *Six Gardens of Paradise*, Mme de Noailles's garden would be . . . the only one where nature reigned supreme, where poetry alone entered.' There is a hint of the great novel to come in his review when he writes: 'She knows that a profound idea which has time and space enclosed within it is no longer subject to their tyranny, and becomes infinite.'

227　*To Madame Straus*

Friday [21 June 1907]

Madame,

Forgive me for bothering you yet again: have you any idea now how you will be placed on Monday the 1st; do you think you could come to dinner or could you only come after? I feel that it's odious of me to pester you. But you see, *six* people have replied to me as you did! So if I'm left in uncertainty until the last moment, I'm threatened either – if I count on people who then don't come – with being left more or less alone with Madame d'Haussonville without her husband[1] and Madame de Clermont-Tonnerre who are definitely coming, which will be very disagreeable for me for their sakes, as I know them only very slightly, or else – if to guard against that I invite more people now and at the last moment the six others come – with having ladies whom I won't know where to seat or at any rate will seat badly. And if you tell me at seven o'clock on the 1st of July that you're coming to dinner, I shall be mad with joy but I'll seat you very badly if the six come and I invite more people now. I invited Mme Aimery de La Rochefoucauld whom I haven't seen for years but whom I'd like you to be friends with because I'm sure it could have beneficial consequences as regards the Princesse de Monaco[2] who is in the process of *killing* her daughter. But unfortunately she wasn't free. My other guests will all be people you know (no Gabriel de La Rochefoucauld). I invited Dufeuille,[3] by the way, telling him the truth, that I was afraid you would only come after dinner. He too sent me a reply making it clear that he wouldn't know until the last minute whether he would be free. But I had to stop there and not tell him that I'd keep a place for him because I needed men and didn't want to have to place him at the last minute if he came.

I already have Fauré who isn't young,[4] Calmette for whom I'm giving the dinner, Béraud who is very touchy,[5] M. de Clermont-Tonnerre who is young but descended from Charlemagne, and some strangers. As it kills me to write and I do it all by telephone, which kills me just as much although I don't do the telephoning myself, I tell them to do it while I'm sleeping, so that a second one is asked when the first has already accepted. In case you should find that you're too tired to come to dinner I've taken pains to ensure that there's something later on that you'll enjoy and that will avoid your having to talk. That's why I got Fauré. They'll play things you like, I think Fauré will play (alas, Reynaldo will be in London), and as we shall be only about twenty, I can still enjoy you a little, your eyes during the music

> Music at times transports me like the sea
> And I set sail . . . towards my sombre star.[6]

But naturally I shall enjoy you more if you come to dinner. I intend to invite Robert Dreyfus after dinner but I shall beg him not to talk about it as I don't even know whether I'll invite my sister-in-law. I want no one to know about it except those invited and I shall send someone round to all the newspapers to make sure they don't mention it. Because of that and also because it would be so tiring for me to have my flat put in order (it's still in the state it was in when I arrived with nothing installed) and to have people smoking there, I think I shall try to find, at the Ritz or rather the Madrid or Armenonville, a private room where one could feel at home; I think I'd find it less suffocating. I had thought of asking M. Reinach whom I haven't seen for a long time. But as M. de Clermont-Tonnerre has accepted which I didn't expect and as it appears that he's very anti-Dreyfusard and very violent, and as M. Reinach is the very incarnation of Dreyfusism, I thought it might be better at such a small dinner party to avoid a collision the first time I was having M. de Clermont-Tonnerre who has invited me so often while M. Reinach never has.[7] If however you would like it, I think it would be possible. I still have three pathetic and hideous little Japanese trees for you.[8] Having seen them advertised in a sale I sent my pseudo-secretary[9] to buy them. What a disappointment when I saw them! But still, they'll grow to be nice, and they're so old and so tiny. It's like when one looks at Mont Blanc framed in an opera-glass and has to tell oneself that it's 4,810 metres high. I wanted to have all this explained to you by the so-called secretary who will bring them round to you one of these days, because writing tires me so and I like writing to you. Mme de Chevigné in accordance with the usual formula asked me to keep a place for her just in case! But since then she has refused. As a matter of fact I'm very sorry; she is so nice. Madame Lemaire doesn't know whether she'll be back from London, Mme de Brantes whether she won't have lost a cousin, Mme d'Eyragues whether she won't be on the banks of the Loire, M. Dufeuille whether his friends from Lower Normandy will have returned (word for word).

My dear little Madame Straus, don't forget this date, the 1st of July. If you decide only belatedly to come, don't blame me if I seat you badly, and if the worst happens and you don't come to dinner, make sure you come as soon as possible afterwards to hear Fauré and see me. I shall let you know the place.

Your respectful friend who would write to you at greater length if he were not so tired

Marcel Proust

I forgot to say that Mme de Noailles is in London and will probably not be back until the 2nd, and Mme de Chimay in Holland and doesn't know whether she'll be back before the 3rd.

No need to remind you that my invitation, either to dinner or after, applies, *with the strongest desire to have him*, to *Monsieur* Straus.

1. Othenin Comte d'Haussonville (1843–1924), great-great grandson of Mme de Staël, Orleanist, anti-Dreyfusard, model for the aged Duc de Guermantes in *RTP*. Proust wrote about his wife (*née* Pauline d'Harcourt: 1846–1922) and her salon in one of his *Figaro* articles signed 'Horatio' (see letter 10, n. 7).

2. *Née* Alice Furtado-Heine (1858–1925), widow of the Duc de Richelieu, divorced wife of Prince Albert de Monaco. In 1894 Proust had met her at Cabourg and she became a model for the Princesse de Luxembourg in *RTP*. Her daughter Odile had married Gabriel de La Rochefoucauld (see letter 130, n. 9).

3. Eugène Dufeuille (1841–1911), a Dreyfusard, former political adviser to the Duc d'Orléans, Pretender to the French throne.

4. The composer was unable to come at the last minute (see letter 231).

5. Jean Béraud (1849–1935), painter, Proust's second at his duel with Jean Lorrain in 1897 (see letter 3, n. 2); in 1920 Béraud was to offer Proust his cross of the Légion d'Honneur.

6. Paraphrased from Baudelaire's 'La Musique': 'La musique souvent me prend comme une mer / Vers ma pâle étoile, / Sous un plafond de brume ou dans un vaste éther, / Je mets à la voile' in *Les Fleurs du Mal*.

7. Not true: he had refused an invitation from Reinach while at Versailles.

8. i.e. Bonsai trees. Cf. letter 16, n. 1.

9. Robert Ulrich (1881–?), nephew of Félicie, the cook.

228   *To Madame de Clermont-Tonnerre*

[Shortly after 20 June 1907?][1]

Madame,

I shall begin by giving you the negative result of my conversation. Loche told me that he didn't want to marry.[2] I believe he thinks very highly of Mlle D.[3] I didn't mention your name or anyone else's (if you want me to name you later I shall do so, I thought it advisable not to without your consent). He first of all insisted that I was mistaken. He said it was a misunderstanding, and this is what made him think so. It appears that his cousin Albert Radziwill (I think that's the Christian name; he has in fact a nickname like Loche, I think it's Aba) is very smitten with the same girl.[4] I told him there was no misunderstanding for the very good reason that I assume that Mlle D. was absolutely ignorant of the plan which some friends of his (I didn't name them, nor did he ask me who they were) had thought up out of affection for him; and that if, which would surprise me, Mlle D. had thought about it, it was because she was attracted to him, as I believe her to be impossible of calculation. I think he had the same idea and I told him there was nothing very extraordinary about it as he was attractive to a lot of women (which is true). He had a captain to dinner and left him for a few moments to speak to me. So we didn't talk for very long although he prolonged the few moments quite a bit. This morning I telephoned him at dawn and he had gone out (it's true that dawn was only half past six but if you knew the state I'm in you'd think it early enough). He wasn't expected back for lunch. I went round this afternoon as I couldn't gather from what was said on the telephone whether he was there or not. He was still out. Then I went round to you (you must have been told) to go into the matter seriously with you since after all we've only exchanged a few words. Now, although I used to have a great deal of affection for Loche and now see him less, through my fault as it happens, I should consider it bad behaviour to influence the destiny of another human creature under false pretences, simply in order to show zeal on your behalf. The more I feel the wish to be agreeable to you – as the sight of the young enchantress whom I once used to see dancing at balls under the name of Mlle de Gramont, resuscitated the other night at Mme de Ludre's,[5] profoundly encouraged me to be – the more reason I have to be scrupulous, to examine my motives, to see if what I'm doing is really dictated by my conscience, and if I would do the same even if it didn't suit you. Otherwise, if I were capable out of obligingness of persuading a friend, or a former friend, to do something which I consider disastrous, you might perhaps think me more satisfactory, but you would be exposed to the possibility that, in a moment of admiration for another enchantress, I might not conceive a loftier idea of the duties which my respectful attachment owes towards you, as well as those which my friendship for Loche owes towards him. However, in this particular case there were no grounds for my scruples, because I think that there are few women capable of understanding, appreciating, getting on with Loche, and Mlle D., with her great and powerful intelligence, seems to me one of those few, perhaps the only one. I told him so, and said that though I knew her very slightly (he too scarcely knows her), she had seemed to me as well as being Beauty itself to have a superior intelligence and a kind and charming character, but that nevertheless I couldn't advise him to marry in order to please his friends. I reminded him that when he broke the news to me of his first marriage, saying he was doing it 'to please his mother', I told him that there was no surer way of causing her a great deal of pain one day and that I thought his mother loved him enough to want him to be happy rather than sacrifice his future. (I didn't add, so as not to appear to be reminding him of my generosity, that I gave him as a wedding present a twelfth-century alabaster figure of Christ inscribed with a sentence from Ruskin: 'You will be happy, but on one condition', which I specified to him and which he failed to fulfil.) I pointed out that if I had said all that to him then I wasn't going to go back on it today by saying, as others said to him then: 'Marry to please us', that this time, if he married, he must try to realize what he was doing and avoid getting into the position after a few years of having to pension off half the girls of Paris whom he would have abandoned; that he alone could decide whether or not he wished to marry; that I simply wanted to inform him that if he did so wish, there was a clear consensus on the part of people who felt very warmly towards him in favour of Miss D., with no possible comparison; that I confessed I had never seen such a beautiful or such a magnificently intelligent girl, or one who seemed more capable of making him happy; that I should be the first to regret, very deeply, his causing pain to Christiane,[6] though perhaps he wasn't considering the possibility of spending the whole of his life with her, that I hoped he wouldn't think my approach presumptuous, and that I had chiefly wanted to warn him and tell him how well I thought of Miss D. He replied by thanking me and saying that he was not in the least offended (he wanted to send his captain away to dine elsewhere so that he could stay with me, or leave him at home and go off with me; at least so he told me), but didn't want to get married. He reverted again to A. Radziwill's infatuation (which he doesn't in the least resent, on the contrary; he isn't absolutely sure that he is the man in question). The long and the short of it is that he doesn't want to get married and my impression is that one couldn't at present persuade him to remarry without exerting on the weaker side of his character (weak-willed and trying to redeem his failings by sudden impulses) the sort of pressure which I feel morally incapable of exerting, the more so since Miss D.'s future also interests me, I should be greatly distressed if she were unhappy, and on the whole, however sympathetic I feel towards this scheme, I'm not certain enough that she wouldn't be unhappy to say to Loche: 'Marry, marry quickly, and we'll see later on.' It's true that if one tried to reason everything out one would never get married or have children. In a word I

P.S. – Madame, I was obliged to interrupt this letter for the whole of the rest of the day, having been completely incapable of writing a line. I'm a little too tired after this attack to go on with it, but it's already terribly long and boring for you. I didn't want to give the impression of boasting of having done more than I actually did, but to show the limits beyond which I didn't go. This scruple accounts for my prolixity in writing to you. This time it's I who must demand the strictest secrecy, especially as regards his cousin's sentiments (which may well be directed towards Mlle D.'s sister), though he, Loche, firmly understood that it was Mlle G.D., whom, by the way, he knows to be entirely ignorant of my approach. In any case, he said he didn't want to get married.

Please accept, Madame, my most sincere respects,

Marcel Proust

1. The only clues to the uncertain date of this letter are Radziwill's marriage (see n. 2) and Mme de Ludre's reception (see n. 5).

2. The first marriage of 'Loche', Prince Radziwill, to Claude de Gramont (a cousin of Mme de Clermont-Tonnerre) in 1905 was annulled in 1906. An 'eligible' young man, he was not to marry again until 1921 when he married his cousin Dolores ('Dolly'), widow of her cousin Prince Stanislas Radziwill.

3. Gladys Deacon. See letter 175, n. 9.

4. In fact, as surmised later in this letter, Prince Albert Radziwill (1885–1935) was to marry Gladys's sister, Dorothy Parker Deacon, in 1910.

5. Probably the reception on 20 June 1907 reported in the *New York Herald*.

6. Loche's mistress, Christiane Lorin.

229   *To Madame Straus*

[Around 26 June 1907]

Madame,

I can't write to you at length because I'm very tired and because, knowing as I do better than anyone what it is to be ill, I don't want to appear to be damaging your health for the sake of my pleasure. But even so! After you telephoned the other day, it never occurred to me that if you couldn't come to dinner you wouldn't come afterwards. I thought it was decided. You told me you were going to the theatre. I'm ashamed to point this out to you. It reminds me of the people who say to me: 'But

you must be well since you're giving a dinner party', not realizing that
it's one day in two months during which I haven't been up even for an
hour. But if you can go to the theatre that evening, would listening to
Fauré tire you as much? I don't believe so, though I don't want to sound
like those unintelligent people who decide for us what's tiring and what
isn't, and who feel that what gives them pleasure can't possibly tire us.
God knows I would forgo all the pleasures in the world rather than that
you should suffer a moment's tiredness! But what if it didn't tire you!
Gabriel de La Rochefoucauld (about whom I do talk to you rather
excessively, considering I never see him) knew the plan of my dinner
party because I had asked him to invite his mother. He came the other
day to bring me his mother's reply and to find out how I was, but before
anything else, knowing it was my dearest wish, he asked me abruptly as
he came in: 'Is Madame Straus coming to dinner?' 'No.' 'Ah, my poor
friend!' I shall always be grateful to him for that remark. Madame Straus,
don't come, don't tire yourself, don't go to bed late on my account,
remember that I'm capable of subordinating my pleasure to yours,
which is another way of saying that I have great friendship for you. It's
just my luck that Monsieur Straus is also unwell and can't come.
Between ourselves, I suspect he doesn't care much for all those people,
and it's perhaps partly because of that that he won't come. You know
what it's all about: I'm giving this dinner party solely as an act of courtesy
to Calmette who is very nice to take my long articles which are little to
the public's taste. And an act of courtesy means inviting people of that
kind. But the greatest courtesy I could pay him was you. When I
mentioned your name, he gave a great exclamation of joy, saying how
fond of you he was and how long it was since he'd seen you. I had gone to
the *Figaro* about that article on Mme de Noailles which gave me so much
trouble. It's a place where one can go at midnight. And not knowing how
to thank him I said I'd like to have him to dinner one day. I thought that
would be the end of it, for how could I have imagined that I should be
giving a dinner party for years hence, or even before my death (less
distant, perhaps). Very amiably he took out his engagement book and
searched for a free date. That is why it was arranged for so long in
advance. He certainly never suspected all the agitation he was involving
me in. For it's only trivial things that agitate one – and love. But I'm no
longer in love. If you see him don't tell him I thought it would be
agreeable for him to invite him with these grand people. Because that
would make it look as though I thought it would flatter him, which
would make him angry and cause a breach between us when he can be so
useful and agreeable to me; and in any case I don't believe it would be

true. It's the sort of thing one feels obliged to do in order to be friendly, like having lots of courses etc., and which nobody cares about. In any case if M. Straus doesn't much like them, he can't really complain, as I met most of them at your house!

Madame, I must say good-bye because I'm all in. I shan't torment you any more, I shan't ask you again to come, I shall simply send you a telephone message telling you the place when I know it. You will know what it means in case you are feeling up to it when the day comes – but *only in that case*.

Your respectful friend
Marcel Proust

230   *To Madame de Noailles*

[Around 26 June 1907]

Madame,

I learn with great delight of your return, in the middle of an attack which will allow me to write only a brief word and will not allow me to telephone you tomorrow I fear. This brief word is this. In order to thank Calmette I have launched (in my present state!) into a dinner party which I didn't even mention to you because I thought you wouldn't be here, and which will take place on the 1st of July (next Monday) I still don't know where but some private, exclusive dining-room at the Ritz or a similar place, and Fauré will make music after dinner. The Princesse de Chimay told me that you wouldn't be here and Ginet[1] said: 'Since Madame la Comtesse has gone to discover London, she is not likely to be so frivolous as to come back on the 1st of July.' But since you have been so frivolous, will you come to dinner? What an unexpected joy, what a unique way of seeing each other again! The annoying thing is that, since I have already invited more elderly ladies or young people I scarcely know than a single table can cope with, Monsieur de Noailles and you will have to consent to being seated in accordance with your kind and longstanding friendship for me rather than with your rank, which is such, did you know, that when the Noailles married or rather mismarried into the family of Louis XIV, they were criticized for this condescension which could only be regarded as *toadying*! Will you, Madame, let me know if you would like to come to this dinner and when I feel less ill I shall write to you on eternal matters. It's a boring kind of

dinner for you but I didn't know you would be here, I was certain that
you would not be and that England the robber of Greece[2] would not give
back so quickly the most beautiful of her Athenian divinities.

Your respectful admirer who is so grateful for your sublime letter,

Marcel Proust

1. Mme de Noailles's manservant, in her own words 'a picturesque and devoted
servant dear to us and to our friends'.
2. Presumably a reference to the Elgin Marbles.

231   *To Reynaldo Hahn*[1]

[Wednesday, 3 July 1907]

. . . arranged on Monday, and my party was due on Monday the 1st of
July. On Sunday at seven o'clock in the evening I learnt through a note
from Hasselmans[2] that Fauré had suddenly been taken gravely ill, that he
would be unable to preside over the Conservatoire competition, etc. I
was ill, I had no time to prepare anything, I wrote Risler[3] a note saying
that I begged him as a favour to come and play, that I'd give him the 1000
francs he asked for, but that even if he was tired he'd be doing me a great
favour by coming, that I would have asked him through you if you had
been in Paris, but that by the time I had wired you in London it would be
too late. He sent me a note accepting. Next morning I sent him a
thousand francs and in the evening he played. There, my Buntchnibuls. I
think it was very nice of Risler. If you think otherwise, I adjure you in the
name of our friendship, in the name of Mama who would have been
deeply upset, not to say a word to Risler that could give him to suppose
that you found his behaviour less than perfect. We'll talk again about all
this but I don't want Risler to think for a second that my gratitude
towards him was not entirely sincere. But that's not all: after having told
me that she was too preoccupied to play, the Hasselmans daughter[4]
changed her mind on hearing 'better news' and came together with
Hayot.[5] So: 1000 + 600 + 700 for the Ritz dinner. Nevertheless it was
perfect, charming. Dinner: Mmes de Brantes, de Briey,[6] d'Haussonville,
de Ludre, de Noailles (Mathieu), M. and Mme de Clermont-Tonnerre
(Philiberte[7] was charming), d'Albufera, Calmette, Béraud, Beaunier,
Guiche, Jacques Blanche, Emmanuel Bibesco. After dinner the Casa
Fuertes, the d'Humières,[8] la Polignac, la Chevigné, Rod, Gabriac

(Alexandre), Berckheim,[9] the young Durfort[10] (not mopschant – said by me to be the son of the Sulamite and the butcher's boy), the young Lasteyrie[11] (ataxia – and romanticism, a moth-eaten Alfred de Musset, a vignette by Tony Johannot so pale that one can scarcely distinguish him let alone recognize him, for I didn't know who he was), Neufville,[12] Lister, Gabriel de La Rochefoucauld, Griffon,[13] Ulrich, Eugène Fould, etc. The absence of Mme Straus, Mme Gaston de Caillavet and especially Miss Deacon saddened me greatly. But the only one I really mished, you know nicens. All these people seemed glad to see me, the dinner was excellent, old mother Brantes was nasty to the others and nice to me,[14] Guiche did the menu and the wine-list:

The thing was exquisite and beautifully arranged.[15]

Programme (I had requested first and foremost some Bunchtnibuls but couldn't have any for reasson I'll splain to you)

> *Sonata for piano and violin* by Fauré[16]
> (Hayot-Hasselmans)
> Risler: *Andante* by Beethoven
> *In the evening* (I think) by Schumann[17]
> *Prelude* by Chopin
> *Overture* to the *Mastersingers*[18]
> *Idyll* by Chabrier
> *Barricades mystérieuses* by Couperin
> *Nocturne* by Fauré
> *The Death of Isolde*[19]
> *Berceuse* by Fauré[20]

For the same price, if it had been in Paris, I could have had the Société des Instruments à Vent in *Béatrice*.[21] Risler had in fact brought some waltzes of yours in his overcoat. But on the pretext that he knew nothing of yours by heart and that anything which had to be played with the music must come at the end, everyone had gone before he finished the other pieces, which he proposed as he went along, for he claimed to know nothing of the things I asked for (*Carnaval de Vienne*,[22] *Soirées de Vienne*[23] etc.).

My dearest Nicens, I think (I may be wrong) I fear that Fauré's indisposition may be more serious than they say. I dare not say what is in my mind. In any case you know that he had to find a replacement for the Conservatoire competitions and hasn't been able to judge a single one.[24] So I intend, having no one left here (Félicie and Ulrich having taken it into their heads to choose the same date of departure – the day before yesterday), to ask Léon[25] to go round every two or three days to inquire after him in your name and mine. If you don't wish your name to be mentioned send *me* a telegram.

Hasdieu tenderness (tenderness is your name in that phrase)

Buncht

1. The first four pages of this letter are missing.

2. Alphonse Hasselmans (1845–1912) was professor of the harp at the Paris Conservatoire.

3. See letter 207, n. 4.

4. Marguerite Hasselmans, pianist, pupil of Fauré.

5. Maurice Hayot was professor of violin at the Paris Conservatoire.

6. Proust borrowed the voice of the Comtesse Théodore de Briey, *née* Amélie de Ludre, for that of Mme Verdurin in *RTP*.

7. i.e. Proust's friend Elisabeth, wife of Philibert, Marquis de Clermont-Tonnerre.

8. Two more of Proust's men friends had married in 1905: Illan de Casa Fuerte to a young Italian widow (*née* Beatrice Focca) and Robert d'Humières to Marie de Dampierre.

9. Comte Alexandre de Gabriac, contributor to *Le Figaro* and *Le Gaulois*; Baron Théodore de Berckheim, diplomat.

10. Probably Comte Bertrand de Durfort, member of the Jockey Club.

11. Comte Louis de Lasteyrie (1881–1955), later member of the Jockey Club. He is compared to the poet Alfred de Musset and to the work of the Romantic painter and illustrator Tony Johannot (1803–52).

12. Baron Alexandre de Neufville, member of the Rue Royale Club.

13. The Vicomte Griffon, doctor friend of Robert Proust, head of the Hôtel Dieu clinic.

14. Mme de Brantes seems to have relented (cf. letter 104, n. 8).

15. Cf. Hugo, *La Fête chez Thérèse* (for which Hahn composed the music for the ballet): 'La chose fut exquise et fort bien ordonnée'.

16. Opus 13 (1876).

17. Robert Schumann, *Fantasiestücke*, opus 12, I, *Des Abends*.

18. The piano arrangement of this overture seems to have been a familiar item in Risler's repertoire.

19. Piano arrangement from Wagner's *Tristan and Isolde*.

20. *Berceuse for violin and piano*, opus 16 (1880).

21. The wind ensemble which had played Hahn's *Le Bal de Béatrice d'Este*. Cf. letter 133, n. 2.

22. Schumann's *Carnival Scenes from Vienna* (opus 26).

23. Nine waltzes after Schubert, by Liszt.

24. Although his deafness began with this illness (cf. letter 227, n. 4), Fauré

recovered to judge the last three competitions which Proust mentions; he died aged seventy-nine in 1924.

25. Hahn's manservant.

232 *To Robert de Montesquiou*

[Saturday, 20 July 1907]

Dear Sir,

You have written (and it's perhaps rather comic on my part to tell you so as though it were something you didn't know) a marvellous letter,[1] marvellous in style, thought, feeling and 'scope'. When your oars are not dripping barcaroles, it wouldn't be nice to receive a blow from one of them! The letter is an exquisite masterpiece; the antithesis which is summed up at the end of the period by 'aristocracy and industry', both of which words are rich and resounding with all the meanings with which you have infused them, the lightweight buoyant statues that one flits past '*inaniaque circumvolitat irundo*',[2] the great mortifying oar-stroke scattering its glistening droplets, and the fine, humane conclusion – all this makes your letter an offering at once 'industrious' and 'noble' which delighted me. Why did you not recall King David and Count Hugo to whom you allude.[3] It's true that everyone remembers them.

Your respectful admirer and friend

Marcel Proust

1. In a *Figaro* article (20 July) entitled 'The Future of the Aristocracy / a letter from Comte Robert de Montesquiou', he replied to an inquiry (in the context of Bourget's *L'Emigré*) about the role of the aristocracy in industry and the arts, at one point using a galley-ship as a metaphor.

2. Cf. Virgil, *Georgics*, I, 375: 'Aut arguta lacus circumvolitatuit hirundo' ('As the shrill swallow flitted round the lake'). Proust adds *inaniaque* (in vain or to no purpose).

3. A reference to an earlier article (1895) nominating several noblemen – Victor Hugo and King David among them – who were artists as well.

233   *To Madame de Caraman-Chimay*

102 boulevard Haussmann
[20 July 1907, or shortly after]

Princess,

Forgive me for not having written before, but I've been so ill recently. Were you told that before your departure I telephoned a number of times to ask if I could come to see you, on one occasion speaking to a kitchen boy who said he couldn't go and tell you 'because he didn't know the apartment'. It was an epic piece of telephoning but the result was that I didn't see you.

Princess, you ask *me* for titles when I haven't read anything for years except Joanne guides,[1] geography books, handbooks of châteaux etc. – anything that enables me to plan journeys, to look up towns, and . . . not go! However I think that this time I will go, to Brittany. If you could recommend any interesting towns or beautiful landscapes there I'd be very grateful. I envy rich people with yachts who can see everything without having to change rooms, but the thought of going to a hotel in Lamballe,[2] then to one in Morlaix, from the one in Morlaix to one in Quimper or Ploermel, delights my imagination but appals my asthma . . .

Princess, it isn't to tell you all this that I'm writing to you but to talk about books. Certain books by Stevenson[3] are the most entertaining and the most delightful that I read years ago. *Treasure Island* is perhaps a bit too much of a children's book (though excellent). *The Dynamiter* is terribly complicated but charming all the same. *The Suicide Club* (perhaps entitled *The New Arabian Nights* or even *The Rajah's Diamond*, from the title of the second story) is also a charming book. But perhaps the most striking is the short novel called *Doctor Jekyll and Mr Hyde* (it isn't in the *New Arabian Nights*, it's a separate volume). The fact that Wells plagiarized, watered down, refashioned it in everything he wrote,[4] notably *The Invisible Man* and *Love and Mr Lewisham*, may have deprived Stevenson's story of some of its thrilling strangeness. But still I can't believe that a book by a man of genius doesn't remain a work of genius even if a man of real but minor talent imitates, vulgarizes and spreads it all over the place.

At all events, Stevenson's story is so correct[5] that one can test its beauty and terror. *The Adventures of David Balfour* is thought, I believe, to be a boring book by Stevenson. Personally it fascinated me – all those Scottish landscapes. Now to come back to France there are at least two novels by Boylesve[6] which are highly enjoyable to read, *La Becquée* (admirable) and *L'Enfant à la balustrade*. I can't say they are entertaining in such a precise sense as with Stevenson, who died, incidentally, of consumption in a railway carriage, anaesthetizing himself with laudanum to kill the pain, and never ceased inventing, with inexhaustible genius, the most entertaining, the most beautiful, the least portentous stories, stories for stories' sake, full of the *joie de vivre* which he communicates to us so powerfully but which he never experienced himself. But still, by reading a good novel by Boylesve your friend will spend his holidays in a delightful garden in Touraine and it will be entertaining too. I have the impression that from his very first books (which I've barely glanced at) such as *Le Parfum des Iles Borromées*, etc. he is groping rather awkwardly to find his way, outside his real province which is not very big but where he is truly the master since Balzac. The last (*Le Bel Avenir*) is a nice book but somewhat pale and rootless, ending up in Paris, a charming creation all the same, by someone who is very gifted but also very limited.

The most enjoyable thing of all would be to settle down to read Balzac (if your friend hasn't read him) or at least one cycle of Balzac, because a single novel can't be read in isolation; it's not easy to get away with less than a tetralogy, sometimes a decalogy. A few of the novellas, truly divine, can be read on their own, this great fresco-painter having been an incomparable miniaturist.

If you'd like some Balzacian advice, I'll write to you but it would be quite a long letter.

Good-bye, Princess, and please accept my respectful regards.

Marcel Proust

Will you please remember me to the Prince.

You wrote me a delightful letter, Princess, with two prodigious phrases (the world is a cocoon, etc.). Are you writing about Holland as you promised? A propos of Holland, I haven't yet answered Paul's charming card, from Amsterdam.[7] But that's because I still thought he would come and see me. I shall write to him one of these days to apologize.

Have you read R. de Montesquiou's letter about M. de Dion?[8]

1. A collection of guides by Paul Joanne published by Hachette. In his letters Proust several times refers to Joanne's *Hollande et bords du Rhin*.

2. Small Breton town, one of the stops on the 'little train' in *RTP*.

3. Referred to by their French titles; as Proust goes on to suggest, two were

published in one volume entitled *Les Nouvelles Mille et une Nuits* (The New Arabian Nights).

4. Cf. letter 79, n. 6. *The Invisible Man.*

5. *Correcte* in the French, but the reading is conjectural.

6. Real name René Tardiveau (1867–1926).

7. Paul Bacart, manservant to the Prince and Princesse Alexandre de Caraman-Chimay; in February he had found a valet for Proust who, in the meantime, had re-employed his parents' old servant Nicolas Cottin (1873–1916).

8. Marquis de Dion, an aristocratic industrialist. Cf. previous letter.

234   *To Robert de Flers*

[Sunday, 21 July 1907]

My dearest Robert,

I can scarcely write to you, as my eyes are blinded with tears, having just read the note in the *Figaro*:[1] I shall never see your dear, your beloved little grandmother again. But what will become of you, my poor boy, where will you be without that soothing awareness of having been the pride, the solace, the gaiety, the life of her life, the breath of her body surviving for you. Dear Robert, the fact that your sweet wife, by her own choice as by yours, so delicately and charmingly avoided taking offence at her immense, jealous affection for you must be a consolation to you at this moment. How I should like to be able to embrace you and weep with you. I shall make an effort to get up, no matter that I've once more been bedridden for a fortnight. My dear little Robert, I am yours with all the immensity of our shared memories, with all the bitterness of my grief-stricken heart, today when you need more than ever the one you have lost to soothe the only sorrow that she will be unable to weep over with you.

Marcel Proust

1. *Le Figaro* had announced the death of Flers's grandmother, Mme Eugène de Rozière. Cf. letter 180, n. 3, letter 236, n. 2. Proust's anguish recalls the death of his mother and anticipates that of the narrator's grandmother in *RTP*.

235   *To Robert de Billy*

[Second half of July 1907]

My dearest Robert,

I think of you affectionately every day, but writing exhausts me I'm so ill. For one day only I got up . . . to give a dinner party at the Ritz! I can assure you that it was really rather nice. After dinner Risler played Wagner, Beethoven, Schumann, etc., Hayot played Fauré's *Sonata for piano and violin* – it was very enjoyable. I had Mme de Noailles, Mmes d'Haussonville, de Clermont-Tonnerre etc. at dinner. Guiche chose the dishes and the wines; unfortunately it was I who paid for them! But still it was nice; Berckheim came for a minute after dinner, but so late I don't think he heard anything.

I have never thought so much about Bulgaria as I do now, and all Ruskin's puns about Sofia, Saint Sophia, Eternal Wisdom and Queen Sophia recur incessantly to my mind, bent under the discipline of that man and my friendship for you.[1]

Write to me, Robert, without asking me to reply because I'm not well. If you can think of splendid journeys for me, recommend them to me; if you have friends in Brittany, recommend me to them.

I had a visit from Bertrand today.[2] He didn't like my beard nor my straggly hair. I very much enjoyed your definition which will endure: I'm a chargé d'affaires but affairs don't charge [burden] me.[3] Remember me to M. Paléologue[4] and try to predispose him to look on me more favourably. I don't think I know your other colleagues. I'm still in Paris but I don't think I'll stay here much longer. Will you be coming here? I've seen Antoine Bibesco again, without his moustache, no longer concealing a lip which is not all sweetness. I've been told that Raoul Johnston's lady friend has – but, dear friend, all this is too difficult by letter. I caught a glimpse of the aforesaid Raoul Johnston[5] the only time I've been out; how pleasing his physiognomy is, how original and luminous. I don't know whether this light comes from the mind:

> But may not appearance alone suffice
> To gratify a heart that flees the truth.[6]

If Madame de Billy is with you, will you ask her to accept my great admiration, my deep and respectful attachment.

Tenderly yours, my dearest Robert
Marcel Proust

1. Robert de Billy had been made First Secretary at the French Embassy in Sofia in January. In his translations Proust had had to elucidate much Ruskinian word-play, in this case of Byzantine complexity.

2. Bertrand de Fénelon. For Proust's straggly hair cf. letter 75 (postscript).

3. Billy's definition of Fénelon.

4. See letter 142, n. 5.

5. A civil engineer and member of the Jockey Club.

6. 'Mais ne suffit-il pas que tu sois l'apparence / Pour réjouir un coeur qui fuit la vérité.' Baudelaire, 'L'Amour du mensonge' in *Tableaux parisiens* (*Les Fleurs du Mal*).

236   *From Robert de Montesquiou*

Pyrenees
[Soon after 23 July 1907]

Dear Marcel,

Thank you for your charming, witty and cordial letter, prompted by an impulse precisely the opposite of that which we have christened *Stullifère*.[1]

As it happens, at the same moment you provided me with an opportunity to write to you in the same sense, with your pretty article which, for us, constructs its variations and its theme entirely to the advantage of the former, since the fact that they are developed in honour of a person for whom I have little liking does not prevent me from appreciating them.[2]

Added to which, the most agreeable part of your commentary gives you the opportunity to talk about yourself, by treating of those valetudinarians whose economy of physical expenditure allows them prodigalities of mind and heart.

Moreover, the type which you describe to us is simply the manifestation, if not perfect, at least complete, of maternal love in its laudable excess. I have experienced and appreciated it myself in Her who served me as Mother, and whom I portrayed in some verses which you like.[3] And what a fine pendant to your article a friend who had your talent could write, speaking of the one you mourn!

As far as you are concerned, it was your father, I remember, who gave me, when I met him one day, that sensation of *the absolute*, in a glorified transposition of La Fontaine's 'My children are pretty'.[4] When I asked him for news of you he replied: '*Marcel is working on his cathedrals.*' And the way he articulated the pronoun made me realize that in his eyes the

Middle Ages, as was befitting, had striven *for you alone* in chiselling and shaping stone.

Another thing that one must acknowledge as one of your merits, after having thought of it as one of your foibles, is the way you displease us, at first, by seeming to put on the same level things and people that are marvellously disproportionate, when you interpose, among your approbatory points of comparison, people whom we are extremely reluctant to admire, and with reason.[5]

But our incipient irritation subsides when we see you yourself smiling in your beard – since you now have one – at the idea that we might well suspect you of confusing what you know to be unequal and only appear (as I sometimes do myself, to the displeasure of some) to be putting on the same level, on the one hand to satisfy the demands of friendship and, on the other, to testify to your power.

Good-bye, dear Friend. Fate has decreed that we no longer see each other. That is better than that it should condemn us no longer to love each other, which is the destiny of those who succumb to the danger of constant meetings . . .

Robert de M.

1. Cf. letter 222, n. 3.

2. Montesquiou means Mme de Rozière (see letter 234), subject of Proust's article 'Une Grand'mère' in *Le Figaro* of 23 July.

3. i.e. his poem 'Ancilla', about the old servant whom he called his 'surrogate mother'. Cf. letter 175, n. 4.

4. 'Mes petits sont jolis' (*The Eagle and the Owl*).

5. Proust takes up this point in letter 250.

237 *To Madame Guéritte*[1]

102 boulevard Haussmann
[27 July 1907]

Madame,

I have been seriously ill (and still am) and had to go away in order to try (in vain) to recover.[2] So it's only now that I have found the delightful letter you did me the honour of writing to me; I have read and re-read it, and am relishing at length the profound if belated joy it brings me. I am truly very proud that such a fastidious admirer of Ruskin should have sanctioned with such warm approbation an effort whose value – and even legitimacy – I have often doubted. 'Let the dead bury the dead', those strange, harsh words from the Scriptures[3] of which sensitive souls alone can understand the imperious and beautiful significance, can be applied to translations, apologias, latria. I was sensitive to your reproach for not having published Ruskin's preface.[4] But I explained in a note that I hadn't even been able to include the third lecture. It isn't my fault but the publisher's. And then, the preface to *Sesame and Lilies*? Which preface? As you know, Ruskin didn't write just one. Whi h to choose? It was difficult. As for mine, it is indeed less literally Ruskinian than the one to *The Bible of Amiens*, but I think in itself it is worth something, and the other not much more than nothing. In a volume of miscellaneous pieces, if I ever recover the strength to assemble those that already exist, I shall certainly include the preface to *Sesame*, and if I give one or two extracts from the one to the *Bible*, that will be all.[5] As for Ruskin, I've ceased to translate him; others are beginning to set about doing so on all sides, not always in what seems to me the most respectful and sensible way, but still the flame is lit and is spreading, and that's enough. I very much regret that you didn't follow through your plans for translating his work. To be interpreted by a woman of such rare and sensitive intelligence would undoubtedly have given him great pleasure. But is it so easy to get translations of Ruskin published as their recent multiplicity seems to suggest? I myself had the greatest difficulty getting mine accepted by the Mercure and they wouldn't have taken a third.

Forgive me, Madame, for interrupting here, so weak am I, a conversation I should have liked to prolong; forgive the muddle of these bits of paper of different sizes – I'm bedridden and haven't my boxes of letter paper to hand – and accept my grateful and enchanted respects.

Marcel Proust

1. *Née* Madeleine Aubry, translator and educationalist; both she and her husband taught in England. She had evidently written Proust an admiring letter about his translation of Ruskin's *Sesame and Lilies*.

2. Almost certainly a white lie to excuse his delay in replying.

3. Gospel according to St Matthew, VIII, 22.

4. Supplanted by Proust's own preface to *Sésame et les lys*, the famous essay 'Sur la lecture' (On Reading). In the collection entitled *Pastiches et mélanges* (Gallimard, 1919), Proust was to include over one hundred pages of his preface to *La Bible d'Amiens*.

238   *To Reynaldo Hahn*

Thursday evening [1 August 1907]

My dearest Reynaldo,

I slightly regret your having shown 'Cydalise'[1] to Cydalise. She will be unable to recognize herself in this mirror which reflects only one aspect (I've changed pens because the other one was awful – and this one is not much better) – perhaps unreal and in any case so fragmentary, so fleeting, so related to me – of herself, which I alone perhaps am capable of finding any truth in, by comparing it with a memory. No, for me, Madame de Reszké is Viviane, the fairy-like apparition at the edge of the Forest of Broceliande or the Lake of Love, whose adorable face and dreamy eyes becharm the legends of Burne-Jones.[2] Figures that seem too conventional in art to be 'believed' by anyone who looks at them in Burne-Jones or Gustave Moreau, but that nature achieves once in a while to show that such 'artistic' beauty can be real. Whence Madame Reszké, and doubtless once upon a time Sarah Bernhardt. To a certain extent Mme Greffulhe. But only Madame de Reszké is the creature of dream, who infinitely surpasses the beauty we have invented for ourselves with Brittany, but who must be the true beauty of Cornwall, the beauty that its poets alone have seen, the beauty of Viviane once more, of Isolda, Isolda who wandered, mournful and disdainful of a princely destiny, until the day when she heard the voice of Tristan. I'm sure that that is the real beauty of Brittany and I shall go as far as Pontaven, as far as Helgoat[3] to see if its lakes are the colour of Mme de Reszké's eyes. Naturally all the people who know her and the pseudo-wits will say that she isn't like that at all, that she is gay, Parisian, worldly, that she would be bored on the moors and in Broceliande and has nothing of the furze flower[4] about her. But that doesn't in the least detract from the veracity of my hypothesis, which Madame de Reszké herself may perhaps find quite erroneous. For all that she herself may be unaware of the mystery in her eyes, her face, that mystery is what a poet must strive to grasp and express – and not simply echo M. de Turenne or M. Bourdeau[5] in the conception they may have of her, even if she shares it. That this conception might be true is immaterial to me. Those lines of Baudelaire

> I know that there are melancholy eyes
> That have no precious secrets in their depths[6]

are wrong. Perhaps the eyes alone reflect these secrets, but at least they do reveal them to those who are capable of reading them. Moreover, does not everything you tell me about her voice, her personality, what you used to call her 'genius', conspire with my reverie?[7] And since we don't need to indulge in false modesty between ourselves, may we not be certain that an idea which we both share has every likelihood of containing more truth than all the ideas of the aforementioned persons put together, even if one added all those whom you can think of. 'Cydalise' was written after coming home from the Princesse Mathilde's where Mme de Reszké (then Mailly Nesle) was in red and talked to Porto-Riche.

Tenderly yours
Marcel Proust

I continue to hesitate between Brittany, Cabourg, Touraine, Germany . . . and Paris.

The young girl of nineteen (Ruspoli) whom 'Just your name, Monsieur, but no thoughts' married this morning is related to *Van Zandt* and *Paganini*.[8] 'How do you spell that?' Guiche will say.

1. A sketch first published in *Le Banquet* in April 1892 and later collected in *Les Plaisirs et les jours* (1896). See letter 30, n. 1. The subject of the sketch was Mme Jean de Reszké, then Comtesse de Mailly Nesle (*née* Marie de Goulaine), wife of the famous tenor and herself a singer. Cf. letter 221, n. 5.

2. In *The Knights of the Round Table* (twelfth to thirteenth century), Viviane learns the secrets of Merlin in the Forest of Broceliande, in Brittany, and later becomes its mysterious Lady of the Lake. *Merlin and Viviane* was a subject painted by the Pre-Raphaelite Edward Burne-Jones (1833–98).

3. Sic. Pont-Aven in south Brittany was one of the stops on the 'little train' in *RTP*. Huelgoat is a town in the central granitic moorlands.

4. *Ulex minor*, dwarf gorse characteristic of Breton heathlands.

5. Comte Louis de Turenne (1843–1907) who was to die later that year. Jean Bourdeau (1848–1928), philosopher and publicist, wrote for the *Journal des Débats*.

6. 'Je sais qu'il est des yeux, des plus mélancoliques / Qui ne recèlent point de secrets précieux.' 'L'Amour du mensonge' (cf. letter 234, n. 6).

7. Reynaldo Hahn declared that her singing voice was 'the most astonishing I have known'.

8. Agénor Duc de Gramont – who made the remark quoted at his son Guiche's engagement party (see letter 50) – had just been married for the third time, aged fifty-five, to Maria Ruspoli, whose cousin was the first wife of a Paganini, whose second wife was a Van Zandt. It is not known whether there was any connection with either the famous violinist or the American singer Marie Van Zandt (1861–1920) of the Opéra-Comique, but Proust possessed a photograph of the latter in male travesty that almost certainly inspired the disturbing portrait of Miss Sacripant which the narrator sees in Elstir's Balbec studio in *Within a Budding Grove* and which triggers off certain discoveries about Odette de Crécy's past as well as Elstir's (*RTP*, I, 906–8, 919–22; II, 275; III, 302, 447).

239 *To Reynaldo Hahn*

[Thursday evening, 1 August 1907]

Nicens,

Herewith in the shape of an idiotic, incoherent and illegible letter[1] a more accurate *view* of herself which you can submit to Cydalise, to correct the bad impression 'Cydalise' must have made on her. To add to the letter something that has no connection with her to make it look more natural, I mentioned the Van Zandt connection in a postscript, but I now remember that I'd already told you about it. I was about to cross it out, but was afraid she might think it was some reservation that you had crossed out so I decided not to. As for starting again, I'm too tired. Do what you think best. If I were to cross anything out it would be the name Bourdeau which has some drawbacks, but perhaps I'll leave it.

Love

Buncht

I'm thinking of the salad you '*composed*', nicens. I'm sure it's rebulsive.[2]

1. i.e. the previous letter.

2. Cf., in *Swann's Way*, the Japanese salad in Dumas's *Francillon* which Mme Cottard dilates upon (*RTP*, I, 279–81) and, in *Within a Budding Grove*, the improbable *salade composée* served to M. de Norpois by the narrator's mother who '. . . was counting greatly upon the pineapple and truffle salad. But the Ambassador, after fastening for a moment on this confection the penetrating gaze of a trained observer, ate it with the inscrutable discretion of a diplomat, without disclosing his opinion.' It followed Françoise's triumphant *boeuf à la mode*: cf. letter 339 (*RTP*, I, 495).

240 *To Robert de Montesquiou*

[Early August 1907]

Dear Sir,

I've been so ill these last few days that I haven't yet thanked you.[1] And now I'm doing so very briefly. But I wanted to tell you that I couldn't love my friends without some pain if they didn't love you, and I want to make them all love you not for your sake because you don't care, but for mine because I do. I have few enough, even if you allow me this satisfaction, for you still to have among the rest of humanity a vast field in which to exercise the 'gentle art'[2] which you seem to enjoy no less than the Master with whom, incidentally, you have so many affinities. Indeed if one had to express an opinion on the matter, I believe that your taste for the 'gentle art' is more irresistible, deep-seated and sincere than his. As far as one can guess from a distance, there was an element of philanthropic compunction in his case, and a certain melancholy. At all events I don't believe the steadfast cult of Enmity procured him that savage and salubrious joy which it seems to inflame in you. One would like to ask you for the sacrifice of an enmity as one asks others for the sacrifice of a friendship. And the difference is that you would refuse it!

You rise above incomprehension, like the seagull above the storm, and you would hate to be deprived of this upward pressure.[3]

Your grateful admirer
Marcel Proust

1. For his last letter (see letter 236). The following character analysis is thus in the nature of a tit-for-tat.

2. An allusion to *The Gentle Art of Making Enemies* by James McNeil Whistler, the 'Master' referred to in the same sentence. See letter 15, n. 1.

3. This remark so delighted Montesquiou that he reproduced it in his memoirs.

241   *To Reynaldo Hahn*

[Cabourg,[1] Tuesday, 6 August 1907]

Binibuls,

I read that Foreign Ministry decorations had appeared and that Birn Ybuls wasn't mentioned. But nevertheless sent letter Thomson[2] (written before) so that it looked as though it had already failed this time and might be more pressing between now and next time and won't delay decorating Nicens.

There's that Englishman[3] or sensible madman here who incidentally speaks French with a facility that bodes no good (imitated – like all me – from Bunchtnibuls). You've guessed.

I travelled with Doyen[4] who said to me after great eulogies of Mme Greffulhe: 'And for all that she hasn't succeeded in creating a salon as brilliant as Mme de Caillavet's!' (the exclamation mark is mine).[5] Did you know that there's a play by Capus called *L'Amant de Léontine*.[6] What a singular.

I was received on my arrival – by chance – by that amiable juvenile lead (a very over-ripe Léandre or Octave) at the Tapirs' who is called Léonce de Joncières.[7] He will soon be playing Truffaldin there – a charming intermediary between old man d'Oncieu and young Bardac (son of Sigismond). You never told me that Faisans[8] in person was doing the watering-places under the name of Emmanuel Arène. I called him doctor. He found me moschant. I'm too tired, moschant. Good-bye. Besides, I'm sure that with your hostess[9] you no longer remember

Fasché[10]

I find that Faisans can't manage to change his voice.

I told him that you'd write to him on the subject of Straram, who seems to me worthy of esteem, etc.[11]

P.S. Someone with a certain discrimination was saying (don't try and guess, it's me) that Lautier[12] with his blue beard and mauve cheeks looked like those figures on posters where the sky is pink and the clothes are silk and golden yellow, deliberately arbitrary. – The juvenile lead at the Tapirs' is very nice (I saw him for five minutes).

1. Possibly because of the death of Flers's grandmother (see letter 234) Proust had chosen to go for his health to Cabourg (cf. postscript to letter 238), the Victorian seaside resort in Normandy which he had visited with his mother and grandmother and was the principal model for Balbec in *RTP*.

2. Gaston Thomson (1848–1932) was a cabinet minister 1905–8. Hahn, who wasn't given French nationality until 1908, received the Légion d'Honneur in 1914.

3. Alfred Edwards (1857–1914), born in Constantinople of an English father and French mother. Proprietor and journalist and founder of the popular daily newspaper *Le Matin*, he was a notorious womanizer. Like Proust, he was staying at the Grand Hotel. Evidently Reynaldo failed to guess who Proust was referring to – see letter 247.

4. Eugène-Louis Doyen (1859–1916), fashionable surgeon, principal model for Dr Cottard in *RTP*.

5. Cf. Cottard on the relative merits of the Guermantes and Verdurin salons in *Cities of the Plain*: 'Mme Verdurin is a great lady, the Duchesse de Guermantes is probably a nobody' (*RTP*, II, 910).

6. Léontine was the Christian name of Mme Arman de Caillavet (see letter 150, n. 12), who was one of the principal models for Mme Verdurin. The play by Alfred Capus (1858–1912) was in fact called *Les Maris de Léontine*.

7. Léonce de Joncières (1871–1952), genre painter. Léandre and Octave must be the two young lovers in Molière's *Les Fourberies de Scapin*. The Tapirs are unidentifiable. Truffaldin is a character in Molière's *L'Etourdi*. 'Old man Oncieu' and 'young Bardac': respectively Comte Victor d'Oncieu de La Batie (d. 1906) and either Jacques or Henri Bardac (see letter 207, n. 9: sons of Noël, not Sigismond Bardac).

8. Dr Faisans (see letter 49, n. 8), said to resemble the writer and politician Emmanuel Arène.

9. Sarah Bernhardt, with whom Hahn was staying in Brittany.

10. Proust had first signed himself Guncht, then crossed it out. Fasché = *fâché*, vexed.

11. As Faisans was a lung specialist, presumably this refers to a singer. Walter Straram was Director of Singing at the Opéra.

12. Eugène Lautier (1867–1935), foreign editor of *Le Figaro*.

242　*To Emile Mâle*

Grand Hôtel, Cabourg, Calvados
102 boulevard Haussmann, Paris
Thursday [6 August 1907]

Dear Sir,

I have just spent an entire year in bed, with attacks so unforeseeable that, not daring to make a plan even an hour in advance, I felt unable to ask you for the favour of a visit – which itself would have been presumptuous enough – which I might have found myself unable to receive because an attack had supervened. Recently, feeling somewhat better, I sent someone round to you to explain this situation, and to ask if you would consent to pay me a visit one evening. But I had chosen a bad moment: you had left Paris two days before and were in the country, and according to your concierge would return to Paris shortly but only to cross it to take another train. There was thus no possibility of seeing you. And now I have something urgent to request of you; the doctors having forced me at a few hours' notice to leave Paris, I should like to take advantage of what may well be the last journey to be granted me, to visit some monuments or sites which you consider particularly striking.

I am for the moment in Cabourg, at the Grand Hotel. But given the extreme uncertainty to which my plans are reduced by my state of health (which may force me to return to Paris from one moment to the next, as it can allow me a longer stay), it would perhaps be just as well for you to reply to me – if you will be so kind as to reply – at 102 boulevard Haussmann, Paris. However, I shall leave my address at the Grand Hotel. Answer me whenever you like. If by any chance you lost both these addresses, you could write to me at my old address, 45 rue de Courcelles, from whence the letter would be forwarded to me with a little delay. Here then are my questions:

1. What are the most interesting things to see in Normandy? I'm

not thinking exclusively in terms of cathedrals or even monuments. Indeed a town that had remained untouched (as Semur,[1] glimpsed from the train some years ago, appeared to me delightfully to have done), or some old port or whatever that you knew, would provide more food for my imagination than a cathedral that wasn't very special – or really sublime – and since we're talking of old towns, may I leave Normandy for a second and ask you if Fougères, Vitré, Saint-Malo, Guérande are out of the ordinary and worth making a journey to? Are there equiv-alents (or better) elsewhere, in other regions? To return to Normandy, are there more interesting monuments (and if I say that I don't want to restrict myself to the churches resuscitated by your pen, that doesn't mean that, restored to their rightful glory by the mere fact of being picked out by you, they won't enchant me) at Lisieux, or Falaise, or Vire, or Valognes, or Coutances, or Saint-Lô, or perhaps in the less well-known places where the surprise of finding them might make them all the more touching, especially if the landscape conspires with them a little? I don't in the least insist on these particular names, and I would joyfully welcome those you give me.

2. If I feel strong enough to undertake a journey to Brittany, have you any special sites that you could recommend to me, beauties either of art, or nature, or history, or legend? And incidentally are Roscoff and Paimpol of outstanding interest and could one of them stand in for the other or both be replaced by a third? Finally, if I couldn't either stay in Cabourg or go to Brittany, could you point out anything nearer Paris, in Seine-et-Oise for instance, as striking to the imagination as in Normandy or Brittany? There are automobile taximeters in Cabourg[2] which, if my present sufferings calm down a little, would enable me to explore quite far afield in Normandy. In general, without entirely sharing current opinion on the matter, I find that unless they reconstruct a very particular way of life, restored monuments don't make the same impression on me as stones that have been dead since the twelfth century, for example, and have remained as Queen Matilda saw them. (I'm thinking of Caen, where I have been and where I can return if you advise me to, and where a few stimulating and guiding words from you would have been invaluable – I should have enjoyed it a hundred times more.)

Is Mont Saint-Michel much restored or is it one of the finest things in France? I've seen it, but only when I was very small, otherwise the difficulty of access would make me give it a miss.

Forgive me, dear Sir, I'm so tired that I'm asking you pell-mell a whole string of brash and boring questions. It's an ignorant layman who is soliciting your advice. But you know that it is also a devotee of your work.

Please accept, dear Sir, my apologies and my grateful admiration.

Marcel Proust

1. Semur-en-Auxois (Yonne), a medieval town in northern Burgundy. See vol. I.

2. During the season in Cabourg, chauffeur-driven motor-cars, a rare luxury, were for hire from Taximètres Unic de Monaco, a business enterprise run by Proust's old friend Jacques Bizet. Proust had three regular drivers of whom one, the young Monégasque Alfred Agostinelli, drove him around Normandy 'like a flying cannon-ball' (see letter 248), shining the headlamps on church façades at night. Proust wrote about this experience for *Le Figaro* (see letter 256) and remembered it for his account of the second visit to Balbec in *RTP*, when the narrator impresses Albertine by hiring a chauffeur-driven car for their excursions. He continued to use the same 'Unic' drivers in Paris (see letter 253) and in Normandy again in 1908. It was not until 1913 that he began the relationship with Agostinelli often identified with that of the narrator and Albertine in *The Captive*; in 1914 Agostinelli died, aged twenty-eight, in an air crash.

243   *To Georges de Lauris*

[Cabourg, shortly before mid-August 1907]

My dear Georges,

Thank you from the bottom of my heart for the sweet token of friendship you gave me by lending me that photograph.[1] Forgive my importunate request, but I needed to study it alone, and in view of the feelings it aroused in me, which surpassed all my expectations, I did not regret having prevented you from taking it home with you. I had no difficulty in recognizing that pure brow and that exquisite face in which I could easily incorporate all the dreams I had built up around your mother. And in comparing those eyes watching over your welfare, those vigilant eyes, and that resolute expression in the lower half of the face of unswerving gentleness, of resignation to suffering and of utter devotion, with the face of your father whom I was delighted to meet at Houlgate, I easily recognized the genealogy, and one after another all the 'patents' of your intellectual, moral and physical nobility. I shall not forget your kindness in letting me see your father and, even more, your mother's portrait.

With my most affectionate thanks

Marcel Proust

1. A favourite photograph of his dead mother, the Marquise de Lauris, which Proust had insisted on borrowing when he and Lauris had stopped at Houlgate on their 'unforgettable outing' to see the Marquis de Lauris whom Proust had also never met (letter 248). Cf. letter 82, n. 11 in which he also solicits a photograph.

244 *To Emmanuel Bibesco*

Grand Hôtel, Cabourg
[Shortly before mid-August 1907]

Dear friend,

Thank you most warmly for your charming card; I was so pleased that you liked the little article.[1] It was very good of you to tell me so. I've been in Cabourg (Grand Hotel) for a week, from where I'm going to look at churches all over Normandy. If you have any landscapes or monuments to recommend to me you would make me very happy. But hurry, because I shall soon be leaving for Brittany. It's true that there too you might suggest a few places that you found really exciting. You'd be amazed to see me on the road every day. But it won't last. I haven't seen only old stones. Guiche, who is my neighbour thanks to the motor-car, introduced me to two ladies, Baroness d'Erlanger and Mlle de St-Sauveur,[2] who greatly disturbed me, so much so that it wasn't until after I'd left them that I realized that I could have mentioned your name. I must have made an incredibly stupid impression on them, but I found them charming and I keep thinking of them. I hope Antoine is well, calm, industrious and happy. Everything that I'm not! I have never been so agitated, so sterile, so miserable. But he has so much future ahead of him that he absolutely must be well and contented, and work. I think I could do him a lot of good, because I know what's good without having the strength to do it, and besides, for me it's no longer of any importance.

Dear friend, I'm very fond of you both and send you my best thoughts – as Victor Hugo put it much better, 'my thought, the best thing I have in me'.[3]

Marcel Proust

1. 'Une Grand'mère': see letter 236, n. 2.
2. Baronne Emile d'Erlanger, *née* Catherine de Rochegude, and Mlle Yvonne de Saint-Sauveur, later Comtesse Léonce de La Celle.
3. 'Ma pensée, la chose la meilleure que j'aie en moi.'

245 *To Emile Mâle*

[Cabourg, soon after mid-August 1907]

Dear Sir,

My deepest thanks for your charming and admirable letter. I have been to Caen, to Bayeux, to Balleroy and to Dives. I shall go to Jumièges if it isn't too tiring, to Lisieux, to Saint-Georges-de-Boscherville, to Falaise, to Saint-Wandrille and, if I can discover exactly where it is, to Cêrisy-la-Forêt. You tell me not to go to the Mont Saint-Michel this time, but you also say that it's a marvellous thing. And that troubles me; if I decide to go there after all, I shall ask you for some more precise guidelines.

I'm so dazed by living on my feet for a change that I don't enjoy anything. I've spent such a melancholy year. I've written a few articles for the *Figaro* which I would have sent you if I had felt bold enough.

Meeting some academicians here, I told them that you ought to belong to the Académie Française; and also to the Académie des Inscriptions et Belles-Lettres.[1] I know it means very little, less than nothing; but in a society that ought to be alienated as little as possible from the hierarchy of minds, it's inconceivable that a leading mind such as yours shouldn't enjoy a leading position.

I can't write to you any more amid the deafening and melancholy tumult of this appalling and sumptuous hotel, but I wanted to express to you my respectful and grateful admiration.

Marcel Proust

I was charmed by the oriental figures in Bayeux cathedral (in the Romanesque part of the nave), but I can't understand them, I don't know what they are. And on the outside it seemed to me that there were some admirable statues on the pinnacles, but difficult to see and identify. I'm looking for an old, untouched, Balzacian provincial town, but haven't found a complete one. I shall try Valognes, but it's very far.

1. Mâle, who became professor of the history of art at the Sorbonne in 1912, was elected to the Académie des Inscriptions in 1918 and to the Académie Française in 1927.

246 *To Antoine Bibesco*

Grand Hôtel, Cabourg
[August 1907]

Thank you, flatterer.[1] I thought it was more like Emmanuel Arène, but still it's very nice of you and I'm enormously fond of you Antoine and miss you a lot. Tell Emmanuel that I'm conscientiously looking at everything he was kind enough to recommend to me and countless other things, that I'm very ill, and that I'm very, very grateful to him.

Fond regards to you both

Marcel

1. Bibesco had written to congratulate him on his article on Mme de Noailles (see letter 213).

247 *To Reynaldo Hahn*

Grand Hôtel, Cabourg
[Second half of August 1907]

My dear little Bunchtnibuls,

It isn't because I have a favour to ask that I'm whriting to you. I was going to hwrite you all sorts of nicenesses but a note from Ulrich telling me he was leaving for Orcival has prevented me from sending him round to my flat as I intended and I'd be very grateful if you could send Léon.[1] This is what for: I'd like him to pick out from the big collected Ruskin (you know the splendid edition of the complete works of which I've occasionally shown you some of the volumes[2]) Volumes VI, VIII, XII and XIV. Are they in the little sitting-room, or in my bedroom, I don't know, in both probably, perhaps even in the cupboards in both. They are red unless they're still in their grey-green paper covers. Some of them are sub-divided, for instance a volume might be (I'm making this up) Volume V, part II of *Modern Painters.* But the numbers I want (VI, VIII, XII, XIV) refer not to such-and-such a work, but to Ruskin's work as a whole. If he can't find Volume VIII, he can take instead from among my little green Ruskins *The Seven Lamps of Architecture* (it's the same work). Finally, I'd like him to take as well (probably in my bedroom, on my chest of drawers or in a revolving book-case) the small stitched volume entitled *Carpaccio* from the collection of painters.

Could he add to the above-mentioned volumes the volume on Turner[3] in the big collected Ruskin (I don't know the number but the title is *Turner*), and would he be kind enough to send the lot to me *as soon as possible* by parcel post. Finally, if it's not asking too much, would he be kind enough to call at Neal's or Galignani's or the third one whose name I've forgotten (Brentano's)[4] and ask the price of Turner's *The Rivers of France* (preferably second-hand) and especially, if it's possible (but I'm ashamed to ask him so many things), copy down the titles of each study in the album (for example, *The Seine at Rouen* etc. – there's a table).

Buninuls, I received the nicens little notes and yet I thought and thought. But I live such an artificial (from the point of view of health) and busy life that I don't know what I think and feel and can't even write a letter. And yet I have all sorts of nicenesses to write to you but all this has already put me into a sweat and since the hotel is full of draughts I try to avoid perspiring so as to avoid catching a cold.

The Englishman thought to be etc. was Edwards.[5] If you are a friend of Defreyn,[6] can you give me some message for him which would provide an excuse for speaking to him. But if it's the slightest bit complicated don't give it another thought as I don't give a damn. I could find all sorts of ways of getting myself introduced to him. But naturally I don't want to.

The hotel looks like a stage set (not squinty[7] considering what I mean) on which are assembled as in the third act of a farce:

Edwards

Lantelme, his mistress[8]

Mme Edwards[9] (Natanson), his latest wife, separated from him

Natanson,[10] first husband of Mme Edwards

Dr Charcot,[11] first husband of the last Mme Edwards (the fourth of the species because before the present one he had already married two Americans, a Frenchwoman and a Greek).

Yesterday evening there was a rumour circulating that Mme Edwards (Natanson) had killed Edwards (the Englishman, who is in fact a Turk) but there was nothing in it, nothing had happened at all.

                         Bonjours, Bonsjours
                              Munchtnibuls

What's wrong with your Mama?[12]

Mme Lemaire writes to say that I could tell her some more interesting things than complaining about the noise in the hotel which is not very exciting it's true.

1. Hahn's manservant. Ulrich had gone to Auvergne.

2. Cf. letter 77, n. 3.

3. *The Harbours of England.*

4. The names of Paris bookshops. The Turner volumes were for Mme de Clermont-Tonnerre (see letter 251).

5. Alfred Edwards. See letter 241, n. 3.

6. Henry Defreyn, Belgian operetta-singer.

7. *Louchon*: see letter 79, n. 2.

8. Geneviève or Ginette Lantelme, a beautiful, reputedly lesbian young actress, had just appeared in a play by Francis de Croisset. In 1911, on a Rhine cruise with Edwards, she drowned (strangled and thrown overboard by him, according to scurrilous rumour). Possibly the model for Léa in *RTP*.

9. *Née* Misia Godebska (1872–1950), intimate friend of Mallarmé, Bonnard and Vuillard. First married, aged fifteen, to Thadée Natanson (see note 10), then to Alfred Edwards. After separating from Edwards, she became a well-known patron, closely involved with Stravinsky, Diaghilev and Picasso. In 1914, she married the Catalan painter Jose-Maria Sert (1876–1945). The Serts both appear in *RTP* in connection with the Ballets Russes, she as the striking Princess Yourbeletieff, he *in propria persona*.

10. The banker Thadée Natanson (1868–1951) was co-founder with his brothers of *La Revue blanche*.

11. Dr Jean Charcot (1867–1936), specialist in nervous diseases like his more famous father, and polar explorer. Formerly married to Léon Daudet's divorced wife, *née* Jeanne Hugo, he was at Cabourg with his new bride, *née* Elisabeth-Mariette Cléry.

12. Mme Carlos Hahn, *née* Elena de Echenagucia (1831–1912), had a fever.

248 *To Georges de Lauris*

[Cabourg] 27 August [1907]

My dearest Georges,

You have written me some delightful letters and I haven't thanked you because the life of a flying cannon-ball[1] that I lead here without even stopping for a second to drink some coffee literally prevents me from writing a single word, a sort of trembling like that of the engine continues to purr and vibrate in me when I've got out of the car and won't let my hand come to rest and obey me. But if I could precipitate, condense and send you all the fond and grateful thoughts of you which constitute the mental climate in which I live, it would make up a whole tender and passionate volume, and if that condensation of you which saturates my inner atmosphere were to be precipitated, I would see the air explode into a thousand little Georges de Laurises like the teeming hosts of angels in Giotto's paintings winging their way after the Virgin or Christ.

I must tell you about the multiplication and diversification of my relations with the Brès,[2] who were really delightfully kind and intelligent and who are touchingly devoted to you. But I have so many things to tell you! Excursions, 'pleasures', yet I've never been so miserable. Sorrow is decidedly not meant to be stirred up; a great deal of stillness is needed to enable it to settle and become a little more serene and limpid again. The fact that I no longer eat anything, or practically anything, luckily induces a mental emptiness which prevents me from being aware of very much.

I much preferred Bayeux cathedral to the churches of Caen. I'm too tired to explain why, but I can tell you that the interior work which impressed you has a lot to do with it. I don't know whether I shall go to Brittany; I still have things to see near here for some days. Sem and Helleu[3] who are *extremely intelligent* have both tried (separately) to persuade me that primitive art was worthless, that nobody knew how to paint before Rubens etc., and one of them took me to see the Boucher tapestries at Balleroy (beyond Bayeux). The château[4] was built by Mignard according to the guide, and decorated by Mansard. The tapestries by Leboucher were brought to the château in 1622 when it was built. I don't need to tell you that all this is wrong, that Mansard is the architect, Mignard the painter, and 'Leboucher', who is none other than Boucher, didn't exist when the place was built. One can also admire in the drawing-room a number of hunting pictures by M. le Comte, the father of M. le Marquis, the present owner. These are not the least diverting ornaments of this mansion. Unfortunately M. le Marquis seems to have inherited M. le Comte's taste and, if he doesn't do hunting scenes, he has framed the Leboucher tapestries in red damask which leads one to think that he could try his hand at synergetic painting with as much success as M. le Comte. There is yet a third kind of painting in the château, namely that with which the walls of the dining-room have been repainted several times, the tapestries receiving each time a little coating of splashes, so that the borders are embellished with a delicate and vibrant stippling that could not, without injustice, be attributed to Leboucher.

I see a great deal of the Guiches, Louisa and the Strauses, a threefold intimacy of which we laid the foundations on that unforgettable outing with you.[5] Since then, Léonce de Joncières has been entirely excluded.[6] Croisset I don't see at all; he is, I suppose, completely absorbed by his play, his mistress, Edwards's mistress, and countless complications.[7]

You know that I found your father delightful, the most noble and charming, the most handsome and imposing looks, the most witty and artistic talk, the most lively and literary choice of expressions and turns of speech. But I was so overcome with emotion on seeing him while thinking of her whom I shall never see that I was unable to appreciate as I should have liked that conversation which would have been so enjoyable had your mother been there. Thus do we continually re-create in our thoughts the family circle such as it would have been, as it still is for us, without the intervention of death. We really do have with us those whom we love. But to think that they cannot realize that we have them, that they cannot feel and enjoy it, that Mama cannot see me up and about, that you will achieve success without your mother knowing of it, all this is more than enough to confirm a longing for death. At least your mother could foresee your success. Mine died believing that I would never get up again.

If you are with Bertrand,[8] give him my truly fondest regards. Above the ruins of our intimacy there often hovers what Chateaubriand would have called the Genius of Friendship and would have depicted so well in a prose at once poetic and funereal. And I sometimes wonder whether I haven't passed over the only friend I ought to have had, whose friendship could have been fruitful for both of us. I don't know if his sister the Comtesse de Montebello[9] is here, I think not, I would have called there if I hadn't been afraid of meeting his mother (Mme de Fénelon) and Mme de Ritter, which would have been presumptuous since I don't know them. On the other hand, if I went to Avranches I might perhaps ask him for an introduction to the aged relatives he has there. But it's unlikely. As regards his sister there's no need to tell him even what I've told you because in any case I wouldn't have the time to go and see her.

Good-bye, dear Georges, and thank you again. With fondest regards
Marcel

1. i.e. being driven around the countryside by Agostinelli (letter 242, n. 2).

2. Dr Pierre Brès and his wife lived in the same Paris apartment building as Lauris's mother.

3. The caricaturist, Sem (see letter 223, n. 2) and the painter Paul Helleu (see letter 19, n. 6); painter of flowers and seascapes among other subjects, he was an important model for Elstir in *RTP*.

4. Proust seems to have remembered Balleroy when he came to describe the Château de Guermantes: Saint-Loup tells the narrator that the tapestries he had supposed to be medieval were in fact by Boucher, bought in the nineteenth century by an art-loving Guermantes and hung 'with a number of mediocre sporting pictures which he himself had painted in a hideous drawing-room upholstered in red plush' (*RTP*, II, 9).

5. To Trouville, via Houlgate. Cf. letter 243, n. 1.

6. Cf. letter 241, n. 7.

7. Cf. letter 247.

8. Bertrand de Fénelon.

9. Mme Louis de Montebello, *née* Marie-Louise de Salignac-Fénelon.

249 *To Reynaldo Hahn*

Grand Hôtel
Cabourg
[1 or 2 September 1907]

My dear little Birnechnibus,[1]

I still have the trembling which prevents me from writing. But how many times a day, how many times a night, does my heart melt at the thought of Buninuls, how many times do I bury myself in him, and indeed always, whatever else I'm thinking of, his dear little muninulserie and face loom up and fill my horizon. I've received envelope from Muninils and inside nothing but a letter *from me*, and an *admirable* one from Lady de Grey[2] who has become more and more of a philosopher. And it hasn't taken away her colour. She has remained an artist nonetheless. And yet she knows that society is sterilizing. What a genius, what a woman! Whistler used to say: 'I find her too handsome a man' and I've never understood what he meant. But I admire her enormously now that I've read her letters. If I had written them I'd die in peace. Yesterday I went to see Vuillard[3] who was wearing a blue workman's smock (slightly too soft a blue, like a worker by Augusta Holmes in a Song of the Guilds.[4]) He repeats with intensity: 'a chap like Giotto, d'you know, or then again a chap like Titian, d'you know, knew just as much as Monet, d'you know, a chap like Raphael, d'you know etc.' But he's no ordinary man, even if he does say 'chap' every twenty seconds.[5]

Do you know the Marchetti Orchestra which is raging here? And can you give me 'Un baiser n'est pas un péché'[6] which they played the other day and which is very beautiful.

I was offended by Léon's letter of introduction to Defreyn (lie).[7] I wouldn't have talked to him about Buninuls. I shall probably ask him to come and sing for the Duc and Duchesse de Guiche who are dining with me on Wednesday and would be happy to hear him. But I have such a sore throat that I think nothing will happen. In any case he will forever remain unaware that I know Buninuls, that Buninuls has heard him sing, etc.

Have seen Tristan Bernard here, charming (lives with Vuillard etc. etc.) and Mme de Maupeou[8] who sings pretty things by the pony[9] which moved me even in her voice, divine things wrapped in the bright blue paper generally associated with bars of Lombard chocolate. Joncières is the most idiotic creature I've ever met. Much talk of Irnibuls with la Soubreuse, la Clermont-Tonnerre. (I amused myself by exaggerating your virtues to such an extent when talking to her that I felt by the end that she had an impression of the *supernatural*. Only don't ever express a vulgar sentiment or a mediocre thought in front of her. My reputation for perspicacity depends on it.) Louisa and Gangnat speak fondly of you.[10] What made you think Mme Ralli is here.[11] Could it be true? Come back quickly to my heart, my Buninuls, come and come and nicenesses. I have a horrifying suspicion that Nicolas drinks.[12] Good-bye and one more kissikins.

Where did you say Clarita[13] lived? If I can get up again I'd like to go and see her.

[Unsigned[14]]

1. Much of this intimate letter is in their private idiom.

2. Lady de Grey (*née* Constance Gladys Herbert: 1859–1917), sister of the 14th Earl of Pembroke, married (1) the 4th Earl Lonsdale, (2) Earl Grey, later Marquess of Ripon.

3. The painter Edouard Vuillard (cf. letter 29, n. 3).

4. Augusta Holmes (1847–1903), composer of songs. Daughter of an Irish officer stationed in Paris, she was a pupil of César Franck.

5. Cf. Elstir in *RTP*: 'The chap who carved that façade, you can take my word for it, was every bit as good as the men you admire most' (I, 900).

6. 'A Kiss is not a Sin', presumably a popular song of the time.

7. See letter 247, n. 6.

8. Comtesse René de Maupeou, *née* Koechlin, an amateur singer who had given a charity concert at the Grand Hotel.

9. i.e. Reynaldo. See letters 18 and 64.

10. Louisa de Mornand was with her new protector, Robert Gangnat (1867?–1910), lawyer, head of the Society of Playwrights; Proust encouraged this relationship despite her acceptance of an allowance from Albufera. Cf. letter 96.

11. Mme Hélène Ralli, yet another of Edwards's former wives (cf. letters 241, 247).

12. Nicolas Cottin, former servant to the Prousts whom Mme Proust had suspected of drinking but whom Proust had re-engaged nevertheless, together with his wife Céline.

13. One of Reynaldo's sisters, Mme Miguel Seminario, *née* Clarita Hahn.

14. Instead of a signature Proust sketched a profile with a long nose.

250   *To Robert de Montesquiou*

Grand Hotel
Cabourg
[Saturday, 7? September 1907]

Dear Sir,

Your magnificent letter, as rich in delicate ornamentation as the most spacious formal garden and as prolific in its perfumes and its blooms, is among other things one of those which will remain most fondly engraved in that corner of my heart reserved for debts of gratitude, because of all the kind and admirable things you say in it about my mother.[1] It is also the one which, through its grave vistas, its dripping avenues stretching towards the invisible, offers the most wide-ranging idea of you, with infinite perspectives. I thank you for it with all my heart and will re-read it often.

I should have been interested to know whom you had in mind when you spoke of people who ought not to be admired and whom I compare to admirable people.[2] A simple word would have cleared up this misunderstanding; what I can tell you is that it must be a want of taste on my part, not of sincerity or seriousness. Perhaps by talking to you I might come to change my opinion on such and such a person if we re-read his works together and tested their worth. But without having an inkling of whom you mean, I can assure you that there is not ᴜne *adjective* of indulgence in anything I've written, at least for some years past, and of course if it's about an artist or a writer. On the contrary it often happens that the hazards of phraseology, or rather the contingencies of one's thought and reasoning, lead me rather to belittle what I most admire, as has happened to me with Maeterlinck for example, but I feel that the fervour and seriousness of the tone are worth more than compliments. A propos of Maeterlinck,[3] I ought to have remarked, if I hadn't been prevented by lack of space, that many of the most splendid presages of his last works often used to figure in your augural conversation. For instance 'It is the most touching miracle of the vegetable kingdom that I know', which I sometimes heard you say a number of years ago. These coincidences (in the geometrical not the adventitious sense), these superimpositions or anticipations of thought, move me greatly.

You were equally perspicacious in your letter in what you said about the physical reserves which I rightly build up in bed. Since I've been here, and for the first time in many years, I've been leading a relatively normal life, up dressed and out almost daily. And I have the sensation of no longer having either heart or mind, only a physical heart that is more exhausted, palpitating and painful every day. But it's an activity that tends to become a habit as bed was. And I feel that I'm using up and wasting what remains of my strength. I find it so tiring and difficult to write, especially these last few days when I've had a constant fever, that I seem to have reached the ultimate end of tiredness beyond which I can go no further, and I haven't yet mentioned your delightful article on *Le Balzac des Enfants*,[4] the end of which is subtly and charmingly Balzacian. I shall try to find a less unfavourable moment to congratulate you on it at greater length. •

I've also heard of an astonishing article of yours on the subject of hats and other things which I've had someone look for in *La Vie parisienne*,[5] a fairly considerable flask, I'm told, of the most delicious ambrosia with which I shall asperge myself at leisure.

Your most grateful admirer
Marcel Proust

1. The anniversary of Mme Proust's death fell in September.

2. See letter 236 dated (soon after) 23 July to which Proust is replying.

3. See letter 28, n. 2.

4. In an article for *Le Figaro* headed 'The Children's Balzac' Montesquiou had expressed his satisfaction that, for once, a statue was to be raised to a deserving person, namely the children's writer Mme de Ségur (1799–1874).

5. In August *La Vie parisienne* published 'Monsieur Monde et Madame Mondanité' (Mr and Mrs Worldly) in which Montesquiou comments satirically on the new fashion for enormous hats.

251   *To Madame de Clermont-Tonnerre*

102 boulevard Haussmann
[Early October 1907]

Madame,

After a night disturbed by all the sabbaths of hell – the hell of Evreux – which the paradise, glimpsed and then lost, of Glisolles[1] made even crueller by contrast, I suddenly decided to return to Paris, so that the taxi, which was taking me back, was unable to deliver the Turners[2] to you. But before leaving Evreux I had a parcel made of them which will, I hope, have reached you as quickly. And I included with them a few volumes of the new complete edition of Ruskin which is in course of publication and in which I think there must be fairly numerous reproductions of Turner. Moreover, as I possess all the volumes already published, I could send you others if you were interested. The volume that will include *The Bible of Amiens* hasn't yet appeared. I shall be very curious to see it and I regret that it didn't appear before the publication of my translation; it would have spared me a great deal of research.

Thank you once more most warmly for the friendly welcome which you and Monsieur de Clermont-Tonnerre were so kind as to give me in your delightful residence. Its bright Norwegian panelling and its old French pictures will remain in the forefront of my memories of the charms of Normandy, together with Gothic churches and Renaissance mansions. I was reluctant to superimpose inferior visions of 'battlefields', however splendid, on the charm, at once primitive and refined, of that forester's or trout-fisherman's lodge, which is also the studio of two artists. And I only went to see, beneath the indifference and opacity of a rain-swept sky from which (by a miracle that might well have been represented in the cathedral among so many others less interesting) they contrived to filch an abundance of light, the stained glass windows of Evreux.

Please accept, Madame, my respectful gratitude and thank Monsieur de Clermont-Tonnerre once more for so charitably guiding my tottering, caffeine-weakened steps down the nocturnal staircase.

Marcel Proust

(In feverish haste, of which the disordered ideas, the style and the handwriting of this letter will provide sufficiently evident proof to serve as an excuse.)

1. The Clermont-Tonnerres' eighteenth-century manor house. Invited to stay at Glisolles, Proust spent four dispiriting days in Evreux instead (see letter 253), possibly in search of his 'provincial Balzacian town' (see p.s. to letter 245).

2. See letter 247, n. 4.

252 *To Robert de Billy*

[Early October? 1907]

My dearest Robert,

I was most touched by your card. It was sent on to me from Cabourg to Evreux and I'm answering it from Paris where I've just arrived exhausted. But I've been asked a favour for which I feel I may address myself to you and I'm only writing you a note as I'm in a state of extraordinary fatigue. Rather than explain to you what it's about I'm sending you the enclosed note[1] which I'd like you to communicate to your father-in-law, who *is* director of the bank, is he not?

My dear Robert, I've spent a summer which would have astonished you. I went to Cabourg where the air was fairly beneficial to me (after having been very ill in Paris for a long time after you left) and where thanks to medicaments which were unfortunately pernicious I was able to get up and dress every day, go for long drives every day in a (closed) motor-car, visit churches (without any pleasure), watch polo, gamble – and lose – at baccarat every evening, etc. – all this among the commonest set of people in the world. When the hotel closed I went to Evreux (by car) and finding it too suffocating I came back (still by car) to Paris where I'm suffocating even more and where I leave you, apologizing for being too tired to express at greater length my feelings for a friend whom I've thought of every day. But the caffeine I've been taking made my limbs tremble to such a degree that I could hardly write.

Your

Marcel

1. Auguste, formerly manservant to Proust's uncle, Louis Weil, had written to ask a favour on behalf of his daughter. Cf. letter 254.

253   *To Madame Straus*

102 boulevard Haussmann
Monday [Tuesday, 8 October 1907]

Madame,

I left Cabourg on the same day as you left Trouville,[1] but not at the same hour! Just before arriving at Evreux (where I spent four or five days) we came down into a valley where the mist was visible from a distance and one sensed the coolness in the air. And from that moment to this (and for I don't know how long in the future) I haven't stopped suffocating and having incessant attacks. That is why, though I think of you more or less every minute of the day, I haven't written to you, I haven't had the heart to pick up a pen. Of course it wasn't the little valley that brought back my attacks; but from that moment I did nothing but make my way back to Paris and at Evreux I was already very ill.

I saw there 'by candlelight' a bishop's palace which is not very beautiful inside, at nightfall a church called Saint-Taurin which struck me as very pretty (Romanesque and Gothic if I'm not mistaken, since you now know the styles) with some rather curious piscinas[2] and fine stained-glass windows. Then a cathedral which you've doubtless seen and which is a mixture of every period, with beautiful windows which contrived to be luminescent even at the twilight hour when I saw them and in dull weather under an overcast sky. From all the dreariness of a day which from the first thing in the morning had resembled the night to which it was about to give way, they managed to steal jewels of light, a purple that sparkled and sapphires full of fire – it's incredible.[3]

I went to Conches, just outside Evreux, to see a church which still has all its sixteenth-century windows, many of them by a pupil of Dürer.[4] It was like a pretty little Renaissance German Bible with illustrations in colour. The windows have captions written underneath in Gothic characters. But the windows of that period don't interest me much, they're too much like *pictures* on glass.

Since I've been back in Paris I haven't left my bed. However, yesterday I went out with Jossien,[5] poor Agostinelli having been obliged to leave for Monte Carlo because of his brother's health. I think there are some grave defects in the way the 'Unic' is run. I shall point them out to Jacques as I'm certain that he could do ten times more business than he does. I myself who had all the information direct from him, and the firm resolve to hire one of his cars, spent more than a day trying to get through to them on the telephone. Some friends of mine to whom I recommended them didn't have the same patience and leisure and gave up. Their garage isn't in any directory. If one telephones Georges Richard one is told that the Taximètres de Monaco are not known there. And finally if through 'personal connections' one manages to get through to Jacques's garage (it's impossible otherwise) one is told that there aren't any cars, etc. The consequence is that when I hired Jossien yesterday it was the first outing he'd made since leaving Cabourg (I had had Agostinelli twice with extreme difficulty). Unfortunately I'm too ill here to be a very faithful customer and anyhow it's ten times too dear for Paris where so many coachmen now do taxis for practically nothing. Anyhow I'll tell all this to Jacques whom I've asked to come and see me.

I saw in the newspapers that Edwards doesn't keep his friends any longer than his wives.[6]

While I was at Evreux I went one evening to see M. and Mme de Clermont-Tonnerre at Glisolles (which is close by) in a very attractive spot. They were supposed to show me some very beautiful things in the neighbourhood, but I was so fed up with Evreux that I left the next morning and so did none of those excursions, nor did I visit Claude Monet's garden at Giverny, near the beautiful bend in the river which is lucky enough to see you through its mist in your drawing-room.[7] I very much want to give Monsieur Straus, if I can find it in a nice edition, a delightful book which I've been lent and which I had in fact read years ago but which is well worth re-reading and studying: the *Dictionary of Architecture* by Viollet-le-Duc. It's a pity that Viollet-le-Duc has spoiled France by restoring, scientifically but soullessly, so many churches whose ruins would be more moving than their archaeological patching together with new stones which don't speak to us, and mouldings which are identical with the original but have kept nothing of it. Nevertheless he did have a genius for architecture and that book is admirable.

Madame, it's so difficult to write to you like this from my bed, with an elbow which refuses to go on putting up with the pain of resting on the table, that I shall say good-bye to you here and now without having tried to tell you how my affection and my admiration for you increased even more during that stay in Trouville, and those blessed weeks in the course of which I saw you again and of which I shall always retain a delightful and tender memory, intact against a background of flowers. Please tell Monsieur Straus that if I've been so outrageous as not to have yet written to either of you, it's neither ingratitude nor forgetfulness, but impotence and pain. Please don't reply; I know – and will try to find out more precisely if I can see Jacques or inquire through Monsieur Straus at the rue de Miromesnil[8] – that you are resting in Switzerland and that writing letters is painful for you. May it not be too painful for you to receive one!

Your respectful and infinitely grateful friend
Marcel Proust

1. Where the Strauses rented the Clos des Mûriers (see letter 148, n. 7).

2. Although the church of Saint-Taurin has none, there are four sixteenth-century piscinas in the cathedral.

3. Cf. the description of Combray's stained-glass windows in *Swann's Way*, 'never so sparkling as on days when the sun scarcely shone, so that if it was dull outside you could be sure it would be fine inside the church' (*RTP*, I, 64).

4. Sainte-Foy at Conches-en-Ouche has windows by Dürer's pupil Aldegrever.

5. Jossien (whose name inspired that of Jupien in *RTP*) was another of Proust's regular chauffeurs from Jacques Bizet's Taximètres Unic for which the firm of Georges Richard (mentioned below) supplied motor-cars. Cf. letter 242, n. 2.

6. *Le Figaro* had just reported a libel case brought by Croisset against Alfred Edwards which resulted in a duel; both men were wounded.

7. An allusion to Monet's *An Arm of the Seine near Giverny, at Dawn* which hung in Mme Straus's Paris drawing-room.

8. 104 rue de Miromesnil was the Strauses' Paris address.

254   *To Robert de Billy*

[Around 9 November 1907]

My dearest Robert,

Although I've been meaning to write to you for the last century, I'm so exhausted that you'll forgive me for sending you just a note. Thank you a thousand times for agreeing to recommend my protégée[1] to your father-in-law. I've learnt that in the meantime Robert[2] had taken action through the Rothschild Bank. It and your father-in-law are two powerful gods and I feel that the effect of the collaboration will be electrifying.

I've been told something very naughty – or rather very graceful – concerning two ladies who I'm sure are members of your intimate circle, Mmes D. and de N. (two sisters-in-law). Did you know about it? It may of course be entirely untrue. What do you think about this homosexuality trial?[3] I think they've hit out rather at random, although it's absolutely true about some of them, notably the Prince, but some of the details are very comic.

Will you remember me to M. Paléologue.[4]

Your affectionate and devoted

Marcel Proust

I'm distressed because the boulevard Haussmann house has just been sold, *very badly*. I have a week to make a higher bid but since the purchaser is my aunt I don't dare.[5] If you had been here I would have asked your advice.

1. Cf. letter 252, n. 1.

2. His brother, Dr Robert Proust.

3. In October a German political and sexual scandal had broken involving Kaiser Wilhelm II's intimate entourage, known to their enemies as the Camarilla, which included two elderly francophiles and pacifists, Prince Phillipp von Eulenburg and General Cuno von Moltke. Accused by the right-wing journalist Maximilian Harden of homosexuality and of having a pernicious moral and political influence on the Emperor, they sued for libel, but in a farcical twist their evidence was turned against them and the editor acquitted.

4. See letter 142, n. 5.

5. The house was jointly owned by Proust, his brother Robert, and their aunt Emilie Weil (whose husband Georges had just died). The three shares were put up for auction in early November and Mme Weil acquired all three, Proust having missed the sale (see following letter, to Montesquiou).

255   *To Robert de Montesquiou*

[Shortly after 8 November 1907]

Dear Sir,

I was touched as well as sorry that your charming thought forestalled the expression of my gratitude and my admiration for Friday's soirée. I shall never be able to forget it; perhaps some day I shall attempt to describe it, and all that splendid recompense I had in the evening (having been unable to bring 'my little folding stool' to the Femina Theatre in the afternoon[1]) in the house[2] which is entirely pervaded with your presence and an absence that is also a presence,[3] and in which the odour of dead leaves from the courtyard and the music of dead lyres from the Court mingled with the scent of massed orange blossom.

I'm very unwell this evening and cannot write to you at length, not having the admirable supremacy which you have over your body, which reminds one at once of Pascal and of Sarah Bernhardt. The sale of the house in which you were kind enough to take an interest went very badly and has halved my income.[4] Perhaps I should have gone to my lawyer rather than to you that evening, to the Cabinet of Ruses rather than the Pavilion of the Muses! I don't even dare say that I have no regrets, for fear of being disdainfully taxed by you with that 'disdain for dibs' with which you reproached the 'marvellous deceased'. But if she hadn't had that disdain, would she have been 'unique among the living'?[5]

Your grateful admirer
Marcel Proust

1. A lecture by Montesquiou on Versailles.

2. Montesquiou's house, Le Pavillon des Muses, in Neuilly.

3. An allusion to the late Gabriel Yturri.

4. On the contrary, he was now rich, thanks to his inheritance from both Weil and Proust estates. The fall in his income from the sale of his share in 102 boulevard Haussmann must have been compensated by the rise in his capital.

5. Montesquiou's dedication of *Les Chauves-souris* (see letter 221, n. 2) to the late Marquise de Casa-Fuerte read: 'A cette merveilleuse Morte / Qui fut Une, entre les vivants, / Mon pieux souvenir apporte / Ces Ex-Voto fiers et fervents.'

256   *To Madame Straus*

Thursday evening [28 November 1907]

Dear Madame,

I can't tell you how touched I was by your charming little note. I'm not used to being so appreciatively read and I'm immensely grateful for everything you were kind enough to say. Incidentally the principal author of the piece was Jacques, thanks to whom I was able to write it.[1] Can you imagine which was the prettiest letter among the few others I received: the one from Agostinelli to whom my manservant had sent a copy of the article.[2]

I've had such frightful attacks these past few days that it was impossible for me to thank you sooner and, of course, to go out. I seldom leave my bed, scarcely once a week, and when I do I never travel except by 'Unic'. Jacques's business success gives me great pleasure.

I have at last received from the Mercure the *Bible d'Amiens* which I ordered for you and I'm sending it to you with this letter together with my respectful and grateful regards.

Marcel Proust

My best wishes to Jacques. Tell him that I've paid my last bills direct to Jossien. Only the last, which is to be sent to me shortly, remains unpaid.

1. 'Impressions de route en automobile', published in *Le Figaro* on 19 November (which he would have been unable to write without Jacques Bizet's hired cars). Describing his summer drives chauffeured by Agostinelli (cf. letter 242, n. 2) it opens with his first excursion to Caen and the play of church spires on the horizon which, prompting a childhood memory, became, in *RTP*, the theme of the narrator's essay on the Martinville steeples (I, 196–8).

2. Cf. the letter of congratulation 'in an illiterate hand and a charming style' which the narrator in *RTP* receives in identical circumstances (III, 604).

257　*To Robert de Montesquiou*

[Shortly before 4 December 1907]

Dear Sir,

A thousand thanks for your letter. I should have liked to write to tell you all the things I admired anew and as though for the first time in *Le Chef des odeurs suaves*.[1] I intended to go back to see you, having twice been unable to go out since my visit to the Pavilion,[2] but I had to go to my dentist who kept me for an unconscionably long time. I had slightly hoped to be able to come to the tea-party to which you so kindly invited me but I've once more caught a chill and I don't expect to have recovered between now and Wednesday. If I have, I would come with Reynaldo, as long as I don't have an attack.

What you are kind enough – in an access, I fear (and should I not rather rejoice) of benevolence rather than justice – to call my talent would on the contrary be no more than an aid in learning to efface myself in front of your work, to interpret respectfully what you might not have had time to explain, to preserve it intact in its entirety. But my health, whatever you may say, alas promises no longevity, my essential organs having now been affected. Besides, you have demonstrated only too well that there are cases where it's only too easy to die. In any case your splendid, irascible old age, which I try to imagine, will still not have exhausted its thunderbolts and will be more and more haloed with lights that are not only lightning flashes when I have long since been reunited with those who left me too soon. I should be lacking, therefore, in the first qualification of the 'depository' which is the probability of survival, since moreover, as you say, you created this character of the 'depository'. And it isn't surprising that, being too occupied – and better occupied – to produce or write theatre, but creating it in your life, you should have established new posts. But I persist in believing, ill-informed and even less perspicacious though I am, but precisely for that reason having a clearer, more detached view, that the two deposed incumbents did not deserve their fall.[3] And I cannot share your reasoning or the contempt for inheritances which I find indeed very silly (the contempt) but where you are concerned it seems to me that there are much better things to inherit than 'dibs'[4] and that 'the forfeiture of familiarity' as you sometimes say, quoting the Prince de Ligne,[5] is a much more tangible form of exclusion.

I sent you a hideous little plaster passion flower[6] which I hope you received.

Good-bye, dear Sir, and please accept my admiring and grateful respects.

Marcel Proust

1. A new edition of an earlier collection of poems (1894). People were quick to apply its title (derived from a character in Flaubert's *Salammbô*), meaning 'Commander of Delicate Odours', to Montesquiou himself.

2. Cf. letter 255, n.2.

3. No doubt the names of dedicatees in the first edition of *Le Chef des odeurs suaves*, several of which were missing from the new edition, including Prince Edmond de Polignac and the Marquise d'Eyragues (cf. letter 209).

4. Cf. letter 255.

5. Charles-Joseph Prince de Ligne (1735–1814), soldier and man of letters, in the service of the Austrian Emperor Joseph II, later at the court of Versailles. Mme de Staël published his *Lettres et Pensées* in 1809.

6. Proust's delicate tribute to a poem in the forthcoming book entitled *Passiflora*, Montesquiou's nickname for his invalid sister-in-law and cherished confidante, Comtesse Gontran de Montesquiou (*née* Pauline de Sénety), who died in 1887.

258   *To Georges de Lauris*

[December 1907]

My dearest Georges,

I miss you so much that I must be really exhausted to resist the desire to tell you so, but I've had the most fantastic attacks almost uninterruptedly ever since I saw you. I didn't come to fetch you on Sunday because I collapsed, suffocating and done in, at Reynaldo's where I stayed for a long time before going home. I drove his sister Madrazo[1] back to her place. What a museum! A really heavenly El Greco with tones as ravishing in their different style as a Vermeer and an unimpaired freshness under its incomparable glaze. A small Tiepolo, drawings by Raphael, compositions by Titian which he never painted but which were engraved in his lifetime, engravings and drawings by Goya of a delightful obscenity, a La Tour – in short a collection as interesting as it could be, each picture long coveted by the master of the house,[2] pursued from sale to sale – delectable fabrics, an eighteenth-century embroidered Italian canopy on which rearing horses are bridled with trappings of such a blue!

Dear Georges, I'm beginning to suffocate again; I simply wanted to tell you that I was thinking of you and wish I could relieve you of a little of your fatigue. Often in the morning, when I feel that it's cold, I think it unfair that I should be in a warm bed while you are obliged to be on the barrack square at a quarter past seven. But tell yourself, dear Georges, that I have other troubles and that all things considered it's still preferable to be doing one's twenty-eight days[3] than not to be in a fit state to do so.

Affectionately yours
Marcel

1. Madame Raymond de Madrazo, *née* Maria Hahn (1865–1948). She and her husband had a magnificent house at 32 rue de Beaujon.
2. Raymond de Madrazo (1841–1920), Spanish painter, like his father, his grandfather, and his son 'Coco' de Madrazo.
3. Lauris was doing his obligatory annual four-week stint in the army.

259   *To Madame Straus*

[Friday evening, 27 December 1907]

Madame,

It's because I still intended to go and see you only an hour later that I didn't write to you first, and then thank you.[1] I don't think more than a day or two will pass before I come to visit you. On that day I shall offer you my New Year wishes. But I feel you are unjust towards last year. It undid the bad effects of those which had exhausted you so much[2] and you've become a good ten years younger. And it doesn't seem to be a temporary improvement but a sustained reorientation in that direction. It seems to me that you are almost at the limit of good health and that you couldn't have more of it without forfeiting all sensibility and intellect. Perhaps you would do so without hesitation. But since one has no choice.

It's too sweet of you to want me to write.[3] It's a change from those people who take infinite pains to avoid talking to me about my articles and not to look as though they are doing it on purpose, because they're afraid of hurting my feelings by admitting that they find them idiotic, and because they attach such asinine importance to their own words and to *sincerity* that they don't want to pay me a compliment out of kindness.

Incidentally I'm astonished to see what even very intelligent people admire and approve of. For instance a speech of Bourget's in which he quotes poems worthy of Legendre[4] and in which there is a profession of faith in anti-semitism which it would have been more delicate of him to keep to himself if it's sincere since he had the misfortune for an anti-semite of having been launched by a Jew, endowed by a Jewess and married to a convert.[5]

I've come across people of taste who found Donnay's speech charming. Personally I've never read anything so stupid and basically so 'academic'. For to find it amusing, shocking, devilish to 'dare' to say to the Academy 'set a teaser' etc.[6] one must be dreadfully academic and have a mind imbued with these same prejudices which one sanctions by the very fact of thinking it extremely daring and witty to challenge them. He and Bourget were like two pious old ladies thrilled with themselves for using a slightly naughty word in front of their parish priest. Really this whole attitude of mind derives from the same idiocy that prompts cretins like Roujon[7] to write: 'I don't give a damn, as Bossuet says.' It's nauseating. And stylistically the whole speech was so fake! What could be less like Honfleur than his Honfleur,[8] to those of us who know the real thing. Poor old Porto-Riche is the one who should have been elected;[9] at least he gave us *Le Passé*. The truth is that people think the love of literature, painting and music has become extremely widespread whereas actually there isn't a single person who knows anything about them and is capable of distinguishing Donnay's speech from a really well-written piece.

Madame, forgive this half-sheet because I've run out of paper, forgive this stupid letter, forgive everything and I hope to see you soon, very soon.

Your respectful friend
Marcel Proust

Did you see the *Figaro* article on poor Cornély?[10] The one in the *Gaulois* (which I ordered specially for it) was, relatively, much more seemly. It's not enough for the *Figaro* to have ditched him after the Affair. They have to reproach him for joining the *Siècle*.

1. Presumably for her New Year wishes.

2. Cf. letter 94.

3. She seems to have encouraged him to write more *Figaro* articles, of which five (signed) had appeared in 1907.

4. Louis Legendre (1851–1908), poet and playwright.

5. Paul Bourget, delivering the customary encomium, had praised a new Academician, the poet and playwright Maurice Donnay (1859–1945), for an explicitly anti-semitic play. Bourget had been the protégé of the Cahen d'Anvers and had married Amélie David, daughter of a rich Antwerp ship-owner.

6. *Pousser une colle*: Donnay used this slang expression in recounting an anecdote about his predecessor, the late Albert Sorel.

7. Henry Roujon (1853–1914), director of the Beaux-Arts.

8. Birthplace of Albert Sorel, Proust's old professor, whom Donnay had succeeded.

9. The dramatist Georges de Porto-Riche, author of *Le Passé* (see letter 137, n. 5), was not elected until 1923.

10. Jean-Joseph Cornély (1845–1907), royalist journalist who campaigned for the revision of the Dreyfus trial.

260   *To Reynaldo Hahn*

[Friday, 3 January 1908]

Hirnuls,

Herewith letterch. Your reply to Marcano[1] is not merely sublime, it's a remark which looks as though it was made by a man of genius to give you, while taking account of your preferences, your idiosyncrasies, etc., the maximum sensation of sublimity. If one had told it to you of someone else, you would have said 'perfection', 'incomparable', etc. And I say so too. I shook with indignation to think that 'the canteen woman who won a million, or rather millions' (that's how I shall nickname La Gustava[2] henceforth) (you know – so that no one understands – that there's a canteen woman who won a million in the lottery[3]) should denigrate, run down, degrade Mme A. de La Rochefoucauld who is *distinction* itself, and who was also beauty itself (see Chaplin's portrait[4]).

Nicens I hope that the purgative (which I wrote to advise you to take as long ago as Sunday, but was afraid of being importunate since my advice rarely seems nicens to you) will be pitiless to your stomach and energetically *drastic*, and afterwards milk and bed, nothing but milk and bed,[5] until the fever has completely gone for two days. Only then will a few restoratives be allowed.

I suppose you read about the Castellane-Elie de Sagan incident[6] (it should be spelt Hélie). In similar circumstances the prophet Elie [Elias] flew up to heaven. Unfortunately for him, Sagan didn't have the same opportunity. Not that he doesn't fly on occasion (if it were Rochefort[7] one would find it charming). But I think for him Gould spells principally Gold.

I read with indignation that *Moret*[8] is now mentioned in accounts of *soirées*. They'll be able to say: 'Stuttering went on until one o'clock in the morning and the party broke up after supper had been served etc.' Moret has just sung a melody. An admirer, anxious for an encore, says: 'Will you sing it for us a *third* time?' It's true that, unfortunately, he doesn't stutter when he sings.

Happy New Year,[9] Puncht, I'm only writing to send you some little kissikins. But I'm too exhausted and if I saw you I wouldn't give you real ones so as not to catch gripppshe.[10] But from afar, yes and yes (and from near too moschant).

Buncht

1. Hahn's doctor.

2. Baroness Gustave de Rothschild.

3. This astounding win, by a Mme Hofer who worked in a barracks canteen, had been a standing music-hall joke for the past two years.

4. Charles Josuah Chaplin (1825–91), society portrait-painter.

5. Exactly what Dr Cottard prescribes for the narrator in *RTP* (I, 536–7)

6. A street fracas after a family funeral that same day between the notoriously extravagant Comte Boni de Castellane (1867–1932) and his cousin Prince Hélie de Sagan. All Paris knew that Boni's wife was about to divorce him to marry the Prince and that he would be penniless: she was the American heiress Anna Gould, hence Proust's pun at the end of the paragraph.

7. Nationalist politician and anti-semitic journalist, Victor-Henri Marquis de Rochefort-Luçay (1831–1913), founder of the daily *L'Intransigeant*.

8. Evidently a singer as well as minor composer (see letter 138, n. 1).

9. The French has 'bonanibuls' = *bonne année*.

10. i.e. the flu (*la grippe*).

261   *To Auguste Marguillier*

102 boulevard Haussman
[Shortly before 8 January 1908]

Dear Sir,

I was deeply touched by your kindness in mentioning my translation of *The Bible of Amiens* in the *Mercure*.[1] I can't tell you how gratifying such praise is coming from you. It gave me great pleasure and I thank you for it sincerely. Together with that post-card you sent me from Amiens which I so appreciated, it constitutes a most precious token of friendship on your part.

I have been solicited from various quarters to write some short notices in the *Gazette (Chronique)*[2] on books about which I fear it will be impossible for me to address your readers:

One is a very interesting translation of Keats which has just been published by *Le Livre*, but I fear that a poet, however much of an aesthete he may have been, doesn't quite fit into the *Chronique des Arts*. What do you think? The translator is the Marquise de Clermont-Tonnerre.

Another of these works is *Le Sens de l'Art* by M. Gaultier,[3] which falls exactly into the jurisdiction of the *Chronique*. But haven't you already mentioned it? It would be a good excuse for me not to bother with it.

Finally Madame Crémieux who translated Ruskin's *The Nature of the Gothic* would, I think, have liked a rather more detailed notice than the few lines that appeared in the *Chronique*. It's too specialist an editorial matter for me to be able to judge whether or not your readers would like me to go back over this work in greater detail.

Finally, while I'm bothering you, would it be at all possible for you to send me *on approval* a few of your English engravings,[4] especially those in which an animal is represented beside the person or persons who are the subject of the portrait.[5] I don't know whether they'll be precisely what I'm looking for but perhaps I'll take one or two if the format isn't too small.

Please accept, dear Sir, all my most affectionately grateful regards

Marcel Proust

1. Marguillier, an editor on the *Gazette des Beaux Arts*, also wrote a column on museums and collections for the *Mercure de France* in which he had merely mentioned in passing the resemblance of a sculpted Virgin to the one Ruskin had so admired and which 'had inspired such fine pages from M. Marcel Proust', a remark he had signalled, and already been thanked for, in a post-card.

2. See letter 177, n. 2.

3. Paul Gaultier (1872–1960), philosopher and art historian.

4. Recently advertised for sale by the *Gazette*.

5. Proust is evidently preparing to write the first chapter of the novel he seems to be gestating, which was to be titled *Robert et le chevreau, Maman part en voyage*. See *Le Carnet de 1908 établi et présenté par Philip Kolb* (Paris, Gallimard, 1976). See also the passage in *Contre Sainte-Beuve* edited by Bernard de Fallois (Paris, Gallimard, 1954) – published in English as *By Way of Sainte-Beuve*, translated by Sylvia Townsend Warner (London, Chatto and Windus, 1957) – where Proust describes his brother caressing his pet kid, the scene bearing 'but a scant resemblance to that popular theme of English painters, a child fondling an animal'.

262   *To Madame Straus*

[Shortly after 10 January 1908]

Madame,

I was so unwell yesterday (and still am) that it really was physically impossible for me to thank you. But I'm overwhelmed by your kindness. I hope, thanks to you, that the Sauphars will be accommodating.[1] Monsieur Straus told me that in the old days, in the synagogue, the Sauphars were the noisy trumpets which awoke the dead for the Last Judgment. Those of today are not so very different . . . At the moment there's a lull for a few days because they're leaving some paint to dry before starting to hammer in the nails again.

I thought you sounded rather ironic about Alexandre de Gabriac on the telephone. I don't know why. First of all, though I may be wrong, I think he has some talent. Then everything he said about the respect and admiration that was owed to the Rothschilds[2] seemed very good on his part (they, who are always so ill-disposed towards me, would care rather less about people speaking well of me). As I felt that he would receive only insults for it and that no one would dream of congratulating him, I sent him a brief note (an exchange of courtesies after his congratulations to me), naturally with enough tact not to appear to find it extraordinary that one should admire generous Jews. He answered me with a letter from which I shall quote a few sentences, which are of no particular interest, to illustrate the general moral tone, since I've taken on the thankless task of making you appreciate Alexandre de Gabriac against your inclination: 'Your warm praise will encourage me to deserve more; I simply wrote from my heart because it had been moved, and I am happy to see from your approval that it was thus regarded among the élite. A few lap-dogs have yapped, lap-dogs from the rue de la Chaise and the rue du Bac.[3] They couldn't understand my having dared to put my name to it. I replied that I should have liked to have done so twice over. How sad, dear friend, to think that while all these nauseating people are bursting with health and going around, one can never meet you etc.' (The last sentence quoted is irrelevant but gives you an idea of the tone.)

Madame, I hope to be able to come to see you one of these days. It seems such a long time now and I miss you so much. Thank Monsieur Straus once more and please accept my respectful, affectionate, admiring and grateful regards.

Marcel Proust

I imagine you've heard of Forain's drawing which the *Figaro* didn't take about the Sagan-Castellane affair:[4] a church full of lights, funeral trappings, a catafalque, priests, etc. In the middle of the nave two men are hitting one another. The verger goes up to them, bows, and says: 'Gentlemen of the family . . .'

1. He had asked the Strauses to intervene in a quarrel with his neighbours, M. and Mme Sauphar, over noise. Cf. the similar story of Mme Katz, letter 194.

2. In *Le Figaro*, Gabriac, an aristocrat, had commended the charitable bequests in Baronne Adolphe de Rothschild's will. Proust later waters down his praise (see letter 264).

3. Socially grand addresses in the faubourg Saint-Germain.

4. See letter 260.

263 *To Robert Dreyfus*

[Wednesday evening, 29 January 1908]

Dear Robert,

I recognize that there's nothing stupider than to quibble about a subject and that to me above all, if I said 'Why Alexandre Weill?'[1] it would be only too easy to reply: 'Why Ruskin?' – Perhaps, if truth were told, there could be said to be a few differences between the two 'subjects', entirely to Ruskin's advantage, first of all because he had some talent. But the differences between the two commentators would be entirely to *your* advantage, first of all because you invented your subject, no one else had ever thought of it, whereas I took Ruskin from La Sizeranne,[2] from goodness knows how many others, from Fame. Your subjects are always new. But in this case perhaps you wanted to take your manner to the fullest extreme, write your *Bérénice*.[3] You said to yourself that it would be more fun if you sailed away and took us with you, come hell or high water, against the most powerful waves of antipathy, towards . . . practically nothing. Monet's last works painted nothing but mist, Rembrandt's nothing but sunlight. That's what I like best in art (not necessarily those painters, but that manner). But even so, this Alexandre Weill! You're the first person to write the *Lives of the Obscure*. Of the *Unknown* has been done; *Imaginary Lives* too.[4] But that was the sort of subject I like, not horrible and antipathetic, arousing an antipathy which isn't even to do with greatness. I shall re-read your study, and I hope I shall understand the man and why you took to him. Up to now

I remain impervious, even to the 'exquisiteness' of Guinon,[5] though I know very little about him, and other judgments of yours which seem to me remarkably generous, but they're friends, no names no packdrill. (I don't mean Beaunier, of course, or Barrès.) In short, you could almost have written a book about Hippolyte Rodrigues.[6] Madame Straus would have given you the material. He was less typical, not much less absurd. I'm sure that when I've read it I shall understand and think differently: about him. About you I never change.

Ever yours
Marcel Proust

1. Alexander Weill (1811–99), writer and journalist, subject of a study with the sub-title: 'Prophet of the Faubourg Saint-Honoré' in Dreyfus's book *Lives of the Obscure*.

2. Robert de Monier de La Sizeranne (1866–1932), art critic on the *Revue des Deux Mondes*, author of *Ruskin et la religion de la beauté* (1897).

3. Racine's fifth play, and one of his most successful, written in competition with Corneille, whose *Tite et Bérénice* appeared in the same year (1670).

4. A reference to the *Imaginary Portraits* written by Marcel Schwob, former editor of the *Mercure*, and by Walter Pater.

5. Albert Guinon (1863–1923), dramatist, author of an anti-semitic play, *Décadence*. The authors mentioned below also treated the subject of Jewishness.

6. Hippolyte Rodrigues (1812–98), stockbroker and amateur historian, especially of the origins of Christianity. Mme Straus was his cousin.

264   To Madame Straus

[Friday, 2 February 1908]

Madame,

I'm enchanted by your little almanacs,[1] and the thought that they come from you gives them an added poetry. In short, I'm delighted and thank you with all my heart.

I'm less well at the moment, which is why I haven't come. And I'm anxious to settle down to a fairly long piece of work,[2] which would make it even more difficult for me to come.

It's so sweet of you to defend yourself for having been ironical about that article;[3] now I'm afraid of appearing to have been not ironical enough. *Talent* is going too far. I meant that it isn't as stupid as society people say. But society people are so full of their own stupidity that they can never believe that one of their own can have any talent. They only appreciate writers who don't belong to their world. However (yet another effect of their stupidity) they appreciate writers only if they express *their*, society people's, mentality. They find Madame de Noailles's books stupid and Bourget's sublime. As for Alexandre de Gabriac, while he has written a few articles in which there really is an amiable impulse to express the sort of things you and I like, the one I mentioned is noteworthy only for its nice intentions and a whiff of virtue. But still it's at least that, a journalist and an aristocrat who doesn't make virtue consist in anti-semitism. And the whole thing adds up to a character on the whole more attractive than his physical appearance, which I have to admit is more worthy of Sem's pencil than our apologias. I realize that in saying *we* I'm being highly presumptuous and that I'm imitating your daughter-in-law without having at least the excuse of being Jacques's wife!

Your respectful and grateful friend
Marcel Proust

I'm desperately sorry to hear about the building work that's being done near your house. I'd much prefer (I assure you this is sincere) that it was happening near me and that you were spared the noise. I shall be thinking about it all the time. Alas, I suppose there's nothing to be done. Would you like us to hire a boat in which there will be no noise and from which we shall watch all the most beautiful cities in the universe parade past us on the sea-shore without our leaving our bed (our beds).

1. Her New Year present of five pretty little notebooks in which Proust was to make notes for his embryonic novel. Cf. letter 261, n. 5.

2. See above note.

3. See letter 262, n. 2.

265 *To Madame de Noailles*

[Shortly after 22 February 1908]

Madame,

This phrase of yours about Michelet ('that noble heart without guile')[1] is the finest thing that has ever been said about him. You are so superior to everybody else that your simplest remarks are the most beautiful offerings to genius.

I'm pleased that my pastiches[2] amused you. It's a facile and vulgar exercise. But still I think I put a certain breadth into them all the same, that they're good 'copies' as they say in painting . . .

Good-bye, Madame
Marcel Proust

1. Presumably in a letter from the poetess that has not been found.

2. Parodies of Balzac, Michelet, Edmond de Goncourt and the critic Emile Faguet (1847–1916) published on the front page of the *Figaro* literary supplement on 22 February. The theme of them all was the Lemoine affair, a current scandal whose comic potential Proust quickly recognized, as he did the possibilities of parody as literary criticism.

Lemoine, an engineer who claimed to have invented a method of manufacturing diamonds, hoping to reap a fortune from the ensuing slump in shares, had been paid over a million francs (roughly £64,000) by the credulous Sir Julius Wernher of De Beers before being taken to court; once his diamonds were shown to be genuine (that is to say, bought), Lemoine fled abroad.

In the meantime Proust had published further pastiches of Flaubert, Sainte-Beuve and Renan (see letter 268), but others already planned were rejected as no longer having news value. One on Régnier was to appear a year later. Most were collected in *Pastiches et mélanges* (Gallimard, 1919). See extracts in notes to letters 272 and 326.

266 *To Paul Helleu*

102 boulevard Haussmann
[Late February 1908]

Dear Sir,

Since this morning, in my joy, my chagrin, my infinite gratitude, I have already made a dozen conflicting decisions. I made up my mind to send back to you this masterpiece which I cannot accept,[1] which in my house, where not even the shutters are opened, would find itself in a disagreeable prison; but then, just as it was about to depart, its divine beauties stayed my hand, I could not do it, I felt that I should have let slip the good fortune of possessing this sublime picture and I couldn't help hearing it say to me: 'My name is Might-have-been . . . No-more, Too-late, Farewell' (Rossetti). And I tell my manservant, who was about to take it back to you, to wait! Will you allow me to wait for a day or two; I'm sure the masterpiece will inspire me with some idea that will enable me to keep it. And the greatest pleasure you could give me would be to allow me to buy it. Would you consent to do so? I know that it would be a great favour, that it is something you would have wished to keep. I'm stunned by the grandeur of your generosity, you have the sublime great-heartedness of a Rubens, of all great artists. But please understand me, please understand that I cannot accept it, tell yourself that if you let me buy it *it would perhaps be even more generous on your part* because you would fill me with joy as well as freeing me from the cruel necessity of not being able to keep it. And truly I cannot. I can say of this marvellous work which contains everything, all the sky, all the trees, all the earth, all the water, all the shadow, all the light, but which is woman too, what a great poet said of a woman:

> We should be unhappy together
> Though both of us are innocent[2]

I should be unhappy with this beautiful object. But if you allow me to pay the ransom for its captivity, never will a marvellously beautiful slave have received greater respect and adoration. There is sometimes a supreme delicacy in yielding to another's scruples. In agreeing to my solution, you would be even greater and more generous than you are. Will you?[3]

Your grateful admirer
Marcel Proust

1. The painter (whom Proust had met in Normandy the previous summer: see letter 248, n. 3) had presented him with a study for a picture called *Autumn in Versailles*. See n. 3.

2. Cf. 'Conseil' by Sully Prudhomme: 'Jeune fille, crois m'en, cherche qui te ressemble, / Ils sont graves ceux-là, ne choisis aucun d'eux; / Vous seriez malheureux ensemble / Bien qu'innocents tous deux.'

3. Ten years later, in a letter to the painter Jacques-Emile Blanche, Proust relates that having seen the picture in question and told the artist how much he liked it, he received it next day. When all payment was refused Proust returned it, but Helleu sent it back again with a dedication.

267   *To Madame Catusse*

[Early March 1908]

Dear Madame,

I've asked Reynaldo to send you two seats for his concert[1] and he promised to do so. So I imagine you will receive them shortly.

Your respectful friend
Marcel Proust

I hope you received the ticket for the box at the Théâtre des Arts the other day. Helleu having made me a present (though I scarcely know him) of a delightful study of Versailles, I've been racking my brains trying to think of something splendid to give him. Do you know how much those old Dutch ships (caravels) in silver (silver-plated) in full sail cost, and if they're easy to find. As he already has one, I think he might like to have its pendant (I've already asked a friend to have a look for one). He also adores eighteenth-century French art, but has such marvels given him by Mme de Béarn[2] that I dare not compete. His craze is for tapestries, but they'd have to be Bouchers at least.

1. A song recital which was a great success.
2. Comtesse René de Béarn, *née* Martine de Béhague.

268   *To Francis Chevassu*[1]

[Wednesday morning, 11 March 1908]

Dear Sir,

I sent the young man who acts as my secretary round to the *Figaro* last night with my completed manuscript of the Pastiches for Saturday.[2] Since you hadn't arrived by one o'clock in the morning, he left it in your name, asking – so he told me on the telephone – for it to be set anyway if you failed to turn up. But the person he spoke to (I don't know who it was) did not feel that he had the authority to do so. I shall send round for the proofs this evening. If they aren't ready, I'll send round again on Thursday morning and return them corrected at midnight; if it wasn't too late I might keep them until Friday morning, but if you prefer I can let you have them at midnight on Thursday. If when I send round for the proofs my manuscript even in bits could be included with them it would make it easier to correct them but isn't essential.

My pastiches are very long. In spite of this I hope you will agree to insert them every third Saturday. Without wanting to appear to attach more importance to them than they deserve, and to go so far as to talk of a balance and a sort of precise measuring out between the three of them which is perhaps evident only to me, I think these pastiches, a little more serious perhaps and referring to a somewhat select group of authors, would gain from being printed simultaneously. Those who don't know Renan very well will fall back on Flaubert and vice versa. Even the last time literary people, if there were half a dozen who mentioned these little pleasantries to me, never preferred the same one. One said he was a better judge of Balzac, another admitted to being less at home with Goncourt. And there were four pastiches that time. This time there are only three. I don't think it's too many. However, so as not to oblige you to put them all in if you don't come round to my view in spite of my arguments, I have revised the original order: there is first of all Flaubert, then Sainte-Beuve's critique of Flaubert, and finally the Renan. That way, if you only want two, you can postpone the Renan to another Saturday. But in that case he would appear alone, as I really cannot do any more for the moment. And I feel that the first combination would be better, unless it is awkward for you in which case it is odious and must be rejected. There is another eventuality in which I would resign myself to seeing only two: if printing all three forced you to drive me off the front page. I must admit that the somewhat secondary nature of the genre needs elevating by being given a fairly prominent position. Between the ornithology, however estimable, of M. Zamacoïs[3] and the weekly theatre of M. Legendre,[4] it would lose a little of its grave significance (!) as literary criticism in action. In this connection allow me to say, at the risk of lengthening even more, by a somewhat tiresome scruple, a letter whose proportions will earn me your malediction and deprive me of your attention, that if the fact of occupying the front page at such length were, as I suspect, to reduce the fees of the other contributors to the supplement, you must not allow this to influence you, for I will joyfully abandon mine, the honour of enlisting in your columns being reward enough. Finally, I should like my titles to figure as complete as possible in the synopsis of the Supplement that appears on the front page of the daily on Friday.[5]

Dear Sir, believe me, I won't often bore you with such long letters, but since I'm too unwell to go out I wanted to be as clear and precise as possible. And now, do whatever you think fit, everything you do will be well done, I leave it entirely to you and beg you to accept my grateful and sincere respects.

Marcel Proust

1. Francis Chevassu (1861–1918), editor of the *Figaro* literary supplement which was publishing Proust's parodies. Cf. letter 265, n. 2.

2. The second series of Lemoine pastiches, those of Flaubert, Sainte-Beuve and Renan.

3. Poems entitled 'Les Canards' (The Ducks) by Miguel Zamacoïs (1866–1955) had appeared on the supplement front page in February.

4. A playlet by Louis Legendre (see letter 259, n. 4) had appeared in the supplement on 25 January.

5. Proust's 'Pastiches' were duly announced on Friday 13 March. Flaubert and Sainte-Beuve appeared on the 14th, Renan on the 21st.

269 *To Madame Léon Fould*[1]

[Shortly after 15 March 1908]

Dear Madame,

Thank you for your letter; time hangs heavy on me without having seen you again and I miss you all very much. I've thought of you *every day*. But don't come; I'm too disorganized to 'receive'. When I am well I go out, but so late that you're asleep. However I shall come round one of these days to bring you the first pastiches. Yes I did indeed receive a charming letter from Eugène who for his part had only seen the first ones. And if I haven't answered him it's because I don't know his address . . . ? I'd be very interested to know who mentioned these pastiches to you. I have in fact had some very flattering testimonials, quite out of proportion with their object, from Lemaître, from France, from Hervieu, from Mme de Noailles. But I wouldn't have thought anyone had read them '*in society*'. I don't believe anyone really reads the *Figaro*, never the supplement, or even notices the signatures. And yet one would like to be read not only by literary people.

May I ask you something discreetly and confidentially (don't bother to reply, you can tell me when the opportunity arises): is a pretty Mlle de Goyon[2] by any chance a friend of Mademoiselle Elisabeth?[3]

Your respectful friend
Marcel Proust

1. Mme Léon Fould, *née* Thérèse Prascovia Ephrussi (1851–1911), mother of Eugène Fould (see letter 115).

2. Oriane de Goyon (1887–1928), daughter of Aimery Comte de Goyon and distant relation of Albufera: her aunt by marriage was his step-mother, the Duchesse d'Albufera, *née* Zénaïde de Cambacérès. Proust's impatient longing to be introduced to this pretty twenty-one-year-old girl, evident from the following letters, was not to be fulfilled until 22 June; a month later she appears to have vanished from his life. She was not to marry until just before her death, eighteen years later. In *RTP*, Proust gave her forename Oriane to the Duchesse de Guermantes (whose cousin, like Mlle de Goyon's aunt, is called Zénaïde).

3. Elisabeth Fould (1881–1952).

270   *To Robert Dreyfus*

[Wednesday evening, 17 March 1908]

My dear Robert,

Allow me to disobey you and write to you. It's extremely nice to receive letters like yours, but thanking you is part of the pleasure, so don't deprive me of it. As you well know, when I write I think of you and wonder what you'll think of it. And I even take some credit for knowing that if I write one thing I shall get a silence full of disapprobation, and if I write another I shall get a line from you. But I don't write 'at will'.

You're wrong to think that I compliment you on what you write out of gratitude for the compliments you pay me. I couldn't. No, if you find something you've done less good, it's for the same reason that I'm sick of my pastiches, it's because we haven't the privilege of our friend, the talisman which for him casts a magic spell over everything he does.[1] (And I fear that, as in the tale I'm alluding to, the spell may be for him alone.) And so, quite naturally, our thought holds no charm for us, and that of someone else holds more.

Although theatres have ceased to be called after their directors, to be given 'emblematic' names, the previous fashion has recently returned with a vengeance with the Sarah Bernhardt, Antoine, Réjane and Gémier theatres. It's true that they are names of actors. How one could philosophize (and if I were in a fit state to write I'd give you my reflections) on the topicality of lyrics on the subject of divorce. I'll talk to you about all this.

As for the pastiches, thank God there's only one more. It was because I was too lazy to write literary criticism, or rather because I found it amusing to write literary criticism 'in action'. But perhaps after all I shall be forced to do so, in order to explain them to people who don't understand them. In which case I shall ask you for all sorts of advice.

Who, by the way, was the gentleman who came and sat down with you and M. Serth[2] at the table at Wéber's where I was with Dethomas,[3] that evening when we saw each other for a second? You had just dined together, you said where but I can't remember.

Ever yours

Marcel Proust

Don't bother to reply!

---

1. Fernand Gregh. The 'tale' alluded to was a 'poetic' article called 'La Péniche évangelique' published in *Le Figaro* in the autumn of 1907.

2. Sic. The Spanish painter Jose-Maria Sert (see letter 247, n. 9).

3. Maxime Dethomas (1867–1929), painter, interior designer, whose 'superb studies' of Venice are referred to in *The Fugitive* (*RTP*, III, 640).

271   To Robert Dreyfus

[Saturday, 21 March 1908]

My dear Robert,

It's enormously sweet of you; I was actually going to write to you to ask whether 'quite charming' was your own phrase or a quotation (nowadays the more 1907 something is the more I think it's 'of its time' so similar does everything strike me as having been; you've confirmed one of my favourite ideas, which it would take too long to explain here)[1] . . .

As regards my pastiches, your reference to my 'technique' made me laugh a great deal. I don't put either as much pretension or as much malice into them. And since you are impenitent about '*uniment*'[2] and even believe '*tout uniment*' that you would have invented it if Lemaître hadn't anticipated you, I might indeed say that I do them *tout uniment*. Only I've had two unwelcome recollections. I'm so upset by them that I'd feel obliged to reprint these little pastiches simply in order to take out the two sentences (only two in the whole series) which seem a bit plagiaristic – I'll show you the originals and you'll see the resemblance – if I weren't so determined not to reprint them. But when I see you I'll show you these appalling blemishes. Fortunately they only amount to three or four lines all told, perhaps five or six out of a thousand. But I can now see nothing else. And I'm terrified of making new discoveries. But I've explored the original works, and haven't found anything else. I might almost say that in another sense it bothers me – one never knows with parody what degree of closeness is the right one. For instance, I talk about 'the (something or other) of the vainglorious and the aberrant'.[3] I find 'aberrant' extremely Renan. I don't think Renan ever used the word. If I found it in his work, it would diminish my satisfaction in having invented it. But if I don't find it, I shall be tempted to omit it (in my own mind, since I shan't make a slim volume of them) because it doesn't belong to Renan's vocabulary. I didn't make a single correction in the Renan. But it came pouring out in such floods that I stuck whole new pages on to the proofs at the last minute, so much so that there are quotations from Mme de Noailles that I didn't have time to check. I had adjusted my inner metronome to his rhythm and I could have written ten volumes like that. You ought to thank me for my restraint. I can't lay my hand on the Bernheim,[4] but can easily reconstruct it if it amuses you. And I've sunk even lower than that. But now it's all over; no more pastiches. What an idiotic exercise.

Ever yours
Marcel Proust

No, it wasn't Régnier, or Leclercq,[5] or Sert. In order to know who – but it isn't exactly important – you'd have to remember where you had dined. Did you see that Marcel Boulenger quoted your book about Weill[6] in *L'Intransigeant* or perhaps *L'Opinion* (herewith Boulenger's article; it was *L'Opinion*).

1. Proust is alluding to an historical article by Robert Dreyfus in the *Figaro* supplement, in which he refers to 'la toute charmante Hortensia, actrice de l'Opéra'.

2. Allusion to another article in which he uses the phrase *tout uniment* instead of *tout simplement* ('quite simply'). When he came to publish his articles in book form, Dreyfus took heed of Proust's objection to this affected usage.

3. In the Renan pastiche he talks about 'les sottes fantaisies du vaniteux et de l'aberrant'.

4. Jacques Bernheim (1861–1914), *Figaro* theatre critic, subject of an unpublished parody the manuscript of which is missing.

5. Paul Leclercq, writer and critic. Dreyfus must have replied to Proust's query in the previous letter.

6. See letter 263.

272   *From Madame de Noailles*

[Saturday, 21 March 1908]

Dear friend,

The Flaubert was incredible enough, and towards the end more intoxicating, penetrating, numbing than the immortal seringas in *Madame Bovary* – Sainte-Beuve I hardly knew, but I started reading him, and I must say he imitates you perfectly – But the Renan seems to me the marvel, with its complicated grace-notes, its delightful meanderings, its solemnities and smiles, in short its ineffable ruses which reinforce one another and which could be defined as 'I offer the ruse to the ruse' and not, as in the *Anthology*, 'the rose to the rose, and the scent to the scent'.

I was thrilled by my lovely passage, and the impossibility of taking a train is the whole secret of lyricism.[1]

However, I think I am departing for Syracuse.

Dear friend, I admire you infinitely.

Anna

1. Proust's Renan writes: 'The Comtesse de Noailles, if she is the author of the poems which are attributed to her, has left an extraordinary oeuvre, a hundred times superior to the Koheleth [*Ecclesiastes*], or to the songs of Béranger. But what a false position it must have given her in society! She seems indeed to have recognized this, and to have lived, not without some degree of boredom, an utterly simple and secluded country life, in the little orchard which habitually served her as interlocutor.' This is followed by a note in which 'Renan' suggests that her exile (like Mme de Staël's) may not have been voluntary, and quotes lines from her poems as proof that she was not free to take the train to Paris.

273 *To Madame de Noailles*

[Shortly after 21 March 1908]

Madame,

Did I not know that the greatest are the simplest and the best, by virtue of a natural and eternal law, I should be overwhelmed by your kindness and the trouble you took to write to me so sweetly. But like its predecessor, this letter has its cruel shaft. Last time it was 'I'm departing for Greece'. This time it's 'I'm departing for Italy'. And it's true that I don't see you in Paris. But nevertheless this news destroys the possibility which every day I hope to realize the day after without the accumulated disappointments in the least impairing my hope:

> And youthful hope forever cries:
> My sisters, let us start again.[1]

Good-bye Madame. At the moment in my meditations about you it is of Wagner that you remind me. I hope this doesn't annoy you and is not insulting to the immensity of your dreams and the omnipotence of your orchestration, you who are Siegfried even more than Isolde and whose verses on the Gardens of Lombardy[2] combine the myriad tones of a multitudinous orchestra

> like a celestial choir
> Stirring a thousand voices that sing within one's heart.[3]

Madame, how I admire you, how I love you, and how harsh my constant separation from you is to me. To have lived in the same age as you, in the same city, and never to see you! I no longer dare think of you because of the shocks it gives me:

> I even issued an express decree
> Against your name being said in front of me.[4]

I told the Princesse de Chimay that I'd tell her – and you – about a conversation I had with Hermant. And how nice I thought his son was. For I refuse to believe in the appalling supposition. Although the solemnity of sacraments of a legal kind like adoption no longer serves much purpose except to add a bit of spice to the banality of irregular situations, I cannot believe that he wanted to disguise a commonplace homosexual adventure with the infinitely respectable trappings of incest. I'm absolutely convinced that he doesn't have those tastes. And, like him, the young man certainly only likes women. Besides, people don't behave as well as that with women. Adopt! But one doesn't marry. It's true that homosexuality shows more delicacy, for it remains under the influence of its pure origin, friendship, and retains some of its virtues.

Madame, burn this letter in heaven's name and never divulge its contents!

Your respectful friend
Marcel Proust

1. 'Et la jeune espérance leur dit toujours: / Mes soeurs si nous recommencions.' Sully Prudhomme, *Les Danaïdes.*

2. A reference to her poem 'Musique pour les jardins de Lombardie' collected in *Les Vivants et les morts* (1913).

3. '. . . comme un divin choeur / Eveillant mille voix qui chantent dans le coeur.' The provenance of these lines is unknown.

4. 'J'ai même défendu par une expresse loi / Qu'on osât prononcer votre nom devant moi.' From Racine's *Phèdre*, Act II, scene 5.

274   *To Louis d'Albufera*

[Thursday evening, 26 March 1908]

My dear Louis,

I'm writing to say good-bye to you since I couldn't manage to call round on you and the time when you said you were due to depart has more or less arrived. Are you by any chance going to the Saussines[1] on Saturday? It's *possible*, if I'm not feeling too bad, that I shall go. But I've just been through several days and nights of such fearful attacks that I daren't make any plans. Our poor Rio Tintos aren't doing too brilliantly. I've a good mind to unload them when they get back to the price at which I bought them (1750).[2] What do you think, great financier? Did you see that in my *Figaro* pastiches I mentioned my discomfiture over De Beers?[3] I almost feel like buying some of those Manchon Hellas of Rochette's![4] But it's too boring to write to you to discuss stock exchange transactions.

Am I dreaming or used you to send letters round to me by the hand of a young telegraph operator who was related to one or other of your servants? If so you could be of service to me because for something I'm writing I need to get to know a telegraph operator. You will tell me that all I need to do is talk to the one who brings me telegrams, but first of all no one writes to me any more, and secondly in my neighbourhood they're all children incapable of giving the slightest bit of information. But in any case information is not the only thing I need; what I want above all is to see a telegraph operator exercising his functions, to get an 'impression' of his life. Perhaps yours is no longer one. In that case he wouldn't be any use but perhaps he has friends. Anyhow I'd be grateful to you and to him for any help, if I haven't got it wrong . . .

I don't know whether your friends are as nomadic as mine, but I have them in China, in India, in Egypt, in Tunisia, in Japan, everywhere, thank God, except Paris! You alone, dear Louis, would be welcome if only we could get together, but alas fate seems to keep us apart. I wish you a less gloomy stay in Nice than last year.[5] Don't you think you'd do well to take some precautions when you get there, in case there are mosquitoes (which I don't know), such as a dose of quinine every day? It's impossible for me to go there in this season of flowers and scents.

I was afraid of seeing your name in the newspapers, for, not knowing whether Marshal Suchet is in the Panthéon or not, I feared you might imitate the Duc de Montebello's initiative.[6] Of course I find the transfer of Zola's ashes to the Panthéon stupid. But in spite of this I don't think M. de Montebello's intervention was a very happy one. It's true that not having any of my own family in the Panthéon I can't really judge.

Good-bye my dear Louis. Give my respectful regards to Madame d'Albufera and believe me, your sincere and ever grateful friend

Marcel Proust

I'm told you were to be seen the other evening at Bourget's *Divorce*.[7] But I don't know who with.

A propos of divorce, has your charming friend Madame de Peyronnet[8] obtained hers yet?

1. A musical *soirée*, which Mlle de Goyon was expected to attend (cf. letter 269, n. 2). It was at similar receptions given by Proust's old friend Comte Henri de Saussine, a composer (1859–1940), that Debussy's early piano pieces and Hahn's songs were first performed in public. Cf. vol. I.

2. Rio Tinto shares had dropped by 45 per cent since March 1907, when they stood at 2700 francs. On 26 March 1908 they stood at 1624.

3. Cf. letter 265, n. 2.

4. Proust is alluding to yet another financial scandal which had just hit the headlines. A crooked financier called Rochette had created a number of bogus companies, including 'la société du Manchon Hella', whose shares, recommended by the newspaper *Le Financier*, had soared.

5. When he had had typhoid fever.

6. The Duc de Montebello, grandson of Marshal Lannes, had written an open letter to prime minister Clemenceau, published in *Le Figaro*, protesting furiously at the transfer of Zola's ashes to the Panthéon: 'Zola, who insulted the French army!' Albufera, like Montebello, was a Napoleonic title; Marshal Suchet was his great-grandfather.

7. A comedy by Paul Bourget and A. Cury.

8. Vicomtesse de Peyronnet (*née* Marguerite Jacob) whose young husband had died the previous June; in July she had married Comte Gaston d'Humières. Late in life she was to marry General Georges Catroux.

275   *To the Princesse Marthe Bibesco*[1]

102 boulevard Haussmann
Thursday, 29 March [1908]

Princess,

The sadness of knowing that you have gone away for a long time is even greater now that I've seen you again, that you have spoken to me at length, and that I've read this book of yours,[2] or looked at this uninterrupted series of wonderful, limpid water-colours – what name can one give to this new work of art which appeals to all the senses at once, enchants them all, and indeed the philosophical intelligence at the same time as touch and smell (I'm speaking to a philosopher), what are called, quite simply, the senses, and in which there is even, if one can apply such an ugly name to beautiful things, some literary criticism. I'm not thinking only of all that oriental folklore which you have invented a bit, I suppose, or of the funerary epigraph, but of those pages deliciously painted in seventeenth-century *turquerie*,[3] in which you create a whole decor, a whole atmosphere around a line from *Bajazet*.[4] Ah! you're not like those women of Isfahan who are unrecognizable in their sack dresses and their masks. From the very first pages one seems to be palpably aware of your body exhausted with insomnia, to feel the ache in your eye-lids which the light has forced open too early; and that childlike gaiety which alone must enable you to bear the weight of your perpetual thought, a gaiety which one would fain see re-enacted if some day you were to mime to us your 'endless salutations' to Azodos-Sultan.

You are a consummate writer, Princess, and that is saying a great deal when, as in your case, 'writer' signifies so many different artists in one, a writer, a perfumer, a decorator, a musician, a sculptor, a poet. At times the uninterrupted beauty produces a slight monotony. It is the fact that each one of them conceals a treasure that makes your paving-stones identical.[5] I have only these criticisms to make: the phrase 'womanly bodies', used thus, has become vulgar, I don't know why. You were afraid that if the gender were not made clear there would be an ambiguity. Nevertheless it seems to me that it could be re-arranged and you are such a clever artist that you could find a solution better than I could. Then I noted (but I don't have the book by me) one or two phrases *à la* Barrès (which are perhaps phrases *à la* Pascal: 'The . . . what . . . for . . .'). Finally, and especially, and I even wrote to her about it, Madame de Noailles has, in my opinion, destroyed for at least the next fifty years the possibility for anyone else of addressing cities etc. in direct discourse. Everything done in this style, under that form, however sincere, however lived and felt, however independent of her, unless we go down deeply and deliberately into ourselves, into our heart's core, or rather our heart's brain, to find a different and entirely individual expression, everything will appear to be an imitation of her, her radiance will absorb all our light. I say 'our' somewhat pretentiously, because I had to burn almost a whole volume on Brittany,[6] written before I had ever read anything of hers, in which the apostrophes to

Quimperlé! . . .

Pont-Aven! . . .

seemed to come from *L'Ombre des jours* or *La Domination*.[7] My sacrifice was necessary but bitter. Perhaps it will not be definitive; literary sacrifices rarely are.

From this point of view I regret the choice for the *Figaro* supplement of passages where the reader could precisely be misled, where I myself didn't immediately find the originality which later enchanted me in the book. But the admirable thought which you quote at the close of the book[8] relates it rather to certain passages of Ruskin which you certainly do not know and which in their case in no way detract from such a different kind of originality.

Please accept, Princess, all my respects.

Marcel Proust

1. Princesse Bibesco, *née* Marthe Lahovary (1887–1973), daughter of a Romanian statesman, wife of Prince Georges-Valentin Bibesco, was a cousin of Antoine and Emmanuel Bibesco and Mme de Noailles.

2. *Les huit Paradis: Perse-Asie Mineure-Constantinople* (Hachette, Paris, 1908).

3. i.e. in the Turkish style.

4. 'Bajazet, écoutez, je sens que je vous aime.' From Racine's *Bajazet*, Act II, scene 1.

5. A propos of unrecognizable Isfahan women, she writes: 'Everyone is perplexed as they pass, like the king in the fairytale who knows that a treasure is hidden under one of six hundred identical paving-stones.'

6. Unlikely: Proust never apostrophized, though he evokes the poetry of the names of Pont-Aven, Quimperlé and other Breton towns in 'Place-names: the Name' at the end of *Swann's Way* (*RTP*, I, 422).

7. Books by Mme de Noailles.

8. 'Do not become too attached to this beautiful land, for each day passes you by. And Time counts each of your pulse-beats.'

276 *To Madame Gaston de Caillavet*[1]

102 boulevard Haussmann
[Tuesday, 14 April 1908]

Madame (Dear Madame?)

I write to you with difficulty because I have a temperature after catching a chill (luckily because as a result I'm more or less free of asthma). But I want to tell you that you are extraordinarily nice, both in yourself and with reference to the Calmann business.[2] For the first niceness (niceness in itself) I love you and admire you, and for the second (Calmann) I thank you and love and admire you even more.

I have just written to M. Calmann.[3] It's unfortunate that it had to be *him*. Because I like him very much indeed, and the consequence is that I wrote a persuasive letter explaining to him that it would give me greater pleasure if he didn't publish my pastiches. No doubt he will have no difficulty in falling in with this solution. And naturally I'll be very sorry. During the time it takes him to acquaint himself with my letter and intimate his refusal I'd be most grateful if you could ask Gaston if there are some less smart publishers with whom it would be quite simple. I shouldn't at all mind paying the publishing costs. What I want is to be rid of it. You know that it was Gaston who put this unfortunate idea into my head.

I found your hanging gardens, your antique columns, the creeper growing up the tree and even, in spite of my affected scorn, the signature of Napoleon, all extremely attractive. But I like even better your daughter[4] and her amazing flashes of intelligence in a look or an exclamation. 'I'm doing my best' (to be nice to one) was sublime. She made me experience something that I never feel: shyness. I think it was the first time, and I now understand what it must be like. I didn't like the concierge but I didn't examine him very closely. And I found the ephebe, who wasn't in the least like the one at the rue de Miromesnil,[5] rather commonplace. I continue to think that you looked a hundred times better with bare neck, the golden apples, and your hair in a compact mass than in evening dress, and I suggest that I accompany you into society like that.

Good-bye Madame, I love you and admire you very much, more than you think. It isn't in any case necessary for you to think it.

Your respectful

Marcel Proust

1. *Née* Jeanne Pouquet (1874–1961), Proust's friend since girlhood, wife of his old friend the playwright Gaston de Caillavet. Jeanne is thought to have been one of the models for Gilberte in *RTP*.

2. Proust had just been to see her to talk about the possible publication of his *Pastiches* by Calmann-Lévy.

3. Calmann-Lévy were Proust's publishers for *Les Plaisirs et les jours* (1896). As Proust assumes, they were to decline the *Pastiches*.

4. Simone de Caillavet (1894–1968), then aged fourteen, a model for Mlle de Saint-Loup in *RTP*. She was to marry the writer André Maurois.

5. The Caillavets' former address.

277 *To Louis d'Albufera*

102 boulevard Haussmann
Tuesday [21 April 1908]

My dear Louis,

Thank you so much for your letter. How do you come to have writing paper watermarked St Marcel? In any case I was most amused to see it. I'm worried at what you tell me about Monte Carlo.[1] If you had told me in time, I could have found out lots of things and resumed the role which used to be my *raison d'être* and for which you know you will always find me.[2] In any case I'll give you in person, since you're coming back soon and express the intention of coming to see me, a few words of advice which you would do well to follow.

As to whether I was hostile to the idea of seeing you again, it isn't, as you seem to believe, from want of affection for you. I'm leaving Paris for good in July, and won't be living here any more. I shall be quite sad enough breaking the bonds of friendship when I leave. Since ours have already loosened somewhat, thereby sparing me some of the pain of the final separation, I felt it was pointless to see you again, thus reviving all the strength of my regret. But since you have decided otherwise, you know that as long as there's any question of my seeing you and being your friend, my heart, always filled with the memories of our old friendship, will be only too ready to surrender!

Thank you for all the information.[3] You're so meticulous that you even reply to me about Saussine[4] (a month later when there's no longer any point). I like this precision and thank you for it. I shall write to Louis Maheux one of these days. Your joke about the kind of relations you haven't had with him was unnecessary and the idea would never have entered my head. Alas, I'd like to be as sure that you don't have such ideas about me in that respect. In any case it would be more explicable since so many people have said it of me. However I imagine that whatever your thoughts are about me in that connection deep down (and I hope with all my heart that they're in accordance with the truth, that is to say exemplary), they wouldn't occur to you with reference to Louis Maheux. I'm not so stupid, if I were that sort of scum, to go out of my way to let the boy know my name, enable him to get me put inside, tell you all about it, etc. Perhaps I'm going on a bit about your joke. But alas it was followed by a 'but' ('but if it's to get an impression of his life') which you wrote involuntarily and thus all the more sincerely, and which gave me much food for thought.

Good-bye to you; I'm about to embark on a very important piece of work.[5] Did you read my pastiches in the *Figaro* supplement?

Fondly yours

Marcel Proust

1. Louisa de Mornand had gone to Monte Carlo in February and her return was overdue.

2. Proust kept Albufera in touch with Louisa's movements; in *RTP* the Baron de Charlus plays a similar role between Swann and Odette. Cf. letter 249, n. 10.

3. About the telegraph operator. See letter 274. His name was Louis Maheux.

4. Proust had hoped for an introduction to Mlle de Goyon at the Saussines. See letter 274, n. 1.

5. Cf. letter 261, n. 5.

278   *To Robert de Billy*

[April 1908]

My dearest Robert,

I hope, I sincerely hope because I think about him continually, that your letter means your father-in-law[1] is better. What profound, what true joy, which I should be prepared to pay dearly for, it would be for me if this improvement were definitive. I think I'd be capable of getting better myself in my happiness in order to go and see your father-in-law and rejoice in his recovery.

Who could possibly have told your friend that I was translating *Praeterita*[2] and the extracts from Ruskin's correspondence and journal his publishers added to it? I don't remember talking about it. In any case it's hardly begun, and I would willingly give way to your friend, as I've already done for M. de La Sizeranne with his *Pages choisies*.[3] My sole consideration in these matters is this: 'What will be more advantageous for Ruskin? Who is capable of making him better known etc.' And in my present state of health it's so improbable that I could bring such a long task to a successful conclusion that it would be preferable to have your friend do it. But is he a poet? *Praeterita* is written in 'faded' colours – what evocative powers are needed to translate it! Has he a bit of magic at the point of his pen? Will he do notes? They're essential. If he's afraid of not knowing enough, he could do what the translators of *Mornings in Florence* did, have someone else do the notes (they even got ten others to do them which is too many). In that case, if he were prepared to entrust me with the proofs of his translation when it's finished I'd gladly annotate it, if they gave me plenty of space, a few months, and complete independence of viewpoint.

My dear Robert, since *The Bible of Amiens* says that present-day Protestantism conceives of the Cross as a raft intended to carry our valuable belongings to paradise intact,[4] don't be shocked if I go from Ruskin to the Pins des Landes,[5] and let me know, please, where one can buy them and whether they're a better prospect at the moment than the Harpeners and the Gelsenkirchens about which you spoke to me.

Affectionately yours

Marcel Proust

1. The banker Paul Mirabaud who was to die in May.

2. *Praeterita. Memories of Childhood* by John Ruskin. A translation by Mme Gaston Paris was published in 1911.

3. A selection from Ruskin introduced by Robert de La Sizeranne (letter 263, n. 2). Cf. letter 28, n. 6.

4. 'The idea [of the Cross representing self-renunciation and hard work] has been *exactly* reversed by modern Protestantism, which sees the Cross, not as a *furca* to which it is to be nailed, but a raft on which it, and all its valuable properties, are to be floated into Paradise.' J. Ruskin, *The Bible of Amiens* (III, 43).

5. Pins des Landes, Harpener and Gelsenkirchen are shares.

279 *To Georges de Lauris*

[End of April 1908]

My dear Georges,

Fondest thanks for your letter. If only I could have guessed that you were ill! It did seem a long time, and there were even humble physical signs of your absence, like the yellowing of plants in poems. Nicolas[1] said regretfully: 'It's a long time since we saw M. de Lauris.' But that 'long time' was only relative to the 'often' with which you spoil me with such preposterous kindness, and it would have been tactless on my part to appear to surmise some extraordinary reason for your not coming, as though I had a right to rely on the sublime 'ordinary' which I owe only to your munificence.

I'm sad to think that you've been unwell, as I myself have been, extremely, aggravated by the removal going on above me which has lasted nearly a month since it was only done in the mornings.

I'm glad you've read some Bergson and liked him. It's as though we had been on a mountain top together. I don't know *L'Evolution créatrice*[2] (and because of the great store I set by your opinion I shall read it at once), but I've read a fair amount of Bergson, and the parabola of his thought is sufficiently describable on the basis of his own generation (?), whatever Creative Evolution may have followed, to know what you mean when you talk about him. And moreover I think I've told you of the high esteem I have for him, and also – which is less interesting but does reveal a moral aspect – the great kindness he has always shown me. Indeed it was he who analysed *La Bible d'Amiens* at the Institut. I saw that Pius X had forbidden priests to attend his lectures or read his books.[3] That told me that they did read them and pleased me because I thought they only read *La Libre Parole*[4] (which, by the way, Pius X doesn't forbid). I remember that one of the first things Sollier said to me when I went to his sanatorium was about Bergson whom he had been obliged to read because he thinks their spheres are the same: 'What a muddled and blinkered mind!' I felt a da Vincian smile of intellectual pride flit across my face. And it didn't contribute to the success of the psychotherapeutic treatment.

Good-bye, Georges, I'm tiring myself writing. I've done nothing in your absence. I intend to start working . . . in an hour's time. But this is just a convenient expression which doubtless means never. The Mercure de France and Fasquelle have refused to publish my pastiches. Don't tell anyone as I've offered them to Calmann-Lévy who will no doubt also refuse.[5]

I just happened to go out on the evening you wrote to me. But I didn't get to the rendezvous in time. I read the name of the young heroine (of your novel), amiably quoted by Flament in an article about Mlle Dorziat.[6]

Affectionately yours
Marcel

1. Nicolas Cottin, Proust's manservant.

2. *Creative Evolution* by Henry Bergson, published 1907.

3. Encyclical on Modernism, 1907. No Bergson work was put on the Index until 1914.

4. Anti-semitic journal.

5. Cf. letter 276, n. 3.

6. Article by Albert Flament on the actress Gabrielle Dorziat in which he mentions 'a strange artist', Mlle Léone Georges, apparently the heroine of a novel which Lauris was not to publish until 1910 under the title *Ginette Chatenay*.

280   *To Emmanuel Bibesco*

[April or May 1908]

Dear friend,

You said to me so many times this evening: 'You must hate me' that I finally came to the conclusion that you found yourself hateful, and if I answered very sincerely 'No, I hate you not' in Corneille's strong sense of the word,[1] it was with one mental reservation, the following, which a stupid need for sincerity prompts me to write to you. I told myself that when, like me, one has been the object of constant accusations of salaism,[2] there's a certain lack of delicacy, more intellectual than moral, on the part of a friend in joking so insistently in front of a stranger about a case (apocryphal moreover) of Josephism, and even worse, as I've recently been told and neglected to complain to you about, to make it the theme of pleasantries which you yourself admit are damaging to me. It's a fine distinction but one which it would be as futile to try to explain to you as the charm of a Vuillard painting to someone who didn't see it. It's the sort of thing that reasoning is incapable of demonstrating to those who haven't first sensed it, since dialectical proofs haven't the power to free us from the family of minds to which we belong. And although it may pain one for a moment to see somebody one prizes remain indifferent in front of a Cézanne or laugh unduly at something a bit coarse, it doesn't prevent one from prizing him in other respects. And there are so many things I admire in you, so many things I like, that any small reservation I may make here or there in no way implies the hatred you mentioned, simply the diminution of a friendship which is only too anxious to prove itself. It would only begin to hurt me if these remarks were made in front of people towards whom personally I'm indifferent but who might form a less exalted view of you than the one I might have given them. But as long as it only 'damages' me, to use your own expression, I shall endure it with a resignation made easy by the thought of all the dear and delightful compensations of your wonderful friendship.

So (unreservedly this time) fondly yours

Marcel Proust

I'd just like to add, in case you thought I was a bit on edge, that my niece had just had a serious appendicitis operation,[3] and as I was expecting a telephone call from my brother, I was afraid it might have been him when you answered the telephone jokily.

1. 'Va, je ne te hais point', in Corneille's *Le Cid*, Act III, scene 4.
2. Private word for homosexuality, from Comte Antoine Sala, an overt homosexual (cf. vol. I). The same applies to Josephism, from the Emperor Joseph II (1741–90).
3. Adrienne (Suzy) Proust, aged four and a half, Robert's daughter.

281 *To Louis d'Albufera*

[5 or 6 May 1908]

My dear Louis,

Try and write to me *at once* to tell me what you want me to say . . . Monte Carlo etc. for I have an excellent excuse.[1] Since I'm *very tired*, forgive me for not replying to you to repeat to you the same things which you no longer wish to believe, to wit that my affection hasn't changed, that I'll see you again whenever you like (if I'm not too ill on that particular day), that I've never suggested that your affection was self-interested or that I was doing you favours. If I thought it I'm tactful enough not to say so. But far from believing it, I consider that you have done me immense favours and (to my great chagrin) I have never done you any at all. No, when I said that it was consistent with the nature of our friendship, I meant something more subtle, more profound and more tender. I was thinking of the beginnings of that friendship, in some sense a dual one, etc. There wasn't the slightest notion of service or utility. I'm much less vulgar by nature than you think. But it's so boring to have to explain what one is.

As regards what you say about my friends, if you're thinking of Reynaldo you're right to believe that he's a friend to me, the dearest, the best, a brother. If I learnt that he had murdered someone I'd hide the corpse in my bedroom so that I was suspected of having done the deed. But that eventuality won't arise! For he's the most exquisite soul. I think I told you that he had planned for years to make a trip to Algeria and finally set off for Marseilles, booked his place on the boat (a fortnight ago) and then, because my valet had forgotten to cable him news of me, he was afraid I might have had a relapse, so abandoned Algeria and came back to Paris.

As for my 'acquaintances', perhaps I have some of whom people speak ill more than of yours. But perhaps (from the point of view you refer to) there's more certainty with regard to yours. I don't want to accuse anybody, especially as I know there are some very nice young men who may have vices, but in your generation apart from a few who are above suspicion and beyond all calumny, which in any case would never be directed at them since they are known to be unassailable, such as you, Guiche (I find my list petering out and I don't know who to add although there are certainly others), I assure you that it isn't only in the world of the theatre or literature that malicious rumour thrives. I mentioned Guiche not at all in terms of his place in the hierarchy of my friends, but because of his fundamental purity in that respect. A propos of Guiche you must have read in the papers that he proposed me for the Polo,[2] which must have made you split your sides! As well it might. I get nothing but letters offering me ponies for sale. I wrote to young Maheux[3] a week ago. He telephoned me later but at an hour when I couldn't see him. He hasn't called since. In any case I'm not sure I won't abandon my Parisian novel.[4]

Have you by any chance – something that's always so interesting – any family photograph albums? If you could lend me one for a few hours (especially if Mlle de Goyon[5] was in it) I should be delighted. It's true that I should be even more delighted if you came here and could tell me the names. By the same token, do you have *your genealogy* in a few lines? It's again because of what I'm working on that all this would interest me. For I have in hand:

> a study on the nobility
> a Parisian novel
> an essay on Sainte-Beuve[6] and Flaubert
> an essay on women
> an essay on pederasty (not easy to publish)
> a study on stained-glass windows
> a study on tombstones
> a study on the novel

Affectionately yours
Marcel

1. i.e. for talking to Louisa de Mornand (cf. letter 277, n. 1.): Proust's pretext was to be a supper party (see letter 289).
2. Exclusive club with polo grounds in the Bois de Boulogne. Cf. letter 361.
3. See letter 277, n. 3.
4. Cf. letter 261, n. 5.
5. See letter 269, n. 2.
6. See letter 84, n. 4. Cf. letter 315, n. 1.

282   *To Louis d'Albufera*

[5 or 6 May 1908]

My dear Louis,

Just after I'd written to you young Maheux called round. This time I was able to receive him. He is very nice, very intelligent, he gave me some information and intends to give me more, so he said. His only drawback is that he's much too well brought up, too genteel, not at all representative of his profession. He resembles Bertrand de Fénelon, except that he's much better dressed. Speaking of telegraphists he said to me, 'They're rather the Grenelle type than the rue Saint-Dominique.'[1] Rue Saint-Dominique in his mind meant *his* type. He would be a perfect model for a picture of society mores. We parted very good friends (at least on my side, and *he* was very friendly).

Good-bye, my dear Louis, I wanted to let you know at once because I'd told you I hadn't seen him and so that my letter shouldn't be in contradiction with reality.

Thanks again and
ever yours
Marcel

1. Grenelle is a working-class district of Paris; the rue Saint-Dominique one of the most elegant streets in the faubourg Saint-Germain.

283   *To Madame Straus*

[Sunday morning, 10 May 1908]

Madame,

The death of Monsieur Halévy,[1] which has grieved me so much, torments me especially today when I think of the repercussions which sorrow and fatigue may have on your health; and also the fear of not being able to risk tiring yourself as much as you would wish. I know that with you energy makes up for strength. But since it is after all your health that has to pay for it, I should prefer you to be less energetic. At all events I am extremely preoccupied about this day.[2] I have been thinking only of you and yours since I read the news; I've also thought a great deal about poor Monsieur Halévy, whom I knew very slightly but whom I admired and liked deeply. And I'm very fond of Daniel and often thought how lucky he was to have such parents and to be able to make them happy. I wanted to write something about M. Halévy, but Beaunier forestalled me, both on M. Halévy and on you, whose conversation is a work of art. I feel I understand your family so well, starting with your father[3] in whom a reading of Sainte-Beuve taught me to revere an all-round mind gifted enough to have been a great writer and a great thinker as much as a great composer. I imagine you know those Sainte-Beuve articles. But he knew nothing of the moral grandeur which makes all of you disdain those frivolous possessions which come to you in spite of yourselves, and which caused Daniel, and his brother, and Jacques[4] to be brought up so simply by you, and by M. Halévy, to whom everyone looked up. I feel that I would have spoken of all that – that unique distinction of mind and character – less well than Beaunier, but with more accuracy, sustained all the time by living memories, and overflowing with sadness and sympathy. Madame, I hope you will not be too exhausted and I think continually of your grief.

Your respectful friend
Marcel Proust

And I haven't even thanked you for your delightful letter!

1. Ludovic Halévy (1834–1908), novelist, playwright and librettist, father of Proust's old friend Daniel and Mme Straus's cousin.

2. The day of the funeral.

3. Fromental Halévy (1799–1862), composer of *La Juive*; Sainte-Beuve devoted one of his 'Lundis' to him on his death.

4. Mme Straus's son Jacques Bizet was a contemporary of Daniel and his elder brother Elie Halévy (1870–1937).

284 *To Robert Dreyfus*

[Saturday evening, 16 May 1908]

My dear Robert,

What you wrote about Monsieur Halévy is of such charm, truthfulness and profundity that I felt I was seeing and listening to him again, that I knew him better. I was extremely moved, and I thank you. I should in any case have written to you to thank you for your visit, but I also wanted to thank you for bringing me those pages.

I meant to ask you if you thought the forbidden article would be as inoffensive (I have such terrible pens that after having laboured over every word I'm taking to a pencil) in the *Mercure* or another review as in a book.[1] But in the meantime my project is becoming clearer in my mind. It will be a long short story instead, and so there will be time to consult you again. But the same reason that makes me think that the importance and the suprasensible reality of art prevent certain anecdotal novels, however agreeable, from perhaps being quite deserving of the status which you seem to give them (art being too superior to life, as we judge it through the intellect and describe it in conversation, to be satisfied with copying it) – this same reason forbids me to make the realization of an artistic project depend on notions which are themselves anecdotal and too directly drawn from life not to partake of its contingency and unreality. All of which, moreover, presented thus, seems not so much false as banal and deserving of some stinging slap in the face from outraged existence (like Oscar Wilde saying that the greatest sorrow he had ever known was the death of Lucien de Rubempré in Balzac,[2] and learning shortly afterwards, through his trial, that there are sorrows which are still more real).

But you know that such banal aestheticism cannot be my artistic philosophy. And if tiredness, the fear of being a bore, and especially *this pencil* have prevented me from explaining it, give me credit if not for its truth at least for its seriousness.

Affectionately yours
Marcel

1. Possibly an idea for an article on homosexuality in the light of the Eulenburg affair (see letter 254, n. 3) which became the basis, Dreyfus later vaguely recalled, of Proust's essay 'La Race maudite' (i.e. 'the accursed race of inverts') in *Contre Sainte-Beuve* (see letter 315), itself the preliminary sketch for the opening chapter of *Cities of the Plain* (*RTP*, II, 623 sqq.).

2. In Wilde's dialogue *The Decay of Lying* (translated into French in 1906), Vivian says: 'One of the greatest tragedies of my life is the death of Lucien de Rubempré', a remark quoted by the Baron de Charlus who remembers only that it was made by 'a man of taste' (*RTP*, II, 1084).

285   *To Madame Straus*

[Saturday, 6 June 1908]

Madame,
I read with amazement and disgust that Shlumberg[1] was putting

himself forward for the Academy. Although I'm afraid of running up against the pusillanimous and time-serving spirit of the *Figaro*, I intend to write (which alas doesn't mean getting accepted and published) an article on that blackguard. I shall substitute insinuation for insult so as to have a better chance of being published.

However, there are certain particulars I should like to have and I'd be grateful if you could ask Reinach[2] for them when you see him. Naturally I shan't compromise him nor mention my sources.

1. Is it true that Shlumberg's works are not by him but by Salomon Reinach?[3]

2. Has Shlumberg any scholarly qualifications whatsoever? Are there other members of the Académie des Inscriptions who have more than he has? (All of them, I suppose, but I want names that I can cite.)

3. Is it true that he was only elected through some sort of fraud and no longer would be today? And in particular that he behaved badly to Maspero?[4]

Finally, if you see Hervieu or any other Academician I should like you to ask him whether Schlumberg has *any chance at all* of ever being elected to the Academy, or will never get more than two votes.

Forgive me for bothering you but you understand my feelings. Reinach can write and brief me without any fear; I shall return his letter to put his mind at rest and I'll compromise only myself.

Your respectful friend
Marcel Proust

1. Historian Gustave Schlumberger (1844–1929), habitué of Mme Straus's salon until he 'deserted' to the anti-Dreyfusards (cf. vol. I). He never succeeded in becoming an Academician.

2. The prominent Dreyfusard (see letter 94, n. 5).

3. Salomon Reinach (1858–1932), archaeologist and philologist, brother of Joseph.

4. The eminent Egyptologist (see letter 170, n. 1).

286 *To François Vicomte de Pâris*[1]

102 boulevard Haussmann
[Friday evening, 12 June 1908]

My dear friend,
It would be easy for me – and you have been so unfriendly towards

me for some time – to leave you with the notion which pleases you, and which is a matter of indifference to me, as to this evening's little incident. But I have such a sad proof to give you of the sincerity of what I said and the stupidity of what you thought, or claimed to think, that I cannot resist the melancholy pleasure of indulging in these memories. My poor Mama whom I lost nearly three years ago – without your ever having offered a single word of sympathy for a misfortune so great that you would have taken pity if you had had the least suspicion of it – my poor Mama thought you had the most handsome face of any man she knew, and she knew that I also admired your looks. If you pretended to find something funny in the way I told you this, it's because I didn't want to embarrass you with a compliment *coming from me* in front of Madame de Chimay. And after you left us, I told Madame de Chimay that Mama used to say to me: 'I find M. de Pâris much better-looking than Lucien Daudet.' This you must not repeat because he's a friend whom I'm fond of and who was very fond of Mama. But I told the Princesse de Chimay because *she* knew very well that I would never link Mama's memory with anything that wasn't *the very truth itself.* You can be sure that everything I say about Mama is gospel truth and I would never commit a sacrilege, I would never profane the memory of the person I loved more than anything in the world, in order to pay you a compliment or relieve you of the bitterness of an alleged criticism, which in any case I wouldn't mind in the least. You are not worth my having tired myself writing you all this. But the truth is worth it.

And moreover you credited me with an idiotic sentiment which I did not have. I stayed at that ball to try and get myself introduced to Mlle de Goyon[2] who is the prettiest girl I've ever seen. And I had to leave before being introduced. And doubtless I'll never be invited to another ball and perhaps will never see her again. But how wonderful she is, how intelligent she looks. How lucky you are to be fêted by all those young girls, and perhaps to know that one. Did you see her dance? I've never seen anything to equal it.

Good-bye, strange creature who inspire affection without being able to establish friendship.

Marcel Proust

1. François Vicomte de Pâris (1875–1958), aristocratic acquaintance of Proust, whose family owned the Château de Guermantes (see letter 334).

2. See letter 269, n. 2.

287 *To Madame Straus*

Monday evening [15 June 1908]

Madame,

Naturally I shan't talk about the Affair.[1] But perhaps it would be as well, so that he can't say that it's a case of political revenge, for me to say that I was delighted by Barrès's election to the Academy and would be equally delighted by Léon Daudet's, that if I find Shlumberg's election absurd it isn't because of his opinions but because of his stupidity – that it's legitimate for there to be a dukes' party in the Academy (I don't think so but it's better to put it that way) but it would be preposterous to have a snobs' party; that if being a duke is sufficient qualification for being an Academician, seeking the company of dukes cannot be sufficient; that if Shlumberg were merely a snob and had been nominated for that, well and good, but that he has written books and there's something offensive in the idea that he might be nominated for his books. However I'd say all this rather better, less seriously, less politely, drawing attention to his feet,[2] without any pretension to good taste or even good faith. But Calmette??[3] The more so because I intend to suggest that Arthur should be nominated instead.[4] But even from a serious point of view and quite apart from all political considerations, it's *shameful* that if they have to search in the related sciences (as though there weren't enough writers), they don't nominate: either scholars who are the outstanding and undisputed glory of those sciences, a Boutroux, a Bergson, a Maspero, a Bréal, an Alfred Croiset[5] and so many others; *or*, even better, a rare and little-known talent in these kindred sciences, whom the Academy would be bringing to light rather than consecrating, an Emile Mâle, an Abbé Vignot, an Abbé Huvelin, a Darlu, a Brunschvicg, and so many others.[6] Or else, quite simply, Porto-Riche, Régnier, Boylesve, Hermant, Francis Jammes, Maeterlinck etc. And if snobbery is to dictate the choice, then let it be a great nobleman who has never written a line, let it be clear that it's the nobleman who's being elected. Or even an aristocrat of some distinction and intellect, Galliffet[7] for instance, or your friend Prince d'Arenberg,[8] or why not Montesquiou? But Shlumberg seems to me a national disgrace, the triumph of everything base and stupid, of knowledge without intelligence and even, unbelievably, without seriousness; that sort of idiotic knowledge whose only excuse is seriousness in his case is frivolous. He also represents baseness turned arrogant, the ugly hiss of a rearing snake, a life supposedly dedicated to science (and in reality a science entailing no more understanding and inspiring no more 'elevation' than the science of bridge) crowned by a devotion to ideas which Louis XVI considered antediluvian and Vauban barbaric, and all this without the excuse of loyalty to principles since he never had any, or to friends whom he only met at sixty and who are not his friends, since they can only despise his shameful obsequiousness towards them; not to mention all his betrayals and desertions. None of this will I say, but it will be the secret driving force, the angry head of steam that may well decide me to concoct an article, although I can't even write a letter without getting a headache. But also the fun of portraying this prehistoric buffalo, with his patriotic whiskers, blushing and fawnir ; in front of all those lady converts of the Haber[9] and Heine families.[10]

Madame, how I must bore you, how I must thank you. And will you also give Monsieur Reinach my warmest thanks.

Your respectful friend
Marcel Proust

1. The Dreyfus affair – à propos of Gustave Schlumberger (see letter 285).

2. Grotesquely large, it seems (cf. following letter).

3. Editor of *Le Figaro*; evidently he was to refuse Proust's article (see letter 285).

4. Arthur Meyer (see letter 64, n. 8), anti-Dreyfusard and snob, editor of *Le Gaulois*, was an ironic suggestion.

5. Emile Boutroux (1845–1921), philosopher; Alfred Croiset (1845–1923), professor of Greek elocution at the University of Paris.

6. Abbé Pierre Vignot (1858–1921) of the Ecole Fénelon; Abbé Henri Huvelin (1838–1910), vicar of Saint-Augustin, Paris; Alphonse Darlu (1849–1921), professor of philosophy at the Lycée Condorcet (cf. vol. I).

7. General the Marquis Gaston de Galliffet (1830–1909), Minister of War 1899–1901.

8. Prince Auguste d'Arenberg (1837–1924), chairman of the Suez Canal Company, member of the Académie des Beaux-Arts, and parliamentary deputy 1889–1902.

9. Allusion to Mme Octave de Béhague (*née* Laure de Haber) and her daughters Comtesse Jean de Ganay and Comtesse René de Béarn.

10. Allusion to Princesse Alice de Monaco, *née* Furtado-Heine, and the Duchesse d'Elchingen, *née* Paule Furtado.

288   *To Madame Straus*

[Monday evening, 22 June 1908]

Dear Madame Straus,

This 'poor man's' paper,[1] as the Baronne James [de Rothschild] used to say, will tell you simply that on arriving at the Murats',[2] before even catching sight of the host and hostess, I saw Shlumberg sitting beside them as they received their guests with the falsely bashful look of someone who for sixty years has been pretending abhorrence of society. It gave me great pleasure to walk past him a dozen times to greet other people, without saying how-do-you-do to him, and when someone asked me 'Don't you know M. Schlumberger?' I answered very loudly: 'I'll say how-do-you-do to him when he says how-do-you-do to Madame Straus, who doesn't care a hoot though I do.' I 'spoke' my article to practically all the guests, so that he must now know my intentions. Madame de Chevigné was sublime and restored all my affection for her. Besides being enchanting about a girl I'm fond of,[3] she told me that she'd said to Shlumberg à propos of the Academy: 'Now you'll pay for all your treacheries.' There's someone really decent.

Good-bye, Madame, it's so late that it seems to me almost indiscreet even to write to you. The bashful buffalo[4] smiled inanely each time I walked past, thinking I was about to greet him, and his enormous hooves left fossil footprints on the carpet.

Your respectful friend
Marcel Proust

On leaving the Murats I went round to the *Figaro* to have another word with Calmette whom I can never find. I was told he was at 'the rehearsal of a ballet at the Henri de Rothschilds'!

1. Written on paper kept by his bed for lighting his anti-asthmatic powders.
2. The Princesse Lucien Murat (*née* Marie de Rohan-Chabot: 1876–1951) had given her annual summer ball that same evening.
3. At this ball he was at last introduced to Mlle de Goyon! Cf. following letter.
4. i.e. Gustave Schlumberger (see previous letter).

289 *To Louis d'Albufera*

[Monday evening, 22 June[1] 1908]

My dear Louis,

In reply to your various questions:

1. I wanted to give a supper-party with Fouquières[2] of which our friend[3] would have been the queen and the secret pretext (Fouquières doesn't know she was the secret pretext, he only knows that I was

inviting her). It would have been less intimidating for me not to see her alone so that I didn't have to vent my reproaches to her, and once I had seen her I would have made an appointment with her. The supper-party couldn't take place because of an access of fever that overcame me and for other reasons. So that she doesn't even know that it was going to happen. I had someone telephone her house the day before yesterday without saying it was from me to find out if she was in Paris before deciding anything. Amazingly enough someone answered at once and said she was indeed in Paris. Only Fouquières and Dorziat[4] were in the know about this supper-party which was to be very large and which I think has only been postponed. This evening I telephoned to ask if I could see her between nine and eleven, or between midnight and one o'clock, because having learnt that Mlle de Goyon was to be at your sister-in-law's[5] I decided to go there, otherwise I wouldn't have gone in your absence, for you would have been the sole attraction for me. And then I hoped to see our friend before or after. But I was told she had gone out, that there was no one there, that they didn't know when she would be home (it was half past eight when I telephoned). Since you're so kind to me I'd like you to know that I have at last been introduced to Mlle de Goyon. It was a hugely emotional moment for me – I thought I was going to fall – but also quite a big disappointment, for she didn't seem to me so nice close to and a bit irritating when she opens her mouth, and more coquettish than amiable. I shall think about her again more calmly: all my ideas are a bit muddled. Unfortunately, as I was obliged before finding an introducer to ask several other people who didn't know her, they all went, Mme de Chevigné and others, to stare rudely at her and I'm afraid it may have ruffled her.

As regards Dinard: I have ideas for work for several months ahead, which moving would interrupt, and I should have liked to go to a place where I could settle down indefinitely without having to switch all the time. The hotel at Cabourg closes at the end of September and isn't convenient for me, but at least I feel well there. Since if you went to Dinard no doubt you'd stay only a month it wouldn't really suit but as it's very mild I might stay on after you left. On the other hand if you were going to rent somewhere near Paris I'd prefer that it was I who did the renting; I would live there all the year round and invite you to spend the autumn. Would that suit you? I'm so grateful to Fouquières for introducing me to Mlle de Goyon that I don't know how to thank him. If ever you have occasion to write to him, tell him how grateful I am. I must say François de Pâris was very nice too[6] and also Mme de Chevigné. But it's incredible how unhelpful people are, especially when there's an element of love involved. However I mustn't exaggerate; it isn't real love; I don't really know what it is. There's also a Mlle Marshall[7] (I don't know whether that's how it's spelt) who is really not bad, and a Mme Barrachin who has a very pretty face.[8]

My dear Louis, do you know how to go about finding a job as gamekeeper – who to approach, whether it's difficult – for a decent youngster, very intelligent, fundamentally honest, very knowledgeable about horses, etc. My manservant is becoming neurotic because of being shut in all day and would like such a job. Should I talk to Guiche about it? Or Loche? Or Robert de Rothschild? There's no great hurry. But when you have a chance, tell me the best way to go about it.

If my Fouquières supper-party can be re-arranged do you think our friend would agree to sup with Cléo?[9] I'm afraid of her baring her teeth at that dancer. Whereas she knows Dorziat who is in any case a workmate of hers, and a very good friend. If by any chance you know where Mlle de Goyon is spending this summer, do tell me. By the way, I was told this evening that she wasn't marrying her Roumanian, that it was a joke. And at the very same moment a man introduced himself to me, saying that he'd met me at Mme de Briey's – a name like Vaugirant – and I understood (but I was so flustered at being in the presence of Mlle de Goyon that I must have misheard) that *he* was engaged to her. So that I lost one fiancé only to find another a second later. But I must have misheard and got it wrong. The other day with a small group of ladies such as Mmes Lucien de Murat and Ludre, I built you up as a first-rate intellectual. Please don't let me down.

Once my work is done I think I shall leave for Italy and if I can't settle near you near Paris I shall buy a little house above Florence and try to tell some Italian beggarwoman about the feelings such and such a woman inspires in me, and in particular Mlle de Goyon. I ought to have got myself introduced to her father but I lost my head. Moreover he would probably have clouted me because of the way his daughter had become the cynosure of everyone's eyes through my fault.

Yours with all my heart
Marcel

Emmanuel swears that I was invited to your dinner party. In any case, as he has invited me to another one shortly with the two Lubersac young ladies, he now sends me invitation cards, claiming that otherwise I don't consider myself invited. But I'm sure not to go, because it's my fate, whenever there are Hinnisdaels at a party, not to be well enough to go.[10]

1. This letter, written during the night on his return from the Murat ball, was addressed to Albufera in the country.

2. André Becq de Fouquières (1874–1959), young man-about-town who provided the longed-for introduction to Mlle de Goyon.

3. Louisa de Mornand.

4. The actress Gabrielle Dorziat (see letter 279, n. 6).

5. Princesse Murat. Earlier Proust had gone so far as to solicit an invitation, via Mme de Caraman-Chimay, to another ball where Mlle de Goyon was expected.

6. Cf. letter 286!

7. Possibly the daughter of a Mr and Mrs Charles Marshall of 20 rue de Magdebourg (XVIe).

8. Mme Pierre Barrachin, *née* Brocheton, referred to as a beauty in Paul Morand's *Journal d'un attaché d'ambassade* (1949).

9. 'Cléo' de Mérode (1881–1966), famous dancer at the Opéra.

10. The Hinnisdaels were the Comtesse Jean de Lubersac, *née* Eliane d'Hinnisdael, and her sister-in-law.

290 *To Reynaldo Hahn*

[Monday evening, 22 June 1908][1]

How harse you, Minusnichant? Will you be crossch if I tell you that I've at last met Mlle de G., at the Murasts. Too long and long, but and but. You should send me notes on dinners and receptions London, at least the most impressive. Because will stagger the crowds here. I can understand how snobs for whom *social pleasure* exists should want to shroud these self-sufficing joys in secrecy. But for you, to whom they are pure tomfoolery, it's necessary to give them a wider context in accordance with the laws of the multiplication of vibrations (press, etc.).

At the Murats the Widow[2] shrilled: 'Suzette, show me the girl Marcel's been talking about.' She looked at her and turned away, saying: 'She's very ugly and she looks dirty.' As for the young widow, she thought her nice-looking but made matters a hundred times worse by staring at her the whole time, making endless remarks, laughing loudly and continually saying to me: 'Come over here; take a look at her from the doorway, etc. etc.' The introducer was Fouquières, to whom my gratitude knows no bounds. But as he was completely drunk, he kept saying to me as he effected the introduction, out loud, not merely out loud but bellowing: 'What d'you think of those little cheeks? You wouldn't mind pinching them, eh? And how about a little kiss? Ah! you wouldn't mind, would you, you rascal? What's that? You say you'd like to scrunch those little cherry pippins' (I was saying nothing at all); 'you're quite right, and besides, you're looking very smart today, trimmed your beard a bit, I like the look of you etc. etc.' The expression 'not knowing where to look' is inadequate to describe the state I was in, not to mention the girl and her present fiancé. Very severe attack yesterday, otherwise well.

Hasbouen
Wrirnuls

Don't repeat any of my old witticisms to La Chevigné, as I stuffed a letter to her full of them. I can't find her neat reply, otherwise I'd shends it to you.

I was introduced to Mme Barrachin whom I find ravishing. And I wasn't introduced to the Duchesse de Morny whom I also find ravishing.

1. This letter was written, like those to Madame Straus and Albufera, during the night on his return from the Murat ball.

2. Mme Lemaire; her daughter Suzette was known as 'the young widow'.

291 *To Robert de Montesquiou*

102 boulevard Haussmann
[Shortly after 27 June 1908]

Dear Sir,

I am writing to you this evening because I am unwell and haunted by melancholy thoughts. You know that I don't 'fish' for 'invitations' and if it had been a party I wouldn't have mentioned it. But how unkind of you not to have invited me to the admirable commemorative reading of which I've been told[1] and which, as you know well, mine would have been the intelligence and the heart most capable of entirely appreciating. When I wrote to thank you for *Sérée* and to express my admiration for that book, which I had perhaps re-read less than some others, and which dazzled me as though I had a new pair of eyes, I asked you if I could come to see you whenever my strength permitted, and you did not reply.

Now my evenings of health are very rare. In the daytime I haven't *even once* been able to get up but I should have tried to do so for that reading, which must have been sublime. Since I don't know which day it was I have no idea whether it was one of the days (nearly all) when it would have been physically impossible. But had I been forewarned of the ceremony by you, even if I had been unable to attend, I should have felt a closer bond between us, one of those bonds that bring the faithful together and that used to inaugurate cults. It seemed to me that this reproach would be a mark of attachment and it is for that reason that I have taken the liberty of addressing it to you. I beg you to accept the homage it reflects together with my faithful and devoted admiration.

Marcel Proust

1. His reading from his latest book, a memorial to Gabriel de Yturri. To Proust's chagrin Montesquiou was to make amends, see letter 293.

292 *To Louis d'Albufera*

[Wednesday evening, 1 July 1908]

My dear Louis,

I telephoned you four times the day before yesterday and twice yesterday (that's to say today, anyhow Wednesday – I'm writing to you in the middle of the night of Wednesday to Thursday) and there was no reply. Both on Tuesday evening and on Wednesday you were said to be dining out (at Emmanuel's dinner party, I suppose). Both times it was to ask you if I could come and see you. Now I shall probably rest for a few days and when I'm recovered I'll telephone again. I should very much like to know whether you received my letter at Charmes.[1] For it's full of proper names and I wouldn't like it to go astray. I don't remember exactly how long it was, but I think it wasn't far off twelve pages (I've really forgotten). Nicolas told me he'd put *Charmes poste restante* on the envelope, then crossed out *poste restante* as they said at your house where I had sent the letter only to put *poste restante* if one was telegraphing. Anyhow I enclose the piece of paper which my concierge brought back.

I had supper last night with a certain number of so-called 'society' people among whom was a very amiable man with a big nose and dreamy eyes who turned out to be the brother of your friend Mlle de Bonvouloir[2] – M. Guy de Bonvouloir if I heard his Christian name aright. If I had dared I would have told him to persuade his sister to demand a large income from M. Martin Lerroy, but I thought the subject wouldn't be very suitable. Ever since you spoke to me about it I've thought of that unhappy lady with much sympathy and affection and respect for her disinterestedness and misfortune.

I must frankly confess that the telegram which you were kind enough to send me, and which dissolves my doubts, made me so happy that I burst into tears. For deep down I still have a great deal of affection for her and I cannot believe the gossip I've been told which is too serious to be true.[3] There's nothing I'd like better than a reconciliation, the more so as I believe that if ever it became necessary I'm one of the few people who know and above all understand her well enough to be able if need be to advise her with affection, clear-sightedness and frankness. Others will perhaps love her as much but understand less well (naturally I exclude you: you are too much above it all to enter into account – but I mean if you weren't there). Others will understand her better but love her less. And others still will understand better, and love her more, but won't dare tell her the truth. And then I have the advantage with her of never having made advances to her, so that she knows how totally disinterested my advice would be.[4]

Have you any plans? Fouquières, for whom I have a great deal of affection and gratitude since he was so splendid over the Goyon affair and others as well, assures me that I could perfectly well work and stay as late as I wish at Dinard since the weather is very mild. One of his Breton friends, M. de Kergariou, recommends a more southerly part of Brittany, Huelgoat for instance. On the other hand the telegram you sent me so sweetly makes me want to go to Cabourg, because I now see that her departure wasn't a lie, which would have put an end to any thought of seeing her again. But Cabourg isn't very sensible because as much for my work as for my health I need a place where I could stay as late as possible, and the hotel at Cabourg closes at the end of September. I shall write to the manager to make sure. Perhaps I shall end up by simply going to Versailles. I've re-rented my flat for another quarter on your advice so as not to have to move out at once. I want to ask your advice about something rather serious that I'm going to write. You know that I can't undertake anything without consulting you. I've spoken again about you to people who are the only ones I'm fond of, though less than of you. But I should like us all to form, even at a distance, an indissoluble band of friends. For you are so often wrong about who is or isn't a friend of mine and you said in a letter only the other day 'He's your friend' about someone who isn't in the real sense of the word.

Good-bye, my dear Louis, and ever yours
Marcel

1. See letter 289, n. 1, which Albufera received on 1 July.
2. Charlotte de Bonvouloir had been deserted by her husband Jacques Martin Le Roy (Proust misspells his name in the next sentence).

3. In spite of Proust's doubts Louisa de Mornand had left for Normandy as she had promised.

4. Perhaps Proust insists too much. Cf. letter 32, n. 3.

293 *To Reynaldo Hahn*

[Early July 1908]

My dear Mintchniduls,

I was so annoysed by your letter loaded with bullion that I wouldn't writes to you were it not for an old orderly habit of acknowledging receipt of moneys and moneys. And I shall tell you by word of mouth that and that. I would have you know that since you behave to me like this I have decided henceforth to give some parties at which I shall pay you a higher fee than La Potter Palmer[1] so that you will be unable to refuse, and I shall approach you via the Dandelot office.[2]

I must tell you that last night, after the exchange of innumerable letters in a pontifical but pressing tone, the fatal Count came to read to me[3] and then present me with the book about Yturri in which we all 'get it in the neck' and in which there are some fine letters from Buncht (Reynaldo), from 'Hahn who doesn't like me'. I should have loved you to hear him at two o'clock in the morning, without pity for the Gageys,[4] stamping his heels on the floor and declaiming: 'And now, Scipio and Laelius, Orestes and Pylades, Horn and Posa, Saint-Marc and de Thou, Edmond and Jules de Goncourt, Flaubert and Bouilhet, Aristobulus and Pythias, welcome me, for I am worthy of it, into your pre-eminent company.'[5]

But how can I confide in a man who is insulting enough to send me 750 francs in exotic notes, the use of which escapes me, and which are not even thin enough to light Legras powders[6] with.

Bye-bye, nicens

Send back letterch

Buncht

In Montesquiou's book there's a letter from Prince von Radolin assuring him of his sympathy for 'your cruel loss' etc. He'd have done better to keep a little of it for Eulenburg.[7]

1. Chicago millionairess Mrs Bertha Potter Palmer, collector, wife of a biscuit manufacturer.

2. Concert agents.

3. Cf. letter 291, n. 1.

4. The neighbours in the flat below.

5. Proust quotes from memory, or rather misquotes. Some of the pairs of bosom friends are ill-assorted: Horn was the friend of Egmont, not Posa (who was the friend of Don Carlos); Aristobulus, King of Judea, had nothing to do with Pythias, whose friend was Damon; Saint-Marc is obviously the conspirator Cinq-Mars, who was executed with his friend de Thou.

6. His anti-asthmatic powders.

7. Cf. letter 254, n. 3. Radolin was the German ambassador.

294   *To Louis d'Albufera*

[8 or 9 July 1908]

My dearest Louis,

Thank you for your letter and the information about Mlle de G. Alas, the fact that she isn't engaged is a purely chimerical balm to me since she will never be mine. But still the fact of having spoken to her, to know that I shall be able to speak to her again – above all to have found her a thousand times less wonderful than I thought – all that has done me a great deal of good and given me a great calm.

You will go to the coastal place you mention if such is your pleasure but I fear that it will cause you sadness if you don't see our friend[1] and trouble if you do. As for the proximity of a forest in which to ride with Madame d'Albufera, the forest of Saint Gatien is a long way from Cabourg and Houlgate (a good fourteen kilometres I think) and the forest of Brotonne is out of the question. Perhaps there are forests near Dinard, I don't know. And then when I think of the frightful coming and going of motor-cars around Trouville etc. I wonder if it's wise to go riding.

I think I told you that I had re-rented my boulevard Haussmann flat for another quarter so that I'm no longer in a hurry to find something in Florence or near Paris. But since horrible alterations are starting in the house on the 15th of August I must be off somewhere before that date. I went out this evening at midnight to see Bernstein[2] but I've come back home with a high temperature and don't know whether I shall be able to get up for some days. Will you go to the Versailles jamboree?[3] I don't think I shall.

With all my fondest regards
Marcel

Burn this letter, or rather send it back to me, as one can't burn letters in a house where there are no fires. I alone can do it!

1. Louisa de Mornand; she was once again with Gangnat (see letter 249, n. 10).
2. The playwright Henry Bernstein (see letter 87, n. 6).
3. Evening charity concert-fête on 11 July organized by the Comtesse Greffulhe.

295   *To Robert de Montesquiou*

GRAND HOTEL
CABOURG
[Shortly after 18 July 1908]

Dear Sir,

As a result of certain incidents of little interest to you, I left suddenly, in two hours, and was unable to thank you for the article in *La Vie parisienne*[1] or call round to see you as I so much wanted to do. I found the article dazzling and true; to think that people have confused you with Oronte when you alone continue worthily to emulate today 'the man with the green ribbons'![2] Ah, you're not one to write for the sake of writing! Every word has a meaning, or two or three. There are one or two locutions quoted whose 'calibre'[3] is indeed stupendous, and whose unconscious humour can only be grasped 'on reflection', that is to say for a lot of people never.

Dear Sir I still find your indifference to the Academy justifiable, but I feel that the time has come. The Lotis, Richepins etc.[4] being your literary guarantors with those who are ... less literary, and moreover your recent incursions into so many different genres having given each doubting Thomas palpable proofs suited to his particular brand of perspicacity, it seems to me that you might piece together a majority. Don't you think so?

Your grateful and devoted
Marcel Proust

1. 'Panaches', collected in *Assemblée de Notables* (1908).

2. Célimène's description of Alceste in Molière's *Le Misanthrope*. In the article in question, Montesquiou had spoken of certain poets in the candid and ironic tones of Alceste speaking of Oronte's sonnet.

3. Montesquiou had used the word *calibre* in poking fun at the solecisms committed in the press.

4. Novelist Pierre Loti (real name Louis Viaud: 1850–1923) was elected to the Academy in 1891, and dramatist Jean Richepin (1849–1926) in March 1908. Montesquiou himself was never elected.

296 *From Robert de Montesquiou*

Neuilly
[Shortly after 19 July 1908]

Dear Marcel,

I learn, with pleasure, of your *escape* (since you were a *prisoner*), and of your *aeration* (since you were *confined*).[1]

May the breath, if not the kiss, of Thetis restore you more completely to your friends and, better still, to yourself!

As for what you say about people capable of confusing Alceste with Oronte, I should like to believe that that form of blindness is no longer very prevalent and that it would appear, to the unenlightened, as one of the oldest manifestations of the Diplodocus of Error!

Salve and *Vale*!

R.M.

P.S. The noble commentaries on my Book provide me with a laurel which is worth more to me than those you wish for me so kindly, but somewhat ill-consideredly in the light of circumstances that it would take too long to enumerate, among which one, not unelevated, may deserve to be mentioned, that of my attaching, sincerely, *no importance whatsoever* to such universally coveted distinctions.

1. Proust had left precipitately for Cabourg (see previous letter).

297 *To Lucien Daudet*

GRAND HOTEL
CABOURG
[Early August 1908]

My dearest Lucien,

I've just received your letter and the *Mercure*. I am so moved and proud to see my name attached to this short story[1] which I've admired a hundred times more on re-reading and wish I had never used the fine words 'proud' and 'moved' and 'admiration' so that you could have the original, virgin proof of them. There are beauties there which I hadn't suspected the first time, and my impression was renewed and enhanced to such an extent that I sometimes felt it was the story that was no longer the same and that you had altered it. From the very first lines, the sunlit wisterias, the remains of a Greek statue and the Balzacian heroes seemed even finer.

But I don't want to talk to you about that today, for being unwell and more or less incapable of writing I want selfishly to talk about my little article.[2] If you can't find a home for it in the *Gaulois,* as you seemed to suggest, will you send it back to me together with Bailby's address and I think he will print it. Alas, the *Figaro'*s reason was indeed, as I realize, the one you first thought of. But what you tell me about *L'Action française* is even more astonishing. Surely your brother must share my feelings about this delightful masterpiece.[3] However, Lucien, something strange is happening. One can write a masterpiece, although it doesn't happen often, and it remains unrecognized. And it's discovered a hundred years later like treasure trove. No one read a book by Stendhal in his lifetime (it's true that even today, d'Haussonville in the *Revue des Deux Mondes* has apparently cited – as proof of Scherer's perspicacity[4] – the fact that he had always despised Stendhal, *L'Education sentimentale* and Baudelaire).

But this isn't the case with *Le Chemin mort.* You say that no journal has mentioned it. But *everyone* has read it. When I brought it to Mme Straus and talked about it to her and Hervieu, she said to me: 'I haven't yet read it, but you've said exactly the same thing as Reinach,[5] although you never think like him. He finds it delightful, etc.' You see I'm not afraid of offending you by quoting the opinion of an 'opponent' since he's a man of great intelligence and taste.

Thank you for your kind intention in 'notifying' Mme d'Eyragues of my presence. But I don't know if I shall be able to go to see her as I no longer budge; I'm ill all the time and am a little less so only by forgoing my long drives of last year. And yet I'd like to go back to Falaise.[6]

I could much more easily have gone to see Mlle Sorel[7] who is staying a few kilometres from here (Falaise is at least sixty). But Lucien, can you see me saying to her: 'Mademoiselle, I know one of your friends slightly, Monsieur Daudet.' She would have rung the bell and shown me the door as in the theatre. I used up most of my cheek 'catching up' Lucy Gérard[8] on the beach and I was ill for two days as a result.

Dear Lucien, in heaven's name answer me about the article, or rather don't bother to answer me if you find it *impossible* to place it better, just send it back to me without taking the trouble to write and I'll send it to Bailby. In any case I haven't given up hope of speaking to the *Figaro* about it again later.

Affectionately yours, and thank you from the bottom of my heart for the greatest honour I have ever received, that dedication which thrills me and would make me insufferably vain if I saw anyone. But I see no one and even my valet is doing his 'thirteen days'.[9]

Your infinitely grateful and admiring

Marcel

1. 'La Réponse imprévue', published in *Mercure de France* and dedicated to Proust.

2. Proust's adulatory review of Daudet's novel, *Le Chemin mort*, which had a discreetly homosexual theme.

3. The article having been rejected by *Le Figaro*, Proust seems to have suggested, however reluctantly, Lucien's brother Léon, editor of the ultra-nationalistic, anti-semitic *Action française*; and then, that having failed, Léon Bailby, editor of *L'Intransigeant*, who used it on 8 September under the pseudonym Marc el Dante presumably because Proust's name was inappropriate in a paper which had been so extravagantly anti-Dreyfusard.

4. Edmond Scherer (1815–89), Protestant theologian and literary critic.

5. Joseph Reinach, leading Dreyfusard (see letter 94, n. 5).

6. The d'Eyragues had the remarkable old fort at Falaise.

7. Céline-Emilie Seurre, known as Cécile Sorel (1875–1966), famous Comédie-Française actress.

8. Lucy Gérard, thirty-six-year-old actress, colleague of Louisa de Mornand.

9. Nicolas Cottin was doing his annual fortnight's auxiliary military service (Proust's own was suspended): cf. letter 362.

298   *To Louisa de Mornand*

GRAND HOTEL
CABOURG
[First half of August 1908]

My dear little Louisa,

Thank you most tenderly for your letter. It only just forestalled the one I was about to write to you to tell you that I was tired and seldom went out, but that I thought a lot about you and will come soon and even write to you beforehand to arrange a little dinner-party. It's the best way of seeing each other because of the late hour I get up at the moment – I'm rarely up by tea-time. If you came at that hour, telephone in advance in case it's one of those days when I have an attack, or one of the days when I go out (which is rare). Moreover there are people whom I fear you would find very irritating who come to see me but this is rare and then perhaps friendship would help you to put up with boring presences. I received a letter from Louis complaining that you never give him any news of yourself and ignore him but as he wrote it some time ago perhaps you have since broken a silence which was doubtless somewhat prolonged and under the influence of which he wrote those disillusioned words . . .

I met Lucy Gérard[1] on the front at Cabourg. It was a ravishing evening and the sunset had forgotten only one colour: pink. But her dress was pink and added the complementary colour of the twilight to the orange sky. I lingered to watch this delicate pink tinge and returned to the hotel, with a cold, when I saw it merge with the horizon, to the utmost end of which she glided like an enchanted sail.

Tenderly yours
Marcel

1. See previous letter.

299   *To Reynaldo Hahn*

[Cabourg, Thursday, 6 August 1908]

Monsieur le petit Binibuls or even Nur-nols,

Your little letterch are excessively mopchant,[1] short, and, alas, few and far between.

But I who cannot write, as though I'd been struck by a paralysis of the hand and the brain, have no right to complain. And thank you for your little messages which I conscientiously apply on hand.

Do you know who Princess Winnie, whom the *Figaro* reports to be at Deauville, is staying with? Might Clothon have got her invited to the Baronne des Vaux's.[2]

There's no hurry about Hopillard for a few months.[3] It's a question of getting from them more cheaply the same parrots (two pairs) as in the place Beauvau.

I can't write to you, because too moschant, no table, no room, no wind, very crossch.

I should like to 'renew' with Lucy Gérard who is here, but she's with a certain Bardac whom I don't dare greet, having seen him only once in the hotel which Mélisande has since abandoned.[4] She was still merely beautiful at that time although she said:

> I am neither a lady, nor beautiful.[5]

I think it was Bu-ni-buls who had got me a hinsvitation, as everywhere nicens I've been.

I believe your little Mama is well from what you tell me and that plizzes me very much.

I should like to be informed about the present and future *material* situation of Nordlingka and about her immediate plans.[6] Can you tell me, if you write.

I went round to Rezka's[7] where a butler full of good nature but well schooled in the ways of the house said to me: 'Madame is *extremely sorry* not to be able to receive Monsieur. It is a real regret for Madame. Madame is really very, very sorry.' As my car drove away I could still hear the Rambouillet de Frontin superlatives.[8] But soon I heard another voice. For, a few minutes later, improbable though it may seem, I was searching Deauville for the Villa Suzaleine, that is to say Georges Lévy's villa.[9] Why was I looking for it? Too long to explain. We asked everyone; no one knew. I saw my driver[10] ask a working woman who said, 'The Villa Suzaleine? I don't know. Wait, we'll try and find out.' She called an old gentleman who was passing by – who stopped and retraced his steps with a cross, dumbfounded, pained expression, and I recognized with horror Gustave de Rothschild.[11] Meanwhile the working woman, not knowing who it was, continued to ask him where the Villa Suzaleine was, while he stooped, cocked his ear, couldn't understand, and after hesitating for a moment put his hand in his pocket and proffered a few coins. She refused; explanations ensued, very bad-tempered on the part of Gustaude; he approaches the car, I hide, shouting to the driver to get away as fast as possible, not at all anxious to clarify the imbroglio, and as we decamped we nearly ran over Robert and his wife[12] who were returning home on foot and doubtless having seen the burgrave in the middle of a group were hurrying towards him out of curiosity.

Hasdieu my Bunibuls, asdieu my Birnibuls, asdieu Fernuls, asdieu Mopchant, asdieu little scrivener, I've ordered *Femina*, Lafitte[13] is 'on our shores', as is also Nozière whom his wife 'has managed to get back'.[14] Alas.

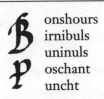

B onshours
irnibuls
uninuls
P oschant
uncht

1. A variant of moschant (*méchant*): see letter 207, n. 1.

2. Princess Winnie is the Princesse Edmond de Polignac (*née* Winaretta Singer: see letter 10, n. 1); Clothon is Mme Gaston Legrand (*née* Clotilde de Fournès: see letter 20, n. 7); and the Baronne was the wife of Proust's royalist acquaintance, the Baron des Vaux.

3. Hopilliart et Leroy, antique dealers in the rue des Saint-Pères, Paris, from whom Proust has asked Hahn to buy some Louis XV porcelain parrots.

4. An allusion to the English singer Maggie Teyte (1888–1976) who had just had a resounding success in Debussy's *Pelléas et Mélisande* at the Opéra-Comique. She had studied in Paris under Jean de Reszké and made her debut as Susanna in Mozart's *Figaro*, conducted by Reynaldo, in 1906.

5. Cf. Gounod's *Faust*: Marguerite, at the end of Act II, 'Non, monsieur! Je ne suis demoiselle, ni belle.'

6. Marie Nordlinger, with whom Proust had been out of touch.

7. Mme Jean de Reszké.

8. Social-climbing valet in the comedy *Turcaret* by Lesage (1668–1747), author of *Gil Blas*.

9. Georges Lévy (1853–1933), economist and Proust's distant relation.

10. Most likely Alfred Agostinelli (see letter 242, n. 2).

11. Baron Gustave de Rothschild (1829–1911), railway millionaire and member of the Jockey Club.

12. Baron and Baronne Robert de Rothschild.

13. Pierre Lafitte (1872–1932), proprietor of the illustrated magazine *Femina* for which Hahn had written an article praising Maggie Teyte.

14. Pseudonym of the *Gil Blas* drama critic, Fernand Weyl (1879–1931).

300　*To Henry Bernstein*

[Cabourg, soon after Sunday, 16 August 1908]

Dear friend,

　　On Sunday I was in the lounge having tea with Mme Edwards, Sert[1] and Mme Forain when I was handed a telegram from you. I wasn't paying much attention, I just noticed 'filthy hotel', and I haven't been able to find it since. But *fear* nothing! The aforementioned people didn't see it. And now I've received 2,200 francs! First of all I shall send you back two hundred francs which mean nothing to me. As for the two thousand, I understand that it's your gambling debt, and it's too nice of you not to have considered it a perfectly natural complement to my invitation to Cabourg.[2] In any case why send it so quickly? If I had had more you might perhaps have won it back, and the fact that the inadequacy of the loan was the cause of your loss ought to forbid my agreeing to your paying me back. We'll talk about it later if you like and in the meantime, for the sake of convenience, I shall keep these two thousand francs although my right of ownership is by no means certain.

I have ascertained since Sunday that the feelings of Hervieu and Mme de Pierrebourg[3] towards you are of the utmost violence. Besides, you are the king for us all, vile metal which you strike brutally with your own effigy, and in that royalist milieu of the Mûriers[4] one would have been hacked to pieces if one hadn't shouted 'Long live the King'.

If you have occasion to write to Mlles Massini and Léone,[5] I hope you will speak to them about me in the manner which seems to you most calculated to imbue them with a respect for me which they would certainly not acquire in the lower depths wherein the love of MM Finaly and Fishoff maintains them. The Marquis de Castellane,[6] that poor but charming Lauzun on casters who swoops with clumsy and vertiginous speed on the armchair, the door or the friend he has targeted, spoke to me about you yesterday for an hour with an affection and admiration which his awkwardness made even more touching. I have already tied round the necks of a few gigolos the chains under the weight of which they must appear in front of you on the appointed day.

Dear friend, I feel it's indelicate of me to accept the two thousand but I'm afraid, if I send them back to you, of starting a futile correspondence and being woken up by a registered letter which has to be 'signed' or having to take the hotel manager to the post office yet again so that he can certify my fitness to pouch the sum which you will have refused.[7] All the hotel servants have left, having gone on strike. 'You don't give a damn but I'm telling you this as a contribution to history.'

Good-bye dear friend, and excuse this letter which is as dreary as befits the acknowledgement of an unwarranted sum which one pockets after having expressed one's scruples, in such a way as to elicit admiration for one's delicacy and to satisfy one's rapacity.

Affectionately yours
Marcel

1. Mme (Misia) Edwards was to leave Edwards and marry Sert (see letter 247, n. 9).

2. The playwright was apparently in Normandy to read his new play *Israel* to the famous actress Réjane (real name Gabrielle Réju: 1859–1920) who was on holiday at Hennequeville. Proust had presumably invited Bernstein to visit him at Cabourg.

3. Paul Hervieu (see letter 150, n. 9) was the *grand ami* of Baronne Aimery Harty de Pierrebourg (*née* Marguerite Thomas-Galline: 1856–1943), novelist, pen-name Claude Ferval (see letter 306).

4. Le Clos des Mûriers was the villa rented by Mme Straus the members of whose entourage were, Proust infers, Bernstein's enthusiastic admirers.

5. Two actresses staying at Cabourg, presumably being 'kept' by men-about-town Horace Finaly (see letter 137, n. 7) and Eugène Fischhoff.

6. Antoine, Marquis de Castellane (1844–1917), dramatist. 'Lauzun on casters' is a reference to Antonin Nompar de Caumont, Duc de Lauzun (1633–1723), soldier and courtier who played a lively and adventurous role at the court of Louis XIV, and married the latter's cousin, la Grande Mademoiselle.

7. Proust later sent Bernstein a postal order for 200 francs accompanied by a note which read: 'Dear friend forgive my inexcusable negligence. And whatever you do don't write to the two lovely ladies who have long since left Cabourg.'

301   *To Madame Alphonse Daudet*

[Towards the end of September 1908]

Madame,

You will surmise that I have been unwell and physically incapable of writing if I have not thanked you for your delightful and undeserved card. I simply crossed Paris and will not go back there finally for another month.[1] But it will be a great joy to talk to you about that admirable book of Lucien's,[2] that unknown river which sets off in a new direction, for an opposing shore, but which springs in its turn from the quadruple Sacred Source. The flattering things which his friend may once have said to the Mother of this beloved son fell far short of what she herself knew. And the entire world is merely echoing what was said then in the drawing-room in the rue de Bellechasse by a bashful young man who was proud to have been the witness and occasionally the confidant of the thoughts that preceded the blossoming, the hours when the sky began to take colour.

Please accept, Madame, my respectful attachment and give my fondest regards to Lucien.

Marcel Proust

1. He had left Cabourg on 26 September for the Hôtel des Réservoirs in Versailles (see letter 164), driven in a taxi by Agostinelli, ostensibly to be near his friend Georges de Lauris, who had broken his femur in a car accident with Lucien Henraux. See letter 303.

2. See letter 297, n. 3.

302 *To Reynaldo Hahn*

[Saturday evening, 3 October 1908]

My dear little Funinels,

My heart sank when I read in the papers the disquieting news about Sarah, but it rose again when I heard that it was a false alarm, that she was very well and had cabled her son.[1] I had at once foreseen the possibility of such a terrible blow to my Buncht, and his future deprived of such a great friendship, that I was shattered. And on learning that the whole thing was untrue, I am extremely moved and thank God for having preserved for you this wonderful friend who loves you and understands you and knows how to comfort your melancholy and fastidious heart. May she live into extreme old age, and continue to distil from the fame of her autumnal years the 'soft and golden ray'[2] which you love to savour.

Hasdieu, my Buncht. I returned home so ill that I don't know when I shall be able to see you and don't think I shall be getting up, but I have the solace of knowing that the heart which is dearest to me has been spared this grief, and your happiness (if I may thus call the cluster of your supportable sadnesses) respected. For the joy of the bad news belied, a double ration of bonsjours for the pony.

Birnibouls

1. A rumour that Sarah Bernhardt, then aged fifty-four, had died during rehearsals at Nîmes, in Provence. She cabled her son Maurice: 'Hunting season as usual the excuse for some idiot to put up a canard [*canard*: game duck or rumour]. Kindly clip the wings of whoever says I am on my death-bed. Am very well, no time to think of being ill . . .'

2. '. . . Goûter, en regrettant l'été blanc et torride / De l'arrière-saison le rayon jaune et doux!' Baudelaire, 'Chant d'automne', in *Les Fleurs du Mal*.

303 *To Georges de Lauris*

[Versailles, Wednesday, 7? October 1908]

My dear Georges,

When I spoke the other day of Moses on the threshold of the Promised Land and yet unable to enter it, I didn't know how apt it was. Twice I have been to Paris and the state of my asthma has suddenly worsened as a result of the difference in altitude (or at least so I suppose, but I know absolutely nothing about it) making it impossible to climb even two steps in spite of all the caffeine in the world.[1] This impotence of my friendship is a terrible thing for me, a mixture of grief and humiliation. I think of my poor Mama saying to me at Evian: 'I'm going back to Paris because I'm helpless and can no longer be of any use to you when you are ill.' I cannot repay the tender care you gave me; I always have to receive from you and never give back. And my friendship is perhaps more unhappy as a consequence of this than of the deprival of seeing you, though this deprivation is all the more cruel just now when, after the shudder of horror and danger, I would so much relish the delight of having you safe and sound. Yes, sound, for your face, your look, your cheerfulness are not those of a sick man. And even more than on *your* face, I could read your health on your father's face in the hall at the rue Washington. If I had some really terrifying attacks on my three returns from Paris, and enormous joy on my first visit (when I saw you), on the other two, when I couldn't have reached the upper floors and remained conscious, I enjoyed some minor pleasures with a girl who is new and dear to me, and a few young friends who are also new.[2]

One of these, who is very intelligent and who is about to leave France for a long period, would have liked to meet you, having heard so much about you from me.[3] I didn't dare send him round to you with an introductory note, for fear that he might tire you at the moment, with the passionate curiosity of thinking youth for whatever is presented for its admiration.

At the moment I'm unable to leave my bed, but I hope to come and see you soon. It is always delightful to see you, but even more gratifying now: each of your limbs so miraculously spared, your beautiful, gentle hands which from time to time, when I express a doubt about your friendship, seek mine in a gesture of persuasive eloquence, your whole body whose natural gait, immobilized now but not altered, is the only one I know that is entirely devoid of conventional mannerisms, swift in its movement towards what it desires or knows itself to be desired by, and above all your eyes, which darken so quickly if a sadness traverses your heart but in the depths of which, in an instantaneous effulgence, magnificent azure flashes pierce the clouds – your whole body, indeed, is what I should like to see and touch now after having too long forgotten that it is the necessary condition of all that spiritual spontaneity which is *you* and which we love and for which we must worship the integrity of this symbol of yourself, this body in which your spirit dwells, those hands through which the force of your grasp runs as through a unique and highly conductive metal. Then one must thank the obscure physiological forces which resisted the shock, the good genii hidden in the depths of the muscles and the nerves which preserved you for us. It seems to me that I have too exclusively loved your mind and your heart hitherto and that now I would experience a pure and exalting joy, like the Christian who eats the bread and drinks the wine and sings *Venite adoremus*, in reciting in your presence the litany of your ankles and the praises of your wrists.

Alas, people have always been so cruel and uncomprehending about me, that these are things which I scarcely dare to say, because of the misunderstandings and misinterpretations which would spring up in others' thoughts. But you who know me and grasp with your infallible intelligence the palpable reality of what I am, will understand how purely moral and reverently paternal is what I say to you.

Thank Lucien Henraux most warmly on my behalf for the exquisite letter he wrote me[4] and say to him that I'm so exhausted that he'll understand why it's to you that I've written. But I'm very grateful to him for that letter (which I hadn't yet received when I saw you).

Your

Marcel

1. Lauris was laid up in his parents' third-floor, rue Washington apartment (round the corner from the boulevard Haussmann) where there was no lift. See letter 301, n. 1.

2. Except for one (see n. 3 below), the people Proust mentions have not been identified, but it seems likely that they were the group of young friends from Cabourg he was to invite to the theatre (cf. letter 354).

3. Marcel Plantevignes, son of Camille Plantevignes, a neck-tie manufacturer.

4. A letter apprising Proust of the accident.

304   *To Marie Nordlinger*

[Versailles, about 18 October 1908]

My dear, dear Mary,

Reynaldo, whom I am too ill to see but who writes to me, put me through every stage of anxiety and grief, as though it were one of my own kin. But isn't whatever is yours to some extent mine? I weep with infinite bitterness for him though I did not know him;[1] I rebel against his death, I hope for a miracle which will be revealed in the letter that arrives, and then I realize the impossibility of it, and I think of you again, and weep again. How can I bear the thought of you weeping in the most terrible distress, when all I wished for you was Joy (as though you were not worth far more than Joy). And your parents, and you all! I bear your bereavement like a heavy cross and can only think of you and grieve.

Your

Marcel

1. Her brother Harry Nordlinger had died in a drowning accident on 2 October. This was to be Proust's last letter to his former helpmeet (but see letter 340) until just after her marriage in July 1911.

305   *To Antoine Bibesco*

102 boulevard Haussmann
[Monday, 19? October 1908]

My dear Antoine,

Will you allow me to recommend to your benevolence my young friend Monsieur Marcel Plantevignes who is spending a few months in London to learn English and business and will be comforted in his exile if you can occasionally give him the solace of your company and your conversation.

I envy him if he sees you, for it's a long time since it happened to me, old boy, and, without flattery, it's still in the last analysis, after the wastage of everything else and after due comparison, one of the very rare pleasures whose value has not depreciated for me, and for which I often feel a nostalgic regret.

Good-bye, my dear Antoine, I hope I'm not being too importunate.

As I have a great deal of friendship for Monsieur Plantevignes and his parents I thought I would give him this pleasure and I'm convinced, charming as you have always been to me, that you will generously oblige.

I have no news of Emmanuel. He wrote to me a few times this summer to disinvite me with some agreeable ladies. But nothing since. In fact I've had no sign of life from the whole family, to whom I must confess I have given none. And the last letter which was perhaps no more than a post-card was from your cousin Chimay,[1] dated from La Baule – you can see how long ago.

You know that poor Georges de Lauris was injured in a car crash with Henraux (a broken thigh, alas, but happily on the mend), that Madame de Billy is dying in Saint-Jean-de-Luz (I hope she'll recover), that d'Albu has built himself a wonderful house in the avenue Hoche, that Loche Radziwill has been decorated by Viviani.[2]

My poor Antoine, how utterly idiotic I am.

Ever yours

Marcel Proust

1. Princesse Alexandre de Caraman-Chimay.

2. At a ceremony where the Minister of Labour, René Viviani, unveiled a monument to Jean-Jacques Rousseau at Ermenonville, where Radziwill lived.

306   *To Madame de Pierrebourg*[1]

[Versailles] Friday [23 October 1908]

Madame,

It's at Versailles, a few steps away from the park which was the scene of the long conversations between David Heriel and Laurence de Kermor[2] and to which I cannot even make a pilgrimage amid the imposing Minervas, the voluptuous Venuses and the wounded gladiators, since I haven't been able to rise from my bed for a month, more than ever tortured by fits of suffocation, it's at Versailles that I have just received *Ciel Rouge*, and for a whole night I was unable to put down this book which gripped me with its passion and life. The marvel is that in spite of this it's a model of even-handedness. M. de Kermor's mother is admirable, and as for him, if we don't espouse his cause, we find it nonetheless defensible – he's all the time marvellously intelligible and reconstructed. And it's perhaps the one with whom you have most affinity that you've treated with the most severe impartiality, knowing that he would easily appeal to our hearts, namely David. A doubly profound insight: the twofold egoism of the lover and the poet laid bare. An objective book if ever there was one, but how delightfully subjective I found it. In the setting itself, in front of the Arc de Triomphe which you describe with triumphant mastery, how could I forget the beloved window, the beautiful eyes gazing out upon it, the delicious evening I spent there while you contemplated it in the ebbing light as the lovely sentences we read today were doubtless forming in the unconscious of your reverie.[3] But where you come alive above all, where you give your feelings free rein, is in your love for Odette. That feeling, or at least Odette's feeling for her mother, is one which some pages I wrote a little time ago[4] about mine will perhaps show you, if I publish them some day, that I am not absolutely unworthy of understanding, and one which, if I conveyed it less well, and moreover very differently, I was well qualified to admire the more moving expression of by another. You would see that 'goodnight' scene by the bedside, though quite different and alas all too inferior. *You're* a novelist! If only I could create characters and situations as you can, how happy I should be! While reading this beautiful and tormenting book I thought, for you make one think a great deal, what I have often said to myself before, that your exaltation of maternal feeling may have found an echo in the widely admired work of Monsieur Hervieu.[5] Madame de La Fayette used to say: 'M. de La Rochefoucauld gave me wit, but I reformed his heart.' You needed no one to give you wit, and much more than wit. Only from his own mother can Monsieur Hervieu have received that sensibility which one feels to be innate in him and so deep-seated beneath the superficial coldness, like a burning cyclamen beneath a fine crust of snow. But it's perhaps permissible to think that close association, constant intellectual intercourse with a woman of your sensibility may have given a larger place in his work to the kind of sentiments which you must inevitably have drawn to his attention. If – however indirectly – you have an obscure share of responsibility for the very theme of that immortal masterpiece *La Course du flambeau*,[6] may you be blessed in the name of French Literature, of universal Literature, which numbers few works as profound and as perfect.

Good-bye, Madame, I hope my health will allow me to see you again soon. Please accept my respectful, admiring and grateful compliments.

Marcel Proust

And yet there are two objections I might make to this fine book which I'm too tired to develop here but will explain to you if and when I see you. In a word, I find that M. de Kermor's rage after the on the whole fairly harmless words 'I know you hate artists etc.' is out of all proportion since these words as yet reveal nothing, and makes the comment a few pages further on, when she confesses the need for this affection: 'M. de Kermor could not get over this. These sentimental expressions etc.' seem a bit cold.

I was thrilled by the delightful dedication, a joy I shall treasure with a great deal of gratitude.

1. The novelist Claude Ferval (see letter 300, n. 3).

2. Characters in her novel *Ciel Rouge* published that week.

3. Here Proust leaves the novel to evoke an evening at Mme de Pierrebourg's house in the avenue du Bois-de-Boulogne (now avenue Foch) in 1902.

4. A scene at the beginning of *Jean Santeuil* which reappears, re-written, as the key 'goodnight' scene near the beginning of *RTP*.

5. See letter 300, n. 3.

6. Play in four acts by Paul Hervieu (see letter 150, n. 9).

307   *To Reynaldo Hahn*

[Versailles, Saturday, 24 October 1908]

My dear little Bugnibuls,

I'm not *hunsgrateful* for Lambert[1] if I don't whrites to you. I can't tell you how pleased I am, what a great favour it is, and how much I realize that it was a very tiresome favour for you to ask. I should have whrittens to you at once but I haven't been well, nicens. But I'm very, very grateful, my Buncht, and thank you a thousand thousand times. Too buninuls.

At the present moment my subtle mind, lulled by the waves, is sailing between the mines of Australia and the railways of Tanganyika and will alight on some goldmine which I hope will really deserve its name.

Bunibuls, I can't whrites to you this evening as I have to work. And yet how many subjects there are to discuss with Bunibuls, since I can say of us two as Ruskin said of Carlyle that 'he was henceforth the only person in England with whom he could agree whether for praise or blame'. How many things to say each day to my beloved ear (you Buncht) and hear from my beloved mouth (you again Buncht) about everything that's being done or said or written, in such a confusion of *styles* and *qualities* that it was impossible the other day, in an article signed B. about a matinée at Mme de Béarn's, to make out whether it was by Beaunier or Mme de Béarn herself. Certain remarks, evoking 'a very ancient tradition', incline me to the latter supposition. It's true that a letter from Mme Raunay[2] next day didn't leave the same room for doubt and that her husband had certainly had no part in it. As for Fauré, I don't know whether he's sincere in everything he says but one's a bit startled to see that after having spoken of Wagner (in an article[3] about *The Twilight of the Gods*) as a 'Titan', he heaps exactly the same sort of praise on Serge Basset, Broussan[4] etc. etc. It's true that when I recall some of the things he has said to me about Wagner, the fact that he speaks of Basset and Broussan (and ... Messager ... valneycharlism[5]) in the same tone of voice doesn't perhaps imply an extreme ... admiration (valneycharlissimo).

Binibuls
Tenderly
B.

(who is not, alas, either Beaunier or Béarn, but your Bunibuls)
Ten bonsjours　　　　　　　　　　　　Ten thousand bonsjours

1. Baron Léon Lambert (1850?–1919), Belgian banker whom Proust wished to consult about his investments.
2. Singer Jeanne Raunay who had recently married the critic André Beaunier (see letter 138, n. 3).
3. The composer Gabriel Fauré had written an enthusiastic review in *Le Figaro* of 24 October of the first full performance of Wagner's *Götterdämmerung* – 'a prodigious masterpiece' – at the Paris Opéra. The conductor was the composer André Messager (1853–1929).
4. The first was theatre editor of *Le Figaro*, the second co-director of the Opéra.
5. Invented word implying mockery, irony.

308　*To Madame Straus*

[Versailles, Tuesday, 27 October 1908]

Madame,

I know nothing of you; as, since I last saw you, four or five hours have rarely gone by without my thinking of you, my thoughts have consisted of memories which I long to revise and re-focus with more recent news. I haven't written to you because I have never yet been so unwell. I am in bed at Versailles, unable to get up at all, with constant attacks, and a state of malaise caused by the medicaments which I take all the time to no avail. But I think if I didn't take any it would be worse. It's all too boring. I came back chiefly to see poor Lauris who had a motor-car accident and I can't leave my bed. I hope it will end soon. I thought of going to Territet. But it's easier said than done if one can't even get up! You mentioned a man who never suffered from asthma there. At Widmer's? At Territet itself?[1] Tell me all this when you come back, for I imagine you'll be coming back soon, with sweet Alpine roses on your cheeks, that fragile good complexion you often have which has none of the vulgarity of high colour and which is so moving to those who have invested part of their happiness in the delicate network of your cheeks.

In my less bad moments I have begun (twice in twenty minutes) to work. It's so annoying to think so many things and to feel that the mind in which they're stirring will soon perish without anyone knowing them. It's true that there's nothing very precious about them and that others will express them better.

Jacques wrote me a delightful letter in reply to a protest, quite justified, he said, that I had made about the car I hired on leaving Cabourg.[2] I shall explain to him when I see him the underlying reason which prompted it. But as I don't know whether or not he leaves his letters lying around, I can't tell him in writing.

I gather that M. Reinach is going to speak out against absinthe.[3] People will doubtless plead liberty against him. He should reply with this little passage from Ruskin: 'You'll send your child, will you, into a room where the table is loaded with sweet wine and fruit – some poisoned, some not? – you'll say to him: "Choose freely, my little child! It is so good for you to have freedom of choice, it forms your character, your individuality. If you take the wrong cup or the wrong berry, you will die before the day is over, but you will have acquired the dignity of a free child!"' (*The Queen of the Air*). I wasn't well enough to go and look up Helleu and Monet[4] in Rouen (nor indeed to do anything) and I was heartbroken. I should like to know (don't tell your daughter-in-law[5]) if Monsieur Straus has any Australian Goldmines, or any shares in Tanganyika, or the Rio de Janeiro Tram Light and Power, or Harpener or Télégraphes du Nord. When I say *if he has* I don't mean it out of curiosity, but 'if he has been told to buy', 'if he has been recommended'. I've certainly been told about Australian Goldmines but I don't know *which*, nor the name of the particular mine.

Reynaldo was here when I arrived and we talked a lot about you; he talks about you better than anyone. He came once or twice to sit by my bed and compose the music of his ballet for the Opéra,[6] while Agostinelli and my manservant played dominoes with me. I don't know how he can compose like that in the middle of all that noise. But after a few days he left and I haven't once been well enough to let him come back, not having had even a bearable evening. As soon as I'm better I shall go back to Paris. I imagine you will not be long too and I shall see you soon.

Please accept, Madame, my affectionate and admiring respects.

Marcel Proust

1. Spa on Lake Geneva where Mme Straus was staying. Dr Widmer had a clinic near Montreux.

2. Cf. letter 253, n. 5.

3. Joseph Reinach made this speech on 27 November in the Chambre des Députés; he later published a book entitled *Against Alcoholism*.

4. Cf. letter 253, n. 7.

5. Madame Jacques Bizet was notoriously avaricious.

6. *La Fête chez Thérèse*, first performed in February 1910 at the Paris Opéra.

309   *To Lionel Hauser*[1]

[Paris, 5 or 6 November 1908]

My dear Lionel,

You wrote me such a nice letter that I must seem very stupid and ungrateful not to have thanked you sooner. But at the moment I am going through a very bad period from the point of view of health. And the very brief span during which caffeine enables me to write gives no inkling to the recipients of my letters of the sufferings I've endured in the interval. So for the moment I am the most execrable and yet the least culpable of correspondents. Forgive me therefore.

I was highly amused by your scruples about the Hamburg America. So you consider yourself responsible for the price of the shares bought by your firm! Dear Lionel, if I weren't so ill I should write a play about the Sentimental Financier who wakes with a groan, saying: 'My God, one of my clients has just lost forty centimes on the Portuguese etc. etc.' In any case if you go to such lengths in your kindness you must be very happy because all the stocks you've advised me to buy have improved so fabulously that if my asthma didn't keep me in bed you would see me every morning driving four-in-hand[2] round the Bois!

I must have explained myself badly in my letter though I thought it was so clear. I wasn't at all asking for the difference between the market prices of the stocks bought by your firm, but, something quite different, the *total income* received by the firm (by Warburgs) on my behalf. By total income I mean the interest produced by each stock, I don't know the exact terms, I mean the dividends, the accumulated interest. From this total I would subtract, or rather your firm would subtract, what it spent on the subsequent purchase of my Mexicans (a purchase which Warburgs didn't pay for in full, since Rothschilds paid part of it) and your firm could send the remainder to me either here, or at Rothschilds or the Crédit Industriel. As for the money accruing from the sale of the Rios (yet another marvellous deal that you arranged for me!), let them on the contrary keep that as I intend to re-invest it shortly at the first brilliant 'idea' that springs up in my brain (or, better still, in yours!).

Finally, as regards the New York City Bonds, if they're really the outstanding investment they're said to be, I don't in the least want to relinquish them. I was only asking you whether the presidential election and the days running up to it hadn't improved the value of this stock to such an extent as to make it less advisable to buy at the moment. In that case one could substitute for it some other investment (at least as remunerative) which occurred to you. And if you have no ideas, perhaps the Banco Español del Rio de la Plata, in which Rothschilds have just bought me quite a few shares, or else United Railways of Havana, if they are still quoted at 72 in London, could receive these 20,000 francs.

But in that case I'd like you to be sure to say to Monsieur Léon Neuburger,[3] who thinks the New York City sounder, that I came round to another idea only because of the rise in New York Cities which made them no longer as favourable as when he recommended them to me (having no American share-list I don't even know whether this rise has taken place, and similarly for the Refinings[4] etc.). Please decide all that as you think fit, without bothering to reply: whatever you do will be well done.

It's only on the German side that, apart from my Harpeners, I don't want to take on anything new, as the future seems to me very sombre. It's possible, though, that these very anxieties will provide scope for a highly favourable investment. You know all that much better than I do. Forgive me, my attack has gripped me again and I can do no more than offer you my friendliest regards.

Marcel Proust

Don't bother to answer; it's unnecessary.

I've received a letter from your firm giving me the figures. I see that you understood me perfectly. I'm about to immerse myself in them.

1. Lionel Hauser (1868–1958), once Proust's childhood friend, now his financial adviser and his intimate. They had recently been exchanging letters about Proust's investments.

2. Smart, dashing equipage drawn by four horses.

3. Léon Neuburger (1840–1932), Hauser's uncle, who had a high position at the Banque Rothschild where Proust had his principal account.

4. The American Smelting and Refining Company.

310 To Madame Straus

Friday [6 November 1908]

Madame,

Thank you enormously for your enchanting, funny, sweet letter. And almost at the same time I've been reading M. Ganderax's article.[1] How I should like to have known you as he did, to be able to call you 'my friend from Bas-Prunay',[2] to have been entrusted with that preface, to know all those things and been able to write them. And then it seems to me that I should have written them . . . rather differently. I don't say this *against* M. Ganderax who has immense qualities, a man cast in a mould that is no longer very familiar, that one will come across less and less, and that I prefer to those of today. But why, when he can write so well, does he write as he does? Why, when one says '1871', add 'that most abominable of all years'? Why is Paris immediately dubbed 'the great city' and Delaunay 'the master painter'?[3] Why must emotion inevitably be 'discreet' and goodnaturedness 'smiling' and bereavements 'cruel', and countless other fine phrases that I can't remember? It wouldn't occur to one if Ganderax himself, when he corrects others, didn't believe that he was serving the French language. He says so in your article: 'the little marginal notes I write in illustration and defence of the French language'. Illustration, no; defence neither. The only people who defend the French language are those who 'attack' it (like the Army during the Dreyfus Case). This idea that there is a French language which exists independently of the writers who use it, and which must be protected, is preposterous. Every writer is obliged to create his own language, as every violinist is obliged to create his own 'tone'. And between the tone of a run-of-the-mill violinist and that of Thibaud[4] (playing the same note) there is an infinitesimal difference that represents a whole world! I don't mean to say that I like original writers who write badly. I prefer – and perhaps it's a weakness – those who write well. But they begin to write well only on condition that they're original, that they create their own language. Correctness, perfection of style do exist, but on the other side of originality, after having gone through all the faults, not this side. Correctness this side – 'discreet emotion', 'smiling good nature', 'most abominable of all years' – doesn't exist. The only way to defend the language is to attack it, yes, yes, Madame Straus! Because its unity is made up of neutralized opposites, of an apparent immobility that conceals a perpetual, vertiginous life. For one 'holds one's own', one cuts a convincing figure, beside the writers of the past only providing one has sought to write quite differently. And when one wants to defend the French language, one actually writes the opposite of classical French. For example the revolutionaries, Rousseau, Hugo, Flaubert, Maeterlinck, 'hold their own' beside Bossuet. The neo-classicists of the eighteenth and early nineteenth centuries, and the 'smiling good nature' and the 'discreet emotion' of all epochs jar with the masters. Alas, Racine's most beautiful lines

> I loved you fickle; faithful, what might I have done? . . .[5]
> Why murder him? What did he? By what right?
> Who told you to?[6]

would never, even in our day, have got into the *Revue de Paris*. Marginal note by M. Ganderax 'In Defence and Illustration of the French Language': 'I understand your thought; you mean that since I loved you when you were fickle, what might that love have been if you had been faithful. But it's badly expressed. It could equally well mean that *you* would have been faithful. As official defender of the French language I cannot let that pass.' I'm not making fun of your friend, Madame, I assure you. I know how intelligent and learned he is. It's a question of 'doctrine'. This man who is so sceptical has grammatical certainties. Alas, Madame Straus, there are no certainties, even grammatical ones. And isn't it happier that way? Because in that way a grammatical form can itself be beautiful, since only that which bears the imprint of our choice, our taste, our uncertainty, our desire and our weakness can be beautiful. Yes, this highly intelligent man has known you all your life. He has already experienced something of life as a whole; he turns and looks back; the diversity of planes ought to multiply the beauty of the varied lighting for him. But grammatical dogma holds him in its chains.

Discreet emotion, smiling good nature. And then this gay little Carmen,[7] is it really you? Isn't there also something of Perdita, something of Imogen, in you? In spite of everything, it was a fine testimonial to a life in relation to other lives, sorrowful and beautiful, in the effulgence of their fame. I read it with a great deal of pleasure. And I found the description of your portrait delightful. Recently he gave a marvellous little talk to some schoolchildren. It was much better.

Madame, what gloomy madness to start writing to you about grammar and literature! And I'm so ill! In heaven's name *not a word* of all this to Madame Ganderax.[8] In the name of heaven – in which, alas, neither of us believes.

Respectfully yours
Marcel Proust

1. *Letters of Georges Bizet*, Louis Ganderax's (see letter 130, n. 2) preface to a forthcoming book of the composer's letters written from Rome or during the Commune in Paris. Proust had read it in *Le Figaro*.

2. Her country home near Bougival before she married Georges Bizet.

3. Elie Delaunay (1821–91), friend of Gustave Moreau, who painted the portrait mentioned at the end of the letter.

4. Jacques Thibaud (1880–1953), famous violinist, twice mentioned in *The Captive* (*RTP*, III, 47, 289).

5. 'Je t'aimais inconstant, qu'aurais-je fait fidèle!' *Andromaque* (Act IV, scene 5). Proust writes *qu'eussé-je* in error for *qu'aurais-je*.

6. 'Pourquoi l'assassiner? Qu'a-t-il fait? A quel titre? / Qui te l'a dit?' Ibid. (Act V, scene 3).

7. Ganderax wrote: 'brown-haired, bouncy and fresh, rather mischievous, olive-skinned, with splendid eyes, a child Carmen'.

8. See letter 130, n. 2.

311 *To Georges de Lauris*

[Paris, Sunday evening, 8 November 1908]

My dearest Georges,

Just a few words to let you know what I haven't yet told anyone else. Having had an attack of incipient bronchitis in Versailles and because of road-mending right under my windows in the rue des Réservoirs I came back to Paris to be plagued by the building work of the dentist who has taken the third floor and literally asphyxiated by an ill-functioning water-heater. As soon as I can get up I shall be off I don't know where, but first I shall come to see you.

Today is the last day of the exhibition I should like to have seen more than any other: the two painters I'm most 'in love' with, Greco and Monticelli,[1] at the Autumn Salon, and I haven't been able to go! Before your accident when I didn't expect to come back to Paris (a lot of good it's done me and it's been extremely helpful to you!) I told myself that no matter where I was, even if it was Venice, I'd come back for forty-eight hours to see this exhibition which may not be sublime but which would have been so fruitful for me because it corresponded to so precise a juncture in my longings. And it was physically impossible for me even to be pushed there in a wheelchair which I would have done without shame.

I think I'm going to get better; I think I'm going to come to see you. Georges, whenever you can, *work*. Ruskin somewhere said a *sublime* thing which we should keep in mind day after day: that God's two great commandments (the second is almost entirely his but it doesn't matter) were: 'Work while you still have light' and 'Be merciful while you still have mercy'. I can assure you that Léon Blum[2] has never said anything as good. After the first commandment, taken from St John, comes this sentence: 'For soon the night cometh when no man can work' (I'm quoting badly). I am already half way into this night, dear Georges, in spite of fleeting appearances that mean nothing. But you still have light, and you'll have it for long years to come, so *work*. Then if life brings rebuffs there are consolations, for the true life is elsewhere, not in life itself, nor afterwards, but outside, if a term that takes its origin from space can have a meaning in a world freed therefrom. And as for mercy, you have had something better than that, a love so sweet that the death of your poor mother sometimes doesn't seem to me so terrible when I think of all the hope she left behind, all the punctiliousness with which you are fulfilling her wishes. As for your life with your father, you know what I think of that and what a delight it is to me.

With me you are very kind, my gratitude tells me, less affectionate than kind although you think the opposite, and kind with reservations arising from your restless character. But nonetheless a thousand times kinder than I deserve.

Good-bye, dear Georges, forgive this evangelical sermon. But accidents such as yours, which I keep picturing to myself, should be a warning. Work, since God has bequeathed you light. No, dear Georges, it's no use searching in Léon Blum, or even, despite what our friends will swear to you, in Claude Anet,[3] you won't find anything so good. Which doesn't mean that Ruskin's works aren't often stupid, cranky, exasperating, wrong, preposterous, but they are always estimable and always great. He was, as you know, greatly admired by George Eliot who, whatever Léon Blum may think, is worth all the Marguerittes and, who knows, perhaps the Rosnys,[4] and who spoke of 'those great works which enable us to reconcile our inward despair with the delightful sense of a life outside ourselves'.[5]

Dear Georges, it's very exhausting to write, but it's pleasant to recall fine thoughts to someone who knows them and understands them and can draw sustenance from them, and for whom the writings of our more or less celebrated acquaintances are of very relative value.

Yours with all my heart and here and now though coughing endlessly and racked with fever, with three windows open at one o'clock in the morning to combat the water-heater. None of that's a patch on love, etc.[6]

Your

Marcel

1. This exhibition at the Grand Palais was the occasion for the rediscovery of the Marseillais painter Adolphe Monticelli, a remarkable colourist. See letter 84, n. 9.

2. The future Socialist leader and Prime Minister Léon Blum (1872–1950) had published his advanced ideas in *On Marriage* (1907), a book admired by Lauris. Proust seems to have had a prejudice against him ever since they had both been contributors to *Le Banquet* (see vol. I).

3. Claude Anet, pseudonym of Jean Schopfer (1868–1931), had just written an anti-feminist book called *Notes on Love*.

4. For the Marguerittes see letter 179, n. 5; Joseph and Justin Boex collaborated as 'J.-H. Rosny' until 1909.

5. An allusion to a passage in *The Mill on the Floss* (book IV, chapter 3), 'A Voice from the Past', in which the reading of the Bible, Thomas à Kempis and *The Christian Year* gives Maggie a new outlook on life.

6. 'Tout ça ne vaut pas l'amour': refrain of a popular song from a new comedy by Proust's friends Gaston de Caillavet and Robert de Flers.

312   *To Maurice Barrès*

102 boulevard Haussmann, please forward
(but above all don't answer)
Monday [9 or 16 November 1908]

Dear Sir,

It is perhaps somewhat presumptuous of me, or somewhat conceited, to inform you that I am reading *Colette Baudoche*[1] with delight.

But after all its publication in a review must mean to you a simultaneous throbbing imparted by you to thousands of hearts. So perhaps you will not be averse to measuring, on an instrument that is not easy to set in motion, the amplitude of the vibration received. The supreme distinction of Metz, the airiness of the countryside, those faces worthy of Chardin and perhaps better, those delightful villages which you designate by a Homeric characteristic, quite naturally and without a hint of pedantry, because one is the village of strawberries and another is distinguished by its fruit trees (as Colonne is prolific in horses)[2] – all this must seem to you well worth the sacrifice of a few colours, in order to obey more readily the sacred instinct which tells us what we must depict. As for the story, you were wrong to warn that it had no surprises in store; it is interesting as life is, and I cannot tear myself away from it. For me the Baudoche ladies are a little lacking in heart (but I like their being so and your being a true portrayer of them and of yourself). I can see Mama very well in Madame Baudoche's place and the similarity of many a trait. But she would never have allowed me to laugh at the Prussian, or else it would have been corrected by a great sympathy for his kindness, his naïvety, for what he had left behind in Germany, his fiancée, etc. But one must appreciate the diversities, the distinctions of human nature, and your ladies of Metz are perfect. I like a little less (in terms of expression) some of the things about hideous German architecture. But everything is delightful and drawn from nature, without ostentation. What an example!

Often I feel a wish, like the Prussian, to interview you, to know the name of the place or the person described. But I asked you if I might do so in connection with *Au Service de l'Allemagne*,[3] and your silence seems to me retrospectively a bit 'Baudoche'. So I no longer dare question you. But in so many respects my admiration has met with a sympathetic response from you that I wanted to tell you that I was reading you slowly, affectionately, my heart attuned to yours, moved by the gentleness which your style assumes at times, as does your voice, and curious, if I ever see better days, to go and visit that noble city of Metz and the wardrobes and the dresses.[4]

Your most sincerely devoted
Marcel Proust

1. Barrès's novel, subtitled 'The Story of a Young Girl from Metz', was being serialized in the *Revue Hebdomadaire*; Proust appears to have read the first two parts.

2. Aware as he was of the potency of names (cf. *RTP*) Proust appreciated what Barrès calls the 'caressing names' of these Moselle wine-growers' villages.

3. A novel (1905) with an Alsatian hero who is obliged to do his military service in the Prussian army.

4. In Part Two of the serial there was a description of an elaborately carved wardrobe made in Lorraine.

313 *From Maurice Barrès*

[Shortly after 9 or 16 November 1908]

My dear Proust,

I'm glad that you like the beginning of this little book. It is, I think, stark and spare like the mental image I have of that military and unhappy town. I'm pleased, too, that you should have distinguished the admirable dryness of these soldiers' daughters faced with transrhenane inferiority.[1]

I hope your health is giving you more satisfaction. With your so well executed pastiches you are on the verge of a delightful form of literary criticism[2] which you should seize and which would readily prove what Buffon knew,[3] that form and substance cannot be separated, that to write in a certain manner is to think and feel in a certain manner: if you added a sharply defined, slightly caricatural portrait of the author interpreted in the sense which your parody has emphasized, without showing your hand, you would say everything.

I shake your hand.

Barrès

1. Barrès had first written 'stupidity' here, but crossed it out.

2. Cf. letter 268, Proust to Chevassu.

3. In his acceptance speech to the French Academy in 1753 the Comte de Buffon said: 'To write well means to use the mind, feelings and powers of expression well; good writing must have at one and the same time wit, soul and taste. Style presupposes the combination and the exercise of all the intellectual faculties . . .'

314 *To Léon Neuburger*

[End of November 1908]

Dear Monsieur Léon,

Even when they're 'official' and not even signed it gives me great pleasure to see your letters, the sort of pleasure one gets from any manifestation of clarity, of light, in the most elevated as well as the most down-to-earth sense of the word. And I admire that conscientiousness in the most minute details. You apply intelligence and virtue even to a stock-exchange transaction! It's amusing to compare with yours the letters of M. Guastalla[1] which inform me for instance that 'I have just remitted 26,000 francs to MM de Rothschild against 75 Banco Español which they transferred to me' when he has done precisely the opposite, received the money and sent the shares, and the other day advised me of the purchase of 35 Santa Fe 4½% when in fact they were 5%. Let's hope he's more accurate with figures than with words! As for Warburgs, if they debited you with 52,000 francs, they only advised me of the purchase of 20,000 New York City and 10,000 Harpeners. I don't suppose there's been an error; perhaps they bought two lots of Harpeners and forgot to advise me of the second purchase. I'll talk to Lionel about it and also about the accounts from his firm when I manage to see him but I've been in a lamentable state for some time and it's even a great effort to write to you, but I couldn't resist the desire to tell you what an agreeable impression your 'letter of enlightenment' made on me.

I've sold my Rios below the high prices they had reached, but I feel it's a healthy thing not always to sell 'at the top' for then one risks waiting . . . for the bottom!

I've been told about a bank with splendid prospects which has just been founded by an Englishman, very well known it seems, called Cassel.[2] The name of the bank begins with something I can't remember and ends I think with Egyptian Bank. The remote possibility of procuring me some shares in it was held out to me as a chance I ought to bless. But I'm not at all sure about it and I don't know whether I ought to want it to happen.

When I heard that the New York City Bonds were at 102⅛ I hesitated to buy them. But I didn't dare write to you again and I bought them since it's a good investment and after all fairly remunerative.

As I was telling you about the frivolities (*confidentially!*) of the Maison Guastalla I couldn't help laughing rather sadly: my poor parents were convinced that I would always be incapable of reading a business letter or taking the slightest interest in money matters. I know it was a real source of anxiety to them. And I think with sadness of the pleasure it would have given them to see what a good accountant I am.

Please accept, dear Monsieur Léon, my affectionate and grateful respects.

Marcel Proust

1. A stockbroker whose mere name was to trigger off two comic scenes in *The Guermantes Way*. Proust, who had just embarked on his genealogical research, discovered Guastalla to be a minor Italian dukedom seized by Napoleon from the Duke of Parma to give to his sister Pauline. For the purposes of the novel he supposed that the title had descended through both families, thereby setting the scene for one of the Duchesse de Guermantes's famous 'audacities': in front of the Princesse de Parme and other guests she enthuses over the magnificent Empire furniture owned by the Iénas – a Napoleonic title invented by Proust – who were, for the Princess, 'rank usurpers, their son bearing like her own the title of Duc de Guastalla . . .' (*RTP*, II, 538, 540–1). In the second scene the Baron de Charlus, provoked by the existence of two dukes of the same name, remarks: 'As for this self-styled Duc de Guastalla, I supposed him to be my secretary's stockbroker . . . But no, it was the Emperor . . . who amused himself by conferring a title which was simply not his to bestow . . . an exceedingly scurvy trick to play on these unwitting usurpers' (*RTP*, II, 586).

2. Sir Ernest Cassel (1852–1921), international financier and adviser to Edward Prince of Wales, was founder of the Egyptian Mortgage Bank.

315   *To Madame de Noailles*

[Mid-December 1908]

Madame,

Will you allow me without preamble to ask your advice. I should like, although I'm very ill, to write a study of Sainte-Beuve.[1] The idea has taken shape in my mind in two different ways between which I must choose; but I have neither the will-power nor the clearsightedness to do so. The first would be a classical essay, an essay in the manner of Taine[2] only a thousand times less good (except for the content which I think is new). The second begins with an account of a morning, my waking up and Mama coming to my bedside; I tell her I have an idea for a study of Sainte-Beuve; I submit it to her and develop it.

Can you tell me which of them seems the better. I ought to apologize to you so profusely that my exhaustion precludes any attempt to do so, but what makes my audacity outrageous also makes it excusable: it's because you are our greatest writer that it's monstrous to bother you with these trifles, but it's also for that reason that your advice is irreplaceable.

Please accept, Madame, my most profound and respectful admiration

Marcel Proust

I can't telephone, otherwise I wouldn't have bothered you with a letter.

1. Charles-Augustin Sainte-Beuve (1804–69), writer and influential literary critic. If Proust's statement that 'Sainte-Beuve misunderstood all the great writers of his day' (in the Preface to *Sésame et les lys*: cf. letter 16, n. 3) was one key to *Contre Sainte-Beuve*, the book he was now undertaking, more vital clues to this complex, hybrid work, part criticism, part fiction, lie in its autobiographical subtitle 'Memory of a Morning' – the morning referred to here – and in Proust's preoccupation with homosexuality (cf. letter 284). In August 1909 (letter 341) Proust explains in detail his aims for this project which by that time had already transformed itself into the embryo of *A la recherche du temps perdu*.

2. Hippolyte Taine (1828–93), philosopher, historian and critic.

316 *To Georges de Lauris*

[Shortly after mid-December 1908]

My dear Georges,

Thank you most warmly for your letter and your telegram.

I never speak to you of the details you give me about your leg, but I can't tell you how acutely I'm aware of them. I have three or four such concrete images in my mind. Now what your leg looked like when you saw it again, and what the photograph showed, have been added to them. Fortunately these images are at once relieved and corrected by the happy knowledge of your health, your recovery and your proximity. Alas, the same cannot be said of another image side by side with those I spoke of, which is of a person I never knew but whom I mourned and still mourn as though I had always known her: the terrible state of your poor mother's gall-bladder after her operation.

Georges, I never know whether you get my letters . . .

Thank you for your advice.[1] It's *the right* advice. But will I follow it? Perhaps not, for a reason you will no doubt approve. What's annoying is that I have again begun to forget that piece on Sainte-Beuve which is written in my head but which I can't put down on paper since I can't get up. And if I have to do it all over again in my head for the fourth time (for already last year . . .) it will be too much.[2]

Having gone without food for almost three days I was unable to answer you, I felt too weak. Did I tell you I went out the other night at midnight but didn't dare knock on your door at such an hour.

Thank you for the Sainte-Beuves. For some time now I've been buying a lot of books, including all Sainte-Beuve. I've already lost a few but what remains is more than enough. I think you'll like my study if it ever gets written. But before writing it, like Musset[3] with truth, I shall already be sick of it.

Georges, as a small inexpensive present I've ordered for you Mâle's new book, and Fouquet's *Antiquités Juives*.[4] Please don't insult me by sending them back to me. Now would you like me to lend you some scandalous works I bought at M. Mirabaud's sale,[5] though I hesitate to do so as they are so filthy. Ah, these Protestant bankers, members of the Evangelical Society!

You never give me any news of your father, who doesn't like me but whom I worship and venerate and whom I think of whenever I think of you, that is all the time. I've got a pain in my hand, in my head etc. and can only send you my most affectionate regards.

Marcel

1. At the same time as he wrote to Mme de Noailles (see previous letter) he had written a similar letter to Lauris.

2. He had clearly had his Sainte-Beuve project in mind for some time (cf. letter 15, n. 3).

3. 'Quand j'ai connu la Vérité, / J'ai cru que c'était une amie; / Quand je l'ai comprise et sentie, / J'en étais déjà dégoûté.' Alfred de Musset, 'Tristesse'.

4. Emile Mâle's massive work on the religious art of the late Middle Ages in France (cf. letter 245) and a short book called *Les Antiquités judaïques et le peintre Jean Foucquet* by Paul Durrieux (1855–1925).

5. Sale of Paul Mirabaud's illustrated books (antiquarian and modern) at the auctioneers Drouot on 17 and 18 December. Cf. following letter.

317 *To Georges de Lauris*

[Second half of December 1908]

My dear Georges,

Thank you with all my heart for your delightful letter which brought a breath of friendship into my melancholy solitude. I didn't mean, Georges, to send you 'disturbing' books but rather the sort of pornography that mortifies the senses. They are two clandestine Verlaines, obscene and stupid.[1] But I have two other books which are merely 'improper' (and not very improper) and which might entertain you for half an hour if you don't know them: *Seven Letters to Stendhal* and *H.B.* by Mérimée.[2] I shall ask for them back for an hour in a few days and then you can keep them. But they're nothing wonderful. Georges, if your innocence is at last beginning to weigh on you I can quite understand that you wouldn't want to bring cocottes to your house, but it's said that

... As for me, I only like (at the moment I don't like anything, as you can imagine) young girls, as though life weren't complicated enough as it is. You will tell me that marriage was invented for such contingencies, but then she ceases to be a girl, you can only have a girl once. I understand Bluebeard, he was a man who liked young girls. Speaking of people who like to complicate things, there are the Rothschilds, who on top of everything else they demand of a future daughter-in-law, require her to be Jewish. This keeps the Halphen family[3] very busy, but that family isn't unlimited. I remember hearing some people say on a train: 'You see, rich as they are, the Rothschilds can't find Catholics to marry them.'

Dear Georges, how do you expect me to work when I can't sleep, can't eat, can't breathe – this letter represents a labour of Hercules.

Good-bye, dear Georges, till soon.

Your

Marcel

1. Among the books Proust bought at the Mirabaud sale and lent to Lauris (see previous letter) were two pornographic works, *Femmes* (Brussels, 1890) and *Hombres* (Paris, 1891), by Paul Verlaine (1844–96).

2. Prosper Mérimée's *H.B.* is in the form of an obituary of Stendhal (real name Henry Beyle: 1783–1842). In both volumes voyeurism seems to have been the vice portrayed; the cover of *H.B.* showed Stendhal spying on his mistress.

3. A large family whom Proust had met at Evian in 1899.

318   *To Georges de Lauris*

[Towards the end of December 1908]

My dearest Georges,

How are you? With each minute that goes by I think how the solemn and to me truly sacred Task of rebuilding, suppling up and strengthening your leg is being accomplished. How nice it is of you to tell me of its progress, though all too rarely.

I am not well. I went out for a moment the other night, but it was a quarter to twelve and I didn't dare come round to the rue Washington. But really I think it was as well, for with the present condition of my heart I wouldn't like your few gentle stairs to be the slope of the hill climbed by the proverbial invalid who died a month later of the lesion he had incurred through breathlessness.

I've not yet started work and have now forgotten everything I've read of Sainte-Beuve, but if I have the strength as I believe I shall, because after all I am a bit better, I shall write it out of curiosity to have your opinion of it when you read it. If I were capable of moving I'd go and shut myself up somewhere to write it, but as I see no one I'm shut up here.

I've stopped reading Chateaubriand (of whom I've done a pastiche)[1] and am deep into Saint-Simon to my vast entertainment. I must say it's quite a change from the *Mémoires d'outre-tombe*. But I'm mainly preoccupied with nonsense, genealogy etc.[2] I swear it isn't out of snobbishness; it amuses me immensely.

Did you get the two Stendhals (*H.B.* and *Letters*) a fortnight ago? Not that I need them in the least.

Georges, I am very fond of you, much more than I say, perhaps even more than I realize, and I have a tender veneration for your father whom in my thoughts I never separate from your mother. I must say good-bye because I have a pain in my eyes, my hand etc.

Your

Marcel

1. 'L'Affaire Lemoine par Chateaubriand', unpublished in Proust's lifetime, was found in one of the little notebooks Mme Straus gave to Proust (see letter 264, n. 1) and published in *Textes Retrouvés* (1968: ed. Philip Kolb).
2. Cf. letter 314, n. 1.

319   *To Lucien Daudet*

[January 1909]

My dearest Lucien,

It's very good of you to think of me again, and I thank you with all my heart. If I feel better I shall write to you to see if we can meet. But at the moment!

You didn't send me those letters from Boylesve and Jammes which would have given me such pleasure, my two favourites among the living. Even without great praise, the fact that they are in friendly correspondence with you seems to me most flattering, if one can apply such a word to someone of your talent. I don't know whether Jammes ever knew what I said about him in an article on Mme de Noailles in the *Figaro*.

Do you have a clear recollection of *Manette Salomon* and do you know in which part of the book there's a eulogy of washerwomen, from the pictorial point of view?[1] I've been very enamoured of washerwomen for the past year and I should like to see it but haven't been able to find it. Don't bother to answer just for that. But if ever you write to me and can remember, tell me then (in several years' time, if I'm still alive, there's no hurry) where to find it.

Good-bye, my dear little Lucien. Is there anything I can do to please you?

Ever yours
Marcel

1. Allusion to a bucolic passage in *Manette Salomon*, a novel by the Goncourt brothers; there are washerwomen, but not depicted as Proust recalls. Washerwomen, or laundry-girls, frequently recur in *RTP*.

320 *To Georges de Lauris*

[Around Friday, 15 January 1909]

My dearest Georges,

Thank you very much for your letter; I'm glad you are beginning to go out. I have just spent some days which, since I ceased to have cause for grief on others' account, are I think the worst of my life, because for the first time I'm utterly despondent: life isn't possible with such constant attacks. I hope I shall get better; it's high time because I'm at the end of my energy which even if it's only a form of inertia is very great, I assure you. I selfishly confide my troubles to you because you are one of the few to whom I can. Reynaldo feels these things too acutely and makes me even iller by writing me furious letters full of stupid advice.

I think the fog of the last few days has been the last straw and also the fact that I haven't been able to eat, because I'm surprised myself by my ignoble cowardice in giving way to this gloom and confessing to you when I've suffered so much and life no longer counts for anything.

Don't have any regrets about not being able to come; you couldn't see me anyhow. No, I haven't yet begun *Sainte-Beuve*[1] and doubt whether I shall be able to, but if only I have a few hours I assure you it will be *not bad* and I'd like you to read it. I may feel better any day, and if I'm at all able I'll get down to it. But I've forgotten everything I'd read. Not that it matters. You can imagine that if I want to do it it isn't in order to write 'criticism'.

I've robbed you yet again: Nicolas has brought me seven volumes of *Port-Royal.*[2] I can't read them at the moment. Perhaps I shall look out the pastiches I scribbled on Chateaubriand and Régnier[3] but they're so utterly illegible that you wouldn't be able to decipher a word; it would be best for me to dictate them and send them to you. But they're less than nothing you know. In any case you already know these exercises.

You still haven't told me whether you received *H.B.* and the *Seven Letters.*[4] I'd be interested to know what you think of them . . .

Good-bye Georges, if I feel a bit better one of these days I'll do *Sainte-Beuve* and as soon as the last line is written I'll send it to you. Perhaps I shall then be able to say (or rather think), scarcely think but anyhow think, like Joubert: 'Behind the strength of many men there is weakness, but behind my weakness there is strength.'[5]

Did you know that when Chateaubriand was arrested in 18-- he was found in bed with two women.[6] How weak the flesh must be if, ill as I've been and still am, I think of that with pleasure.

Affectionately yours
Marcel

1. See letter 316.

2. Sainte-Beuve's history of the Jansenists; evidently Lauris had lent him more books which Proust had had Nicolas Cottin collect. Cf. letter 316.

3. For Proust's parody of Chateaubriand, see letter 318, n. 1; that of Henri de Régnier was to appear in the *Figaro* literary supplement on 6 March.

4. See letter 317.

5. The quotation from the French moralist Joseph Joubert (1754–1824) is found in Sainte-Beuve's *Portraits littéraires.*

6. Proust must have read this probably apocryphal story in the Goncourt *Journal* of 7 October 1866.

321   *To Reynaldo Hahn*

[Thursday evening, 21 January 1909]

My Tuninels,

No one, neither Saint-Victor,[1] nor Gautier, nor Berlioz, nor the overrated deceased,[2] nor Sainte-Beuve, nor Wagner, *could* have written that. It's *magnificent, exquisite, sublime.*[3] To quote France, it's a witches' brew in which there are flowers as well as tigers' blood. What a pity that a remark about Ronay makes it impossible to send it to Bosnier for the review of the press! Perhaps, though I would have been too incapable of writing these pages to dare to criticize them, a slight alteration could nevertheless be introduced in readiness for the inevitable and glorious day when your writings are collected in book form. The image of catching fire from the mere *reflection* of a ray is admirable. But I feel that it shouldn't be a volcano, whose combustion, if I'm not mixing up my geology, is purely internal, aboriginal. You could use two successive images, the volcano for the second only, with its cascades of lava. And I think you could lighten the sentence (carried away into an incandescent vortex) by putting in its place: one is carried away, one abandons oneself, one revolves in the incandescent vortex. Perhaps 'into an' is unnecessary since the vortex is designated.

Hasgouen and agdsmiration from
Bunchnt-Nidoulss

1. Paul Bins, Comte de Saint-Victor (1827–81), literary critic.

2. Ernest Reyer (1823–1909), music critic (as Berlioz had been) on *Le Journal des Débats*, composer of the operas *Sigurd* and *Salammbô*, who had died on 12 January.

3. An article by Hahn in *Les Nouvelles* about a performance of *La Sulamite* by Emmanuel Chabrier (1841–94) and Ernest Reyer. Hahn was critical of the principal singer Jeanne Raunay – whom Proust spells Ronay – who was married to the critic Beaunier – whom Proust spells Bosnier. Proust goes on to criticize certain metaphors employed by Hahn.

322   *To Lionel Hauser*

[1 February 1909]

My dear Lionel,

You give me the impression of the greatest power a man can have on this earth: I have just received from Messrs Warburg a letter[1] which I don't know whether to describe as worthy of Voltaire or of Mérimée, not knowing which of these two writers you prefer, but anyhow written in a French that is not only impeccable but of the most refined elegance. Accordingly, I think I may infer therefrom (to borrow one of their locutions) that your authority, more powerful and above all more rapid than that of Berlitz, has given them in three days (without deducting the despatch and return of your order) not only the most thorough command of the French language and the usage of all its refinements, of all the 'I think I may infer therefroms' which form the superiority of our beautiful tongue, but also an elegance of expression, a purity of style which thanks to you will make French culture in Germany resemble that which reigned in Voltaire's time from the Spree to the North Sea and of which you are the new Frederick the Great. Thank you most warmly for having performed this friendly philological miracle for me and please accept my fondest regards.

Marcel Proust

Since you have such great thaumaturgical power could you use some of it to ensure that I am spared the dozen attacks of asphyxia a day which are killing me at the moment. Never have I been so ill! Needless to say I didn't in the least intend to mock Messrs Warburg's former French. I should be only too glad to write German as they wrote French hitherto! As for the way they write it after your thaumaturgy, if I wrote German like that I should be Schiller, Novalis or Hoffmannsthal!

1. He had begged Hauser to ask Warburgs, a finance house based in Hamburg, to find someone able to write to him in comprehensible French.

323 *To Robert de Montesquiou*

Tuesday [16 February 1909]

Dear Sir,

My illness, which has recently taken a turn for the worse, is now complicated by the fact that I can no longer write the slightest thing without instantly developing a headache. Hence, so as not to write unnecessarily I was waiting until I could go and see you to tell you that I was thinking of you a great deal, that I had read your enchanting articles on El Greco, on Leonardo and on the painter of the Engadine and that had I not been ill I should have congratulated you on them.[1] The other day I read *Les Paons*,[2] and although reading tires me, at least I read some beautiful things there. And even in the simplest pieces such as the one dedicated to Dr Robin,[3] or the one to the Duchesse de Rohan, there is true greatness, flowers that are stylized but irresistible.

I regretted that my name didn't appear above any of them. I don't deserve it as much as André Maurel. But Victor?[4] However the dedication makes no difference. The pieces belong to whomsoever is capable of appropriating for himself the thought that they enshrine.

And in this respect I am arrogant enough to believe that I am their predestined, or one of their predestined readers.

If by any chance you still have the pastiche of Saint-Simon, 'Fête chez Montesquiou',[5] and don't need it, could you whenever you have the opportunity – there's no hurry in the least, any time – have it sent to me at 102, boulevard Haussmann, please forward. As long as I'm unable to work, I intend to dispose of another few pastiches which I did last year, and perhaps collect them all. While having some Saint-Simon read to me recently, I wondered whether that one would be accurate enough to go with the others. Being full of Saint-Simon at that moment,[6] I could easily have corrected it. I'm no longer in the same mood, but if I read or get someone to read me some more, I should be able to make any necessary adjustments, if I have some possible hours; I have just been through some terrible ones.

As a matter of fact the pastiche that would most amuse me to do when I can write a bit (without prejudice to more serious studies) is one of you! But in the first place it would probably annoy you, and I don't want anything of mine to annoy you, I'm too fond of you for that, and secondly I feel that I should never be able to, never know how to![7]

I hope soon to be once more visible and approachable, and shall then try to see you. In the meantime I send you all my affectionate and admiring respects.

Marcel Proust

1. All published in *Le Figaro* in autumn 1908. The painter of the Engadine was the Marquis Gaspare de Vitelleschi degli Azzi (1875–1918).

2. Special edition of collected essays (October 1908) illustrated with Montesquiou's portrait by Philip de Laszlo (1869–1937).

3. Professor Albert Robin (1847–1928), doctor of medicine and writer.

4. Journalist André Maurel (see letter 153) who had written flatteringly on Montesquiou, and opera singer Victor Maurel (1848–1923).

5. Probably Proust's earliest published parody, see letter 25, n. 7.

6. Cf. letter 318 to Georges de Lauris.

7. Montesquiou replied that he trusted to Proust's talent, not to mention his delicacy of feeling; as for the Saint-Simon, he was delighted that Proust had at last admitted paternity!

324   *To Robert de Montesquiou*

Tuesday [2 March 1909]

Dear Sir,

Thank you very much for your letter.[1] I'm too exhausted at the moment to write any new pastiches. And yours is not even begun. I only know that if I did it I should use as an epigraph a phrase of Saint-Simon's on the Regent: 'a diamond as big as a greengage',[2] if, that is, the phrase has never been quoted by you. For a pastiche must never quote anything. And I've just begun to wonder if, though I know it myself from Saint-Simon (at the time of Peter the Great's arrival in Paris, if I remember rightly), I haven't read it also in one of your works. Please don't reply to me just for that! I shall search for it and ask you about it in person. In any case I'm not thinking of new pastiches at present.

Please accept, dear Sir, new admiring thoughts (entirely fresh ones, as I read you all last night), always confirmed, renewed, and indeed assured.

Marcel Proust

1. He had written to suggest that as the pastiches Proust had mentioned in his previous letter had not yet appeared, he was no doubt concentrating on his, which would be the most amusing!

2. Diamond bought for the young Louis XV by the Regent, the Duc d'Orléans, with the encouragement of Saint-Simon, who described it as being 'de la grosseur d'une prune de la Reine Claude' (i.e. a greengage). It became known as 'the Regent'.

325   *To Henry Bordeaux*[1]

102 boulevard Haussmann
[Shortly after 3 March 1909]

My dear friend,

I am touched that you should still remember the invalid and recluse, and that you should send him those beautiful Savoy grapes 'with all their leaves'. I tasted them as soon as they arrived, that is to say all through last night. I'm keeping some for tomorrow but I didn't want to delay sending you my affectionate gratitude. Your Madame de Charmoisy enchanted me.[2] Ah, how nice it would be to travel with you through that country from which you generate, like mystic corn, this harvest of splendid stories. A few names linked in my memory to our meeting down there filled me with melancholy poetry as I read your pages in which such wide horizons unfold, dominated by a significant summit. But should I regret my secluded life this evening? Have I not been walking with you just now?

You are very hard on Madame de Boigne.[3] Aristocratic prejudice I can't deny, since after all she confesses it in her book. But it didn't blinker her. A liberal of today wouldn't judge the events of that time in a more liberal, indeed, more liberated spirit than this lady brought up before the Revolution in the lap of the Queen. As regards her 'genius for names', don't forget that M. d'Osmond's Christian name was Raynulphe. What rankles with you is her attitude to M. de Boigne, if I understand you correctly; and that has inspired you, very nobly, to write some fine pages and also all that delightful annexe to the *Mémoires*, your exquisite landscape of château, park and church.

You surprise me by quoting that fine sentence of Chateaubriand's, which I know well and about which I've written, as having been addressed to Mme de Duras.[4] My books are far from my bed, I'm not in a fit state to get up, and I don't want to ring (it's four o'clock in the morning) but I shouldn't have thought it was addressed to her.

Your Mistral is magnificent and charming.[5] That's as far as I've got for the moment (and the *Inconnue de Sainte-Beuve*: is it an artifice by which I've been taken in, or a priceless reality in which I detect malice?[6]).

Since reading and writing give me headaches and are forbidden me, I shall leave it at that and send you, together with my thanks, my most sincere regards.

Marcel Proust

1. Novelist Henry Bordeaux (1870–1963) whom Proust had met in Evian in 1899.

2. Sixteenth-century subject of a series of studies entitled 'Portraits of Women and Children', published in February 1909.

3. The *Mémoires* of the Comtesse de Boigne also inspired Proust (cf. letter 195, n. 1). Her father, to whom Proust refers in the same paragraph, was Rainulphe-Eustache Marquis d'Osmond.

4. Proust was mistaken here. The Chateaubriand quotation was indeed from a letter addressed to the Duchesse de Duras at Dieppe and runs, as first published: 'Tell the sea of my tender feelings for her, tell her that I was born within the sound of her waves, that she witnessed my first games, nourished my first passions and my first rages, that I shall love her to the end of my days and that I beg her to let you hear one of her autumnal storms.'

5. He had written on the childhood of Frédéric Mistral (1830–1914), Provençal poet and linguist.

6. One of Bordeaux's essays tells of an anonymous letter Sainte-Beuve found addressed to him by a woman who objected to his animadversions on Benjamin Constant.

326   *From Robert de Montesquiou*

Engadine[1]
[Tuesday 6 March 1909]

Dear Marcel,

What an *astonishing* thing your latest pastiche is![2] I am not in the habit of reading authors with whom I have had differences.[3] I don't in the least wish them *not* to have talent. Nor do I desire that they should have *too much*. In consequence, I abstain. So I must take your word for its *accuracy*, and I grasp very well the method which you seek to dismantle (or demonstrate), moreover very gracefully.

But . . . the *ending*![4] Should it *worry* me, or *reassure* me, when my turn comes? The second rather, I feel. Amen!

R.M.

1. A postcard.
2. On Régnier. See letter 320, n. 3.
3. Montesquiou's first duel, in 1897, had been with Régnier.
4. 'Régnier on Lemoine' ends with a description of a dewdrop fallen from Lemoine's nose on to his frock-coat and gleaming there momentarily in the sunlight like 'a diamond still warm, as it were, from the furnace whence it had emerged, and of which this unstable, corrosive, living jelly . . . seemed, by its false and fascinating beauty, to present at once the mockery and the emblem'. Cf. letter 265, n. 2.

327   *To Georges de Lauris*

[Shortly after 6 March 1909]

My dearest Georges,

Just a brief note to thank you for your letter. I don't know what Régnier's feelings are and your doubts disturb me. And yet there's nothing that could offend him. He must know perfectly well that he mixes up his pronouns since this neo-Saint-Simonian syntax is deliberate with him, and as for his repeating the same thing several times over, he must know that too.

I can't publish either Chateaubriand or Maeterlinck because they need a bit of polishing and I'm not in a fit state to make the slightest effort.

What has the best chance of appearing some day is Sainte-Beuve (not the second pastiche but the study) because that full trunk in the middle of my brain hampers me and I must decide whether to set off or to unpack it. But I've already forgotten a great deal and although I oughtn't to read at all I read a lot of stuff of a quite different kind. Nevertheless if I'm still alive this autumn there's a good chance that *Sainte-Beuve* will appear and I think you'll like it.

Affectionately yours
Marcel

Did I tell you about the very intelligent letter I received from Barrès about my pastiches, a really pretty letter in which he begged me to go on with them because I had hit on a formula for criticism based on Buffon's notion that form and substance shouldn't be separated, but Nicolas has gone and lost it for me.[1] But I don't agree with him and don't want to do any more of them. However I'd like to put the Maeterlinck into shape some day because there are two or three little things in it that might make you laugh though it's nothing much 'to write home about'. I feel, contrary to what you say, that when we see each other we shall in no way be able to show our pleasure in doing so, but that's of no importance. Gestures matter less than what one says, what one says less than what one writes – reality is elsewhere.

This letter, dear Georges, was written two days ago. In the meantime I've received a very nice note from Régnier. He doesn't seem displeased, or at least he doesn't show it. He says he finds it a good likeness.

1. It was found: see letter 313.

328   *To Jules Lemaître*

[Shortly after 6 March 1909]

Dear Sir,

Your letter made me very happy, and I thank you warmly for it. I should have expressed my gratitude sooner had I not wanted to enclose a piece which I've been unable to unearth. Allow me to stick to my opinion on your pastiches. The length of a pastiche matters not at all as long as it contains the generative features which, by enabling the reader to multiply the resemblances *ad infinitum*, dispense the author from piling them up! I confess moreover that the slightest *jeu d'esprit* of yours is enhanced and sustained for me by my admiration, which developed in my youth and has retained something of its charm. I often delight in recalling those first evenings when I read those phrases, which were like nothing I had come across, about Baron's voice, for instance, and the arias that emerged from it, or about Banville.[1] I remember little family scenes because Pozzi[2] had offered to introduce me to you and my poor parents thought I was too young to 'go out'. And later, that afternoon when I heard you read *La Fin de Satan* at the Desjardins'.[3] Beginnings that seem almost legendary today, so much do they pertain to things long dead, the cherished mythology of an admiration which has become more conscious without repudiating that first flower of which it is still redolent.

Thank you again, dear Sir, and please accept my admiring and respectful regards.

Marcel Proust

1. It had been ten years since Lemaître's review praising the actor Baron (cf. letter 130, n. 11) who was known for his trembling voice, and fifteen years since his article on the Parnassian poet Théodore de Banville (1823–91) whose principal aim, according to Lemaître, was 'never to express a single idea in his verse'. Cf. Bloch in *Swann's Way*, quoting Racine and Musset as each having once in his life composed a line 'which is not only fairly rhythmical but has also what in my eyes is the supreme merit of meaning absolutely nothing' (*RTP*, I, 97).

2. Dr Samuel Pozzi (see letter 134, n. 4).

3. Parents of Paul Desjardins (see letter 149, n. 2.). *La Fin de Satan* is a posthumously published verse epic by Victor Hugo.

329   *To Robert de Montesquiou*

[Shortly after 6 March 1909]

Dear Sir,

I'm so grateful to you for your beautiful card and your kind and witty encouragement. How pretty it is, that post-card[1] from a country of which you have written so well and so often, quite recently indeed à propos of M. de Schickler's house[2] and also the painter of the Engadine,[3] and the lesson of the peaks to the valleys and the towns. Thank you for distinguishing my intention and my design in these pastiches. But I confess that for a pastiche of you I'm not sure what the design should be, unless I interwove the threads into a veritable tapestry, but it wouldn't be easy. Extreme complication and extreme nakedness make pastiches difficult. Monsieur Lemaître, who was so benevolent at that little *soirée*,[4] more than benevolent in fact, asked me to do a Voltaire and a Mérimée. I don't know quite what they would be like . . . Yes I do, of course, but it wouldn't be easy.

Forgive me for talking so complacently about myself, scarcely a very interesting subject, and accept my respects as an affectionate and devoted admirer.

Marcel Proust

1. From the Engadine (see letter 326).
2. Cf. letter 215, n. 3.
3. See letter 323, n. 1.
4. Proust had recently met him at Mme Alphonse Daudet's.

330 *To Georges de Porto-Riche*[1]

102 boulevard Haussmann
Saturday [20 March 1909]

Dear Sir,

Forgive this stupid little note. I learnt from my sick-bed, through the newspapers, of the recent elections to the Academy. I am too well aware of what the secret and vital ambition for an artist is not to realize that the man who wrote the immortal *Passé*[2] has already fulfilled it, and that his place is between Racine and Marivaux and not between M. Mezières and M. Costa de Beauregard.[3] If the only enviable form of happiness for an artist is to breathe eternal life into what is most individual and profound within himself, you are certainly not to be pitied, and you would be the envy of all, had not the death of your poor child intervened to lacerate your proud heart.[4] In spite of all this, these elections have caused me real pain as well as a great deal of hilarity. It is not good that the real City should be so profoundly at variance with the true and ideal City, that Fauré should have had so much trouble getting into the Institut,[5] that a complete *imbecile* like Shlumberger should dare to offer himself as a candidate,[6] that Brieux,[7] none of whose plays I know but who I suspect is an illiterate didact, should get there before you.

I imagine you will be elected next time, but this first rebuff will leave a scar on my pride as a Frenchman and a man of letters.

Your devoted friend and respectful admirer
Marcel Proust

1. Georges de Porto-Riche (1848–1930), highly respected playwright. Cf. letter 137, n. 5.

2. Cf. letter 259, n. 9.

3. Two second-rate Academicians, Alfred Mezières (1826–1915), literature professor at the Sorbonne, deputy and senator; and the Marquis Costa de Beauregard (1835–1909), historian and member of the Jockey Club.

4. Porto-Riche's son Marcel died in February 1905 (see letter 137, n. 5).

5. Fauré had only just been elected to the Académie des Beaux-Arts, one of the five branches of the Institut de France.

6. Cf. letter 285.

7. Eugène Brieux (1858–1932), didactic playwright much admired by George Bernard Shaw.

331   To Robert Dreyfus

102 boulevard Haussmann
[21 or 22 March 1909]

My dear Robert,

I've been meaning to write to you for a long time to congratulate you on your articles in the *Figaro*.[1] Right at the beginning there were one or two I didn't like, because I hadn't understood that insignificance in an insignificant subject is the mark of true originality. But since then there have been several delightful ones. And now I always look at the *Figaro* with confidence and pleasure as though it were the window to a well-lighted drawing-room where one knows one will find a witty and charming friend. And I enter . . . and, to change the metaphor, I go and make my devotions between the ninth and tenth columns, where in his niche stands the smiling Sage who in the Middle Ages recited the Wisdom of Works and Days and from whose lips issued balloons bearing mottoes certainly more naïve than your charming Comments: 'Shed a thread in April, be sure to catch a chill.'[2] It must be very exhausting for you to give us such frequent pleasures. But it must also be very agreeable to collect thousands of listeners for minor reflections that are so personal and echo your tone of voice, with that smile between putting out one cigarette and lighting another.

I don't ask to see you because my hours now begin at two o'clock in the morning. So no? No, that's my opinion too.

Fondest regards
Marcel Proust

1. Dreyfus was writing the column 'Parisian Notes', signed D.
2. The French equivalent of: 'Ne'er cast a clout till May be out.'

332 *To Fernand Gregh*

[28 or 29 March 1909]

Dear friend,

Just a word to thank you most warmly for your letter. These pastiches are in fact little exercises which don't require talent but which are only *addressed* to people of talent (or genius! I don't set limits to my audience in that direction) because they alone are capable of understanding the jokes. So I'm very pleased that they've come to your attention.

Thank you for your advice about Dubois.[1] I hope to be able to talk to you soon if I feel better.

Ever yours

Marcel Proust

No, to publish these pastiches in book form would be excessive. Like that, in a newspaper, well and good. Perhaps I shall include a few of them in a collection of pieces if I ever get round to it; but not the pastiches alone. However I mustn't make myself out to be more sensible than I am. A number of very kind friends, no doubt happy to have at last something on which to be able to congratulate me, were so insistent that I should collect my pastiches in a book that I accepted the idea for a few days. Luckily the publishers took a more rational view of the matter, and one by one, within a few days, the Mercure, Calmann, Fasquelle and I don't know who else who had been approached by deluded third persons,[2] declined the honour! But that's as it should be. You read one,[3] you find it accurate, you tell me so, that is the real and rare pleasure they can give me, I can't think of anything better.

1. Dr Paul Dubois (see letter 71, n. 2); he had successfully treated Gregh in 1904.

2. For instance, the Caillavets. See letter 276.

3. The Régnier parody; for reasons of health this was Proust's first contribution to *Le Figaro* since the pastiches of a year ago.

333 *To Max Daireaux*[1]

[Around May 1909]

My dear friend,

I shall be happy to try and get your *fantasies* into the *Figaro* if I can and I thank you for giving me this pleasure. But I am the man least qualified perhaps for the purpose, since I scarcely get up more than once a month. And I'm in such a difficult situation vis-à-vis the whole of Paris (that's to say the five or six people I know) that at moments the prospect of being interned in a nursing-home seems to me a 'solution' that would at least cut short my excuses. Which is to explain that to send them your *fantasies* I shall have to precede my letter with innumerable circumlocutions giving the impression that your pieces are the only thing that matters to me in life. But my friends are forbearing and they will be kind. The drawback is not being on the spot. The useful man is the man who has one of your *fantasies* in his pocket. For some nights running he keeps it there. Then the moment comes when Fauré has failed to deliver his article etc. and he takes it out and inserts it. After five or six times you are 'on the strength' and you do it yourself. In this respect Caillavet whom I believe you know has the advantage over me (among countless others) of being 'of the *Figaro*' and being able to do what I've just described. As for Chevassu, he's the *editor* of the Supplement. Which means that in relation to him I'm like a flea in relation to the Eiffel Tower. Nevertheless in my capacity as a flea I jump with joy at the opportunity of proving my zeal and I'll do all I can to get your *fantasies* published and to transform this accidental contribution into a definitive one. But I remember when I used to take articles round to Cardane[2] and every time a new 'topical' subject would oblige him to say to me apologetically: 'Alas, dear friend, you understand, don't you, that with this Moroccan business we can't possibly fit your splendid article in. We're chock-a-block.' Morocco died down but Mme Le Bargy[3] was getting a divorce: 'Our legal correspondent is at the end of his tether.' Then things began to stir in the South, and even in the North, at Courrières.[4] Cardane was more and more desperate. Alas, that excellent man is dead, dead before me, which seemed against all probability.

Did you receive the letter in which I sent you some idiotic and obscene verses about Cabourg?[5] As long as they haven't gone astray!

Do ask me for the box at the Théâtre des Arts whenever you like.[6] And send me a *fantasy* whenever you like. The ones I read in *Comoedia* were very witty. Thank you for sending them to me, and warm regards.

Marcel Proust

1. Max Daireaux (1884–1954) was one of the group of young people whom Proust took up with at Cabourg in the summer of 1908 (cf. letter 303, n. 2). Proust had known his elder brothers in the 1890s and used to go to tennis parties at Neuilly with him, Jeanne Pouquet (see letter 276, n. 1) and her future husband Gaston de Caillavet. Cf. vol. I.

2. Cf. letter 185, n. 3. Cardane had just died, aged forty-nine.

3. See letter 182, n. 2.

4. A mine disaster in 1906.

5. This letter has not been found.

6. Proust was a subscriber to weekly literary debates at the theatre known as the Mardis des Arts.

334   *To Georges de Lauris*

[Sunday evening, 23 May 1909]

My dearest Georges,

You will have to be good enough to tell me what *kind* of position your protégé wants, and his age and his circumstances.[1] Then, if it's all right with you, I shall wait a few days because those people have just done something (by omission) which I found so lacking in warmth considering my niceness to them that I want to launch a few thunderbolts and not mix them up with a request for favours.[2] I shall then take advantage of the clearing of the air that ensues to launch into a clement sky an appeal which will be heard. However, since this brighter spell may come sooner than expected, let me have the necessary information without delay.

Do you happen to know whether *Guermantes*, which must have been the name of some people as well as of a place, was then already in the Pâris family, or rather, to put it in a more seemly way, whether the name of Comte or Marquis de Guermantes was a title used by relations of the Pâris family, and whether it's entirely extinct and available to an author?[3] Do you know any other pretty names of châteaux or of people? What was your family place called?

Affectionately yours

Marcel

Régnier continues to write articles which are, it seems, enormously successful. This evening he talks about the *admirable* book by *Monsieur* le Vicomte Eugène Melchior etc.[4] It's true that he has also spoken of the *admirable* work of M. René Doumic,[5] of the incomparable writer known as Paul Hervieu (here, by the way, I agree), of the great poet who signs himself Richepin,[6] etc. etc. etc. Only Bataille and Augier[7] have been roughly handled. By contrast, Ponsard,[8] Sardou,[9] *The Danicheffs*[10] etc. etc. have been praised to the skies (no, Ponsard merely praised). Ecstasy was reserved for that delightful poet by the name of Georges de Porto-Riche; it was for his verses.

I should add that each of Régnier's pieces contains a number of excellent puns. 'There is indeed the *beginning* of an ex-quisite thing in *L'Ex* by M. Gandillot.' *'L'Impasse* by M. Xanrof is indeed an impasse.' 'I don't know whether the revival of *L'Honneur et l'Argent* will bring much money to the Comédie Française, but it will bring it much honour.' I forget all the good ones.

1. An army friend of Lauris, called Nogrette.

2. Apparently Camille Plantevignes, the tie manufacturer, and his son Marcel (see letter 303, n.3).

3. A significant inquiry! François de Pâris's aunt, the Baronne de Lareinty, *née* Puységur, had inherited the Château de Guermantes near Lagny, Seine-et-Marne. The first reference in Proust's correspondence to this most resonant of all names in *RTP* occurs in 1903, when Proust suggests meeting Pâris at the château.

4. An over-elaborate, obsequious way of referring to the Comte de Vogüé, the historian and expert on the Russian novel (see letter 172, n. 6).

5. René Doumic (1860–1937), critic on the *Revue des Deux Mondes*, had just been elected to the Academy.

6. Jean Richepin was elected to the Academy in 1908 (see letter 295, n. 4).

7. Henri Bataille, poet and dramatist (see letter 209, n. 3), and Emile Augier (1820–89), playwright, who is mentioned in the same breath as Dumas fils.

8. François Ponsard (1814–67), neo-classical playwright.

9. André Sardou, a young playwright, son of Victorien Sardou.

10. A successful play (1876) by a Franco-Russian, Pierre de Corvin-Kroukowski, who wrote under the name Pierre Nevsky.

335   *To Georges de Lauris*

[Shortly after 23 May 1909]

My dearest Georges,

I was desperately sorry because on the previous days I could have seen you so much earlier. What a pity!

No, Georges, I'm not writing a novel; it would take too long to explain. But if Guermantes is one of the names of the Puységur family it comes to the same thing as if it belonged to the Pâris family.[1] I only want to annoy unknowns, who aren't related to people I know; I haven't Balzac's nerve.

This must sound as though I'm doing a novel. First of all I'm not *doing* anything. But I should like to. I want my château not to belong to the family whose name it bears (for example Dampierre and the Luynes) and, if the present owner is alive, the name of the château to be extinct and not related.

But Georges that wasn't at all the reason why I was writing to you; it was to ask you how much time you can allow me before I make the approach on behalf of your protégé which I thought would be more efficacious if I delayed it a bit. However it also depends on your convenience. Say. – And it was above all to express my regret about yesterday, my sadness, my affection, my gratitude.

Marcel

1. See letter 286.

336 *To Max Daireaux*

[Around May 1909]

Dear friend,

As I'm very tired at the moment, I occasionally experience a weird phenomenon which you may perhaps understand (because you are very intelligent!). Opening my correspondence absent-mindedly, it sometimes happens, if a letter has fallen off my bed and I've been unable to find it again, that I don't know whether I dreamt it or whether such a letter really did arrive. Now for the past few days I've seen in my mind's eye a card reading

The Baroness d'Eichtal
will be at home . . .

Have I dreamt it? Did I receive such a card? I incline towards the dream but I've no idea, and as I remember your mentioning this lady to me I'm appealing to you idiotically to tell me if I dreamt it. If so, *in heaven's name* don't start thinking this is a ruse to get myself invited to this lady's parties. I get up about once every two months and I couldn't possibly go to her house. But if by chance it was true tell me what I ought to do. Cards?[1] (Or?) A letter? *La Bible d'Amiens?* Flowers (I suppose not!). Anything, except to have to go there, since I'm ill.

What about the *Figaro*? And Caillavet? And Cabourg?[2] (I'm very much afraid I shan't be able to go there. It does me good. But having to leave!) What's become of the young Berthier ladies?[3]

Your
Marcel Proust

1. i.e. to leave his visiting card at her address. Proust mentions a M. d'Eichtal in a letter to Mme Straus (see letter 118, n. 4).

2. See letter 333, penultimate paragraph.

3. Marguerite and Marie Berthier, each married to a Comte de Maleville, whom Proust had also met at Cabourg in 1908 and was thinking of renewing the acquaintance.

337   *To Georges de Lauris*

[Shortly before 23 June 1909]

My dearest Georges,

Don't think I'm forgetting your protégé. Twice since the expiry of my time limit I've asked for an appointment with my industrialist.[1] On neither occasion was he free: it's true that I asked in the morning for the same evening. Perhaps I'll be able to try again today. On reflection, I think German would be useful because I remember that the son, who is to take over from his father, spent several years in Germany in the house of a professor. At the Foulds, on the other hand, it's Spanish. You ought to have asked not Eugène but his father.[2]

Georges, I'm so exhausted from having started *Sainte-Beuve*[3] (*I'm hard at work on it*, though the results are execrable) that I don't know what I'm saying to you. I literally cannot write. I'll write to you again this evening if I'm calmer on the subject of the coming autumn which will be impossible or at least difficult.

Affectionately yours.
Marcel

1. Camille Plantevignes. See letter 334, n. 2.

2. See letter 115, n. 1, and letter 269.

3. Although he continues to refer to it as *Sainte-Beuve*, Proust has now begun to write what will become his novel.

338   *To Georges de Lauris*

[Shortly after 2 July 1909]

My dear Georges,

I've been extremely unwell since I last saw you and was so ill yesterday that I can't write to you at length. I intend to try and resume work on *Sainte-Beuve* as from tomorrow. So if you're not in too much of a hurry for *Ginette*[1] which I've awaited so impatiently for so long I should prefer to leave it until a bit later, as I fear I may be too tired to be a very useful reader, if one can ever be such. But if it's urgent to hell with *Sainte-Beuve*.

Georges, I can't tell you how unwell I am. Did I tell you I had received letters from Plantevignes of all ages telling me of the impression of determination and decisiveness, in other words the excellent impression made by M. Nogrette.[2] In spite of this I think you would be wise not to neglect other recommendations. But M. Plantevignes seems anxious to do something for him after the holidays.

I have lots of things to explain to you about myself and various other matters, but I'm really too tired. Did I reply to your note that I had never met anyone you knew?

Affectionately yours, dear Georges.
Marcel

Where did Gide's novella appear?[3] (Tell me when we meet.)

1. The manuscript of Lauris's novel *Ginette Chantenay* which Proust had first read in June 1908. It was to be published in 1910.

2. See letter 334, n.1.

3. *La Porte étroite*, André Gide's third novel, was first printed in three successive numbers, April–March, of the recently founded *Nouvelle Revue Française* which was to become Europe's most influential literary periodical.

339   *To Céline Cottin*[1]

[12 July 1909]

Céline,

I send you my warmest compliments and thanks for the marvellous *boeuf mode*. Would that I might bring off as well as you what I am going to do tonight, that my style might be as brilliant, as clear, as firm as your *gelée*, that my ideas might be as succulent as your carrots and as nourishing and fresh as your meat. Pending the completion of my work, I congratulate you on yours.

1. Proust was working on his novel. In this unsigned note he congratulates his cook Céline, Nicolas's wife, on a classic French dish: larded beef in jelly. He would

remember it when he came to describe the dinner party in *Within a Budding Grove* (*RTP*, I, 493–4) at which Françoise's famous *boeuf à la mode* is set down before the ambassadorial figure of M. de Norpois: 'The cold beef with carrots made its appearance, couched by the Michelangelo of our kitchen upon enormous crystals of its own jelly, like transparent blocks of quartz.' (The effect of the salad which follows is less gratifying, alas: see letter 239, n. 2.) And at the very end of the novel the narrator sums up: 'Just as, in a book, individual characters (whether human or not) are made up of innumerable impressions . . . might I not make my book in the same way that Françoise made that *boeuf à la mode* which M. de Norpois had found so delicious, just because she had enriched its jelly with so many carefully chosen pieces of meat' (*RTP*, III, 1091).

340   *To Reynaldo Hahn*

[17 or 18 July 1909]

Bonjour, Metmata.[1]

Bunibuls, since nothing is more convenient and 'close at hand' as a library than a bed when one can't get up, I constantly re-read 'for want of anything better' (no, Buncht, from choice) your divine articles whose wit intoxicates me, whose scurrilousness gives me the shivers, and whose talent fills me with jealousy, a jealousy that is infinite but without rancour.[2]

Buncht, strange dealings this week with 'l'Elisabeth'.[3] The other day I receive two letters in the same envelope. The first from the said lady's secretary saying that 'Madame la Comtesse has the greatest admiration for my talent (!) and were I to write a few lines on Bagatelle, thinking people would be highly gratified etc.' The second letter from l'Elisabeth herself saying that . . . the same thing and that I should write a few lines, just as I felt, in other words 'exquisitely poetic'. Unaccustomed as I am to praise, even self-interested praise, I repeat all this to you, even adding to it a bit. Whereupon, heartbroken refusal from me . . . state of my health, have had to refuse other people etc. What do you think l'Elisabeth does? Insist? Not at all. She understands my reasons and sends me a *vine*, a magnificent vine dripping with grapes. And tells me that if I'm still unwell she'll come and see me whenever I like, just give her a day and a time, 'hoping to find you in the highest spirits' (?). – The letter was very literary, full of words like 'speaking symbol'. But speaking of the vine she said: 'Acceptez-là' with a grave accent which seemed to me principally grave for her and which is in itself also a 'speaking symbol'.[4] As for the vine, since the most beautiful things here below have the worst fate, as Malherbe says more or less,[5] I'm sending it to Marie Nordlinger.[6] She'd probably appreciate a post-card far more. I was going to include a few roses to go with Gérard's line about 'the vine intertwined with the rose'[7] but I reflected that Mallarmé's 'When I have sucked the clarity of grapes'[8] would be just as effective and more economical since it doesn't need any roses and the grapes are there.

Nicens,

> I rather fear my novel on Sainte-Veuve
> May not be very pleasing to the Beuve[9]

But never mind. Nicens, you're going to send me back this letter and not breathe a word about it *to anyone at all*. I have a great liking for l'Elisabeth (less than for Metmata) and besides she has been extremely nice and I should be very sorry if it got back to her that I'd made those jokes the more so as I thanked her grovellingly. The Winaretta[10] would find this story just the sort she most likes to repeat and would do so, and the new Marchioness of Ripon[11] to whom I've sent books though I don't write to her because I don't know what to say, who detests l'Elisabeth and loves Montesquiou would regale him with it and within five minutes l'Elisabeth would be informed because nowadays when anyone blurts out something about another person Montesquiou *writes to them instantaneously*. So mum's the word. I unbosom myself to you as I did to Mama. But she gave *nothing* away.[12]

1. A new, and unexplained, pet-name for Hahn.

2. Hahn was reviewing opera regularly for *Le Journal* among other papers.

3. Comtesse Greffulhe, president of an annual concert-fête to be held this year at the charming eighteenth-century Bagatelle pavilion which still stands in the Bois de Boulogne.

4. The grave accent would, of course, transform 'la' from 'it' to 'there'.

5. Allusion to 'Stances à monsieur du Périer' on the death of his daughter by François de Malherbe (1555–1628): 'Mais elle était du monde, où les plus belles choses / Ont le pire destin, / Et rose, elle a vécu ce que vivent les roses, / L'espace d'un matin.'

6. See letter 304, n. 1.

7. 'La treille où le pampre à la rose s'allie.' Gérard de Nerval (1808–55), 'El Desdichado'.

8. 'Ainsi, quand des raisins j'ai sucé la clarté, . . .' Stéphane Mallarmé, 'L'Après-midi d'un faune'.

9. This coded couplet, with its transposed initials, strongly implies that Proust has turned his attention from Sainte-Beuve to the novel in which one of the major characters will be based on 'the Widow', i.e. Mme Lemaire (see letter 290, n. 2), the principal model for Mme Verdurin in *RTP*.

10. Princesse Edmond de Polignac (see letter 299, n. 2).
11. Formerly Lady de Grey (see letter 249, n. 2).
12. Unsigned.

341   *To Alfred Vallette*

102 boulevard Haussmann
Confidential   ıd fairly urgent
[Around mid-August 1909]

Dear Sir,

It is not from a natural inclination, a sort of perverse compulsion, that I come to bother you so often.[1] Believe me, I don't enjoy the thought of pestering you. And the memory of past failures with you does not give me much courage to make up for native euphoria . . . This said, of course, I shall proceed to bother you: otherwise it wouldn't have been worth the trouble to say it!

This letter is moreover peculiar in this way, that although alas quite unimportant, it is also extremely confidential. Whether or not my proposals meet with your approval, I beg you to keep them secret at least on one point. You will see why. I am finishing a book which in spite of its provisional title: *Contre Sainte-Beuve, souvenir d'une matinée,*[2] is a genuine novel and an extremely indecent one in places. One of the principal characters is a homosexual.[3] And this I count on you to keep strictly secret. If the fact were known before the book appeared, a number of devoted and apprehensive friends would ask me to abandon it. Moreover I fancy it contains some new things (forgive me!) and I shouldn't like to be robbed by others. The name of Sainte-Beuve is not there by chance. The book does indeed end with a long conversation about Sainte-Beuve and about aesthetics (if you like, as *Sylvie*[4] ends with a study of popular song) and once people have finished the book they will see (I hope) that the whole novel is simply the implementation of the artistic principles expressed in this final part, a sort of preface if you like placed at the end.

I should like this book to appear in January or February. However strong my preference for the Mercure, I wouldn't even have mentioned it to you, knowing that you are not very keen to publish me, and having an excellent publisher for the book, less good than you, because there are none better, but the second best.[5] But I very much hoped that the work might appear first of all as a serial in the *Mercure*, and then if you agreed to that, perhaps by the same token you would agree to publish the book. But what chiefly interests me is the review. Of course many parts of the book are perfectly proper, even pure. And those parts, if the *Mercure-Revue* doesn't want me, I could give to the *Figaro* to which the grave state of my health has forced me to suspend my contributions for over a year, and which would publish them twice a week. But in the first place the suppression of the obscene sections would upset me considerably, and in the second place it's a book of episodes, of events, and the reflection of events on one another at intervals of several years, and it can only appear in large slices.

So, to sum up, would you consent to give me, from the 1st or the 15th of October, thirty (or more – that would be all the better) pages of the *Mercure* in every number until January, which would amount to roughly 250 or 300 pages in book form. The novel part would thus have appeared. There would remain the long discussion on Sainte-Beuve, the criticism etc. which would appear only in the book version. The whole would be about the same length as *La Double Maîtresse*[6] (425 pages) and would appear under your imprint if you wished. If it could be at my own expense that is what I should like best, with your permission to choose my own paper, organize my own publicity etc. But this condition is not a *sine qua non*. The word condition is in any case rather comic since it's something that I want and that you don't and that you will only do out of kindness if you do it at all. The question of *dates* is all-important for me. My crippled health makes the timing essential. Perhaps, before giving me a definitive answer, you would like to have a sample of this production. I could in a few days have the first hundred pages copied for you very legibly, or even typed. But they are of the greatest purity. If the thought of the others frightens you and you would like to be reassured on the point (there isn't a hint of *pornography*) I can have a few passages copied for you but the text is not absolutely definitive. But I must ask you not to talk about them to anyone. The hoped-for but alas improbable reply is this: 'The *Mercure* offers you from 30 to 50 pages (or more) every fortnight from October to January and in January or February the book will be published by us.'

That is what I wish but dare not hope for.

Please accept, dear Sir, my grateful and devoted respects.

Marcel Proust

1. In April 1908 he had asked Vallette, as head of Mercure de France (see letter 25, n. 5), to publish his collected *Pastiches*; Vallette had declined.

2. Cf. letter 315 to Mme de Noailles.

3. The Baron de Charlus, the great comic creation of *RTP* (though Proust had not yet settled on the name). Proust imagined that Vallette was unshockable in this regard: his own wife (*née* Marguerite Eymery: 1860–1953) had published pseudonymously between 1888 and 1897 three novels with homosexual overtones for which she had been tried and sentenced in Belgium but which Mercure had published in France nevertheless. But Vallette's attitude had changed, and 'daring' passages were now censored.

4. An allusion to *Sylvie, souvenirs du Valois* by Gérard de Nerval.

5. Presumably Fasquelle, whom he was to approach later.

6. Novel by Henri de Régnier, published by Mercure de France in 1900.

342   *To Georges de Lauris*

[Shortly after 14 August 1909]

My dearest Georges,

Just a line from Cabourg,[1] to thank you, to express my affectionate and grateful admiration and to ask you to convey my deep respect and attachment to your father. I'm not surprised by the uncouthness of your companion since the genius of the family, the best of them, expresses the aesthetics of Winckelmann in the language of Sergeant Pitou.[2]

Georges, I wanted to keep my arrival in Cabourg a secret, but I met Papa Plantevignes in the train. Two hours' conversation about Nogrette. I was able by chance to do the Plantevignes a great favour and I know they would like to do a lot for me. But Monsieur Nogrette seems such an excellent person that I shan't even have the pleasure of thinking that he'll owe his job to me because he'll owe it to himself. I'll tell you all the nice things Monsieur Plantevignes (who saw him again) said about him . . .

Vallette, who had already rejected my *Pastiches, Collected Articles*, etc., has now rejected *Sainte-Beuve* which will doubtless remain unpublished! But I shall *read* it to you. And besides, Calmette is in the neighbourhood and is so charming that he may perhaps undertake to get it published, but precisely because he's so nice I hesitate to ask him. Yes, it will be very long, four or five hundred pages.

Your affectionate, and exhausted
Marcel

I think your title is perfect, your subtitle perhaps superfluous.[3]

1. He had arrived in Cabourg on 14 August.

2. Lauris's companion was Lucien Henraux; by the family 'genius' Proust means his younger brother Albert, an aspiring art historian writing in the manner of the distinguished German historian and archaeologist but the vocabulary of the barrack-room. Sergeant Ange Pitou (1767–1842), hero of a novel by Dumas *père*, was a counter-revolutionary ballad-singer who was deported.

3. See letter 338. Lauris abandoned the subtitle.

343   To Madame Straus

[Cabourg, around 16 August 1909]

Madame,

Your delightful card causes me as well as immense pleasure a little pain because I've been meaning to write to you for several days, my letter has now been transformed into a reply, and you will no longer be aware of the spontaneity of my desire and the impulse of my heart which reaches out to the Mûriers twenty times a day and tires me perhaps more in my anxious, wandering immobility than would a journey which at least would feast my eyes with the picture my memory ceaselessly paints.

The fact is that I left Paris ill and have remained so here. I do however get up for a while at about half past nine in the evening, though I have nasty damp rooms where it won't be easy to recuperate. I go from the hotel to the Casino, which is two minutes away, and that seems so exhausting that I wonder how I could get to Trouville. But I shall. I'd prefer that to seeing you in Cabourg, where the excitement of your arrival, the fear that you might catch cold, the deafeningness of the music and the crowds, the shock of seeing you without the preparatory pilgrimage during which I gradually make myself worthy of you in the course of my meditations on the way – all this almost prevents me from realizing you are here, thinking as I do only of the draughts which must bother you, of the chocolate which M. Straus has gone off secretly to pay for, and it's only when your motor-car has gone, when I'm left alone, that I say to myself: 'It was Madame Straus!' and become sorrowfully aware of a happiness I haven't felt. And I try to identify all the thoughts of affection, beauty and admiration that your name means to me with the vanishing image I have looked at for an hour without seeing.

I shouldn't want you to take too seriously threats often uttered in the past but perhaps this time more real, but I think you'll see quite a lot of me in Paris this year. And before that you will read me – more of me than you will want – for I've just begun – and finished – a whole long book.[1] Unfortunately my departure for Cabourg interrupted my work, and I'm only now about to go back to it. Perhaps a part of it will appear serially in the *Figaro*, but only a part. Because it's too improper and too long to be given in its entirety. But I very much want to get to the end of it. Once everything is written, there'll be a lot of things to recast.

I'm very pleased, being unable to see you or Helleu, that you are seeing him. I know it makes him very happy and I imagine you must like him. Even physically he is charming, and Montesquiou quite rightly said that very few descendants of those who were guillotined had as much 'breeding' as this descendant of a guillotiner.[2] It's absolutely true. I wish you could see his establishment in Paris. We feel blasé about pretty houses because they all seem alike and they bore us. But it's just as with novels: one is tired of them until the day when an original book appears which restores all our freshness of impression and desire to read. I'm sure you'd very much like the whole arrangement which is in exquisite taste and he would greatly enjoy showing it to you. For he has the very simple nature of the true artist, and if very few people give him pleasure, of those few everything does.[3] You know La Bruyère's remark, 'To be with the people one loves, to speak to them, not to speak to them, nothing matters as long as one is with them'[4] (I'm quoting very inaccurately). That is unfortunately a pleasure I lack, and I'm never with the people I love. At least in Paris I have the consolation of not being with those I don't love either, and I haven't that in Cabourg.

I need hardly tell you that I've seen no one on the coast since I haven't seen you. If you run into the Guiches, for I see from the newspapers that you're going about a bit, I don't even want them to know I'm here. Madame Greffulhe sent me a superb vine before my departure, but I feel that if I begin that saga the damp in my room will soak the pages I've already written of this letter to such an extent that I shan't be able to send it to you. A sheet of paper is only fit to throw away after a quarter of an hour and the walls are covered with damp stains; I can't think where it comes from.

Good-bye, Madame. I shall come to see you soon, and you are always in my thoughts.

Please accept, and convey to Monsieur Straus, my respectful regards and my grateful and profound attachment.

Marcel Proust

1. Perhaps the earliest draft of *RTP*, consisting of its long overture, *Combray*, and its conclusion as then envisaged.

2. Montesquiou, in a study of Helleu, had referred to him as the great-grandson of Le Quinio, member of the Convention which condemned Louis XVI to death.

3. These remarks about Helleu foreshadow the character of Elstir in *RTP*.

4. A shortened rendering of a maxim from 'Of the Heart' in *Characters and Morals* by Jean de La Bruyère (1645–96). Charlus quotes the same remark in *RTP*, (I, 819).

344 To Reynaldo Hahn

[Cabourg, shortly after 25 August 1909]

My dear little Bunibuls,

You don't write excessively to your Binibuls. But as he is tireds he won't write either. Anyway he doesn't write, it's boring to write.

The other day I went to hear an act of *Werther* and in spite of the immense inferiority of this music compared to that of my Buncht, I thought to myself all the same that these well written, well prepared, well seasoned works will always whet the appetite, as I attested by going to hear it although I was ill, like Voltaire, like France, like others you will dispute, like Reynaldo indeed who has countless other merits besides Massenet's but whose least ambitious music is sapid and edible. Beside it all other music is caca and makes me run away.

This evening I asked the tziganes if they knew anything by Buncht and when they started *Rêverie* I began to cry as I thought of my Bunibuls in the big dining-room surrounded by a score of dismayed waiters who put on long faces! The head waiter, not knowing how to commiserate with me, went to fetch a finger-bowl.

I seemed very musical to Dr Roussy (more distinguished than Griffon and Lubet-Barbon)[1] by saying, of a tune they were playing, that it was reminiscent of the overture to *Carmen*. It was in fact the overture to *Patrie*.[2] What do you know of a conductor called Firance? Is he better or less good than Vizentini?[3]

Ulrich is here, without however my being able to say 'Ulrich, your eye has probed the ocean's depths' as Musset did to that Guttinguer who lived near here. See Léon Séché.[4]

Good-bye Reynaldo and bonsjours

Marcel

They call the Baronne Maurice de R. 'Mouton Rothschild' and Mme de Maupeou the Merry Widow. Feeble.

What a nice remark of Talleyrand's: 'One should be in imaginary good health'.[5]

I hope your Mama is well and everyone and everyone.

1. Samuel Roussy (1874–1948), leading pathologist. Griffon was a colleague of Dr Robert Proust, Lubet-Barbon an ear, nose and throat specialist.

2. The overture to Victorien Sardou's drama *Patrie* is also by Bizet.

3. Albert Vizentini (1841–1906) conducted in the theatres of Paris, Lyon and London, and also composed operettas. Firance is unidentifiable.

4. Robert Ulrich was acting as Proust's secretary. Alfred de Musset wrote a poem dedicated to the poet Ulrich Guttinguer (1785–1866); it opens 'Ulrich, nul oeil des mers n'a mesuré l'abîme . . .' Séché was Musset's biographer (see letter 123, n. 9).

5. 'Il faudrait être un bien portant imaginaire.' Cf. *The Guermantes Way* where Dr du Boulbon says to the narrator's grandmother: '. . . if ever you have a slight indisposition, a thing that may happen to anyone . . . your nervous energy will have endowed you with what M. de Talleyrand astutely called "imaginary good health"' (*RTP*, II, 317).

345   *To Robert Dreyfus*

[Cabourg, towards the end of August 1909]

My dear Robert,

Will you allow me to ask you the following somewhat vague favour. Calmette whom I've seen here asked me very nicely and with great insistence if he could publish serially in the *Figaro* a novel I'm in the middle of writing. I may add between ourselves that I don't think I shall give this novel to the *Figaro* or to any newspaper or review and that it will appear only in book form.

In the meantime I had mentioned to Beaunier a critical study I had written (which won't appear in a newspaper either). And I'm afraid that if by any chance Calmette told Beaunier that he'd asked me for a serial for the *Figaro*, Beaunier might think it was my critical study and advise against it (not out of hostility towards me – Beaunier is charming to me – but in the interests of the paper). It wouldn't be *very* serious if he did so, since I don't *think* I'll give the novel to the *Figaro*. But still, in case I changed my mind I'd prefer to keep my options open and not have Beaunier, on the basis of a misunderstanding, prevent its publication as a serial. I should like you therefore, if it isn't in any way a nuisance for you (because if it is I can write to him myself), to say to Beaunier that if by chance Calmette mentions to him a serial by me (which in any case won't happen) it's a question of a novel. I should emphasize that I'm not asking Beaunier to say anything in my favour, or anything at all, to Calmette. I don't need any support with Calmette who is enchanting to me, and as regards the serial it was he who offered it, begged me etc. If I do decide to publish the novel as a serial, then let Beaunier advise against it if he finds it bad. All I want is to avoid a muddle.

I felt it would be less tiring to explain all this to you, with whom I'm on more familiar terms, and I thought you would be seeing Beaunier as soon as he returns. But if for one reason or another, perhaps having too many things to discuss with him, you would prefer me to write to him, tell me frankly. It would be too silly for you to do something that embarrasses you when I can so easily do it myself. And please forgive me for asking you so bluntly, so unceremoniously, without telling you how apologetic I feel or how full of affectionate friendship for you. But I'm exhausted.

Marcel Proust

I haven't been able to see the Strauses even once!

346 *To Robert Dreyfus*

[Cabourg, 1 or 2 September 1909]

My dear Robert,

I'm so sorry to hear that you're still depressed. Alas, I fear from what you say that I can't be of any help to you. However you know that I'm entirely at your disposal if there is anything I can do. I'm confident in your brother's complete recovery since Sollier told you so.[1] He's not generally an optimist. But it's a long, slow business for those who suffer, and even more so for those who love.

It's excessively nice of you to have given me that good advice about B[eaunier].[2] You're quite right, don't tell him anything. At most you might say to him 'in the course of conversation' if my name crops up, that you know I'm working on a novel. Alas I'm scarcely working at the moment.

I sometimes think that the impression of extraordinary distinction one gets from what you write, however unimportant, derives from the fact that it contains and presupposes things that you know will not be seen – the opposite of meretricious art being that of the cathedral builders who did carvings that were just as delicate *behind* statues, or at a height where no one would see them, an impression reinforced for us by all the elaborate symbolism which means nothing to us unless we're scholars. (I don't mean to say that people don't read your articles, *everyone reads them*, I mean that there are allusions to things which escape the notice of unlettered readers.) Thus without mentioning the disproportion between your thought and the particular news item – and perhaps there isn't any disproportion since the news item is the whole point – all your refined culture, even in frivolous matters, is lost on readers who don't understand your allusions. How many are there who even know who 'the man with the carbine' is?[3]

It's because I'm very tired that this may seem like a criticism. But it's the supreme compliment.[4]

Ever yours and again many thanks

Marcel Proust

1. Dreyfus had been very worried about his brother Henri, who was gravely ill in Dr Sollier's sanatorium at Boulogne-sur-Seine.

2. See previous letter.

3. An allusion in one of Dreyfus's *Figaro* articles to a poem by Victor Hugo entitled 'Guitare', of which the first line is 'Gastibelza, l'homme à la carabine'.

4. Dreyfus was later to comment: 'It was excellent criticism, a tactful reminder that one should try to make oneself clear to everyone . . . With friendly perspicacity, he slipped in a judicious admonition under a sheaf of splendid flowers.' Cf. letters 131 and 271, where Dreyfus accepts and heeds Proust's criticisms.

347 *To Georges de Lauris*

[Paris, early October? 1909]

My dearest Georges,

Just a note because I'm in the middle of the usual home-coming crises. Your letter was forwarded to me in Paris, as the hotel wouldn't keep me and I didn't want to go back to Versailles after the time I had there last year.

I'm glad that scarcely having finished your book you're thinking of a new one. I liked 'affair of the heart, not very happy' in your letter. You mean there are happy ones? Yes, there are lucky people who assert as much and whose affairs of that kind are very happy. But I wonder if they are really of the heart.

I'm pleased by what you tell me of Bertrand's possible return. I often think of him and would be glad to see him again, as long as he doesn't harm our friendship.[1] For now the situation is reversed: yours is the principal friendship in my life and his is secondary. And if he had as bad an influence on your friendship as you had on his friendship for me, I should feel the same sort of resentment against him as I harboured for so long against you but which has now been entirely dispelled.

Georges, I shall probably not try to see you before the beginning of November. But while I'm recovering a little from my home-coming, I'll have a copy made from my untidy scribbles of the first paragraph of the first chapter of *Sainte-Beuve* (it's almost a book in itself, that first paragraph!) and as soon as it's copied will you give me an evening and come and read it with me? Even if I can hardly speak, what I've written will speak to you.[2] At least I should like it to.

Affectionately yours

Marcel

I'm delighted by what you say about Fasquelle. As your book is 'respectable' and you have a connection with Calmann,[3] he seemed eminently suitable, but I think Fasquelle is even better. So that would be perfect.

1. Cf. letter 248, final paragraph. Fénelon had been *en poste* in Washington.
2. See letter 356.
3. Calmann-Lévy, leading publisher (see letter 276, n. 3). Fasquelle was also a leading publisher, to whom Proust was to submit his novel in 1912.

348 *To Constantin de Brancovan*

[Early October 1909]

My dear Constantin,

After three journeys in pursuit of me, the splendid letter bringing tidings of your happiness has arrived.[1] I do not have the honour of knowing your fiancée. But I have heard her spoken of as a rare, noble and delightful person. All my friendship for you, filled with all the thoughts which have so often gone out to you when, during a separation from my dearest friends, I have had to invent a real presence for them,[2] all my profound friendship rejoices in conjuring up for you the fine, happy and fruitful life of which you are both so worthy.

Please accept my tender affection and lay at the feet of the future Princesse de Brancovan all my deepest respects.

Marcel Proust

1. Mme de Noailles's brother Prince Constantin was to marry for the first time; his fiancée was a divorcee, Eugénie Antoniadis (1874–1917).

2. Proust often uses the phrase 'real presence' figuratively, by allusion to the divine presence in the Eucharist.

349 *To Madame de Noailles*

[October 1909]

Madame,

If ever you have occasion to see M. Barrès, would you tell him that I have thought, that I constantly think of him, with great distress ever since that terrible death. I 'knew' so little that I didn't dare write to him. But it happens rather strangely that, having felt for a long time, together with immense admiration for Barrès's talent, a profound antipathy for what I believed to be his lack of heart, today my admiration is mixed with feelings of human sympathy, deep, cordial, I might almost say tender; I feel bitterly aware of his pain and long to tell him so, but I dare not write to him, and I beg you, if you happen to see him, to tell him that for me.[1]

Madame, I have had a dreadful year, in the emptiness of which the whole hierarchy of regrets took up tangible residence, and during which, in spite of the intensity of the real presence[2] you have in me, no privation was more terrible than that of your cherished face, your voice, yourself.

I have begun to work. And until my work is finished, followed through, despite this formidable obstacle of adverse health that constantly interrupts it, in the desire to put enough of myself at last into something for you to be able to know me a little and esteem me, I do not want to risk the slightest fatigue, which in my now precarious condition is highly dangerous. But if I can continue and finish it, if I don't die beforehand, I shall have such a need, such arrears of accumulated desire to see you, that you will have to allow me to take my fill of you in interminable visits of mutual contemplation.

The Princesse de Chimay may perhaps have told you of my enthusiasm for those pages of yours I read in *Les Marches de l'Est* in which the roofs glistened like the tunny-fisheries on the shores of Catania.[3]

In Cabourg, where (in spite of a decline in health so marked that I can no longer do the thousandth part of what I was still able to do a year ago) I have a slight renewal of life and can see a few people 'at the Casino', I tried out your *Eblouissements* on some adolescents similar to those you write of. And by the time I left them great loves for you had sprung up in those young hearts. You would have laughed to hear a student of mathematics[4] say in the middle of a conversation 'fierce and gentle as the Persian Garden',[5] or else 'just now on the golf course there were some fiery butterflies which seemed like winged jasmine blossoms'.[6] Another, in the name of Optics, asked whether the sun can really insert its prism in a stained-glass window.[7] Your statue is in all their hearts, above that of Victor Hugo.

Good-bye, Madame. Remember me to M. de Noailles and please accept my admiration and respect.

Marcel Proust

1. Proust cannot have known how tactless a suggestion this was. The 'tragic death' was the suicide on 21 August of Barrès's nephew Charles Demange, a young writer. According to Henry Bordeaux, he had shot himself after having been first encouraged then thrown out 'like a squeezed lemon' by Mme de Noailles in order to spite Barrès with whom her long relationship was ending. When the letter first appeared in print she pretended it referred to the death of Barrès's mother and had been written in 1903 despite the later date of the poems quoted.

2. See note 2 to the preceding letter.

3. A description of Strasbourg in a piece called 'Fragments' in the review *Les Marches de l'Est*: 'quand ses toits écailleux étincellent comme une pêcherie de thons sur le rivage de Catane'.

4. Probably Pierre Parent (1883–1964), who was to become a distinguished mining engineer. In the summer of 1908 Proust jotted down various observations about him in his notebook, comparing him to Bertrand de Fénelon in his 'bourgeoisism' and his pronunciation. In the following year he noted down impressions or expressions of Parent which he would use for Saint-Loup or Albertine in *RTP*. See *Le Carnet de 1908*.

5. 'Ce bruit qui donne soif et rend féroce et doux', from her poem 'Jardin persan'.

6. 'Aux fougueux papillons qui, sur la paix des blés, / Se poursuivent pareils à des jasmins ailés', from her poem 'L'Eblouissement'.

7. 'Où, comme un chaud vitrail, le soleil met son prisme', from her poem 'Paganisme'.

350 *To Antoine Bibesco*

[Tuesday, 2 November 1909]

My dear Antoine,

As my life does seem to be rather ill-fated, your letter arrived a few hours after the end of a period during which I could have seen a lot of you, and just at the moment when the switching on of the central heating (1st November), the chimney of which backs on to my bed, has given me terrible attacks of orthopmoea[1] which time will attenuate but which at the moment are totally incapacitating. I did however have one tolerable hour yesterday from nine to ten, and telephoned you (it's no longer 514 00!), but there was no reply. Are you staying much longer? Are there any evenings when you're doing something (I mean when I may be sure that you're not putting yourself out for me, but when *mecum* or *sine me* you are in any event at such and such a theatre or restaurant, or at home? in any event whether I come or not)?

However if I can see you here I'll telephone. Unfortunately it will be at an hour when no one will answer.

At all events, after the completion of a major piece of work which I've undertaken, I want to devote myself this summer, before leaving this earth, to seeing some of the 'companions that have given me all the best joy of my life on Earth' (for I'm very much afraid that in the next world I may no longer be able 'to meet their eyes again and clasp their hands')[2] and there is nobody, my dear Antoine, of whom I have a more affectionate memory than of you. My solitary life has enabled me to re-create in my thoughts those I have loved, and I have always close by me the dear Antoine of the days when he was so good to me. But you, do you still remember me after such a long time?

I see with alarm from the papers that Constantin's marriage has already been celebrated, because I haven't yet sent him my present. Has he left on honeymoon? For long? I ask so as to know when I should send it to him. I hope at least that he received my letter of congratulation which I sent off as soon as I heard of his engagement.

Good-bye, my dear Antoine; I know that you laugh at my long and effusive letters. Good-bye.

Your

Marcel

Give my fond regards to Emmanuel.

---

1. A form of asthma in which breathing is possible only in an upright position.
2. Proust quotes in English from Ruskin's *The Bible of Amiens*.

351 *To Lionel Hauser*

[13 or 14 November 1909]

My dear Lionel,

You can imagine that if I weren't so unwell and above all in the throes of a work in 3 volumes (!), begun, promised, not ready, I wouldn't have left you without news.

Thank you for thinking of explaining to me about the Pennsylvania. You're a real friend! In any case with regard to those shares (the Pennsylvanias) as to all the others, I'm quite prepared for you to arrange for them to be sold whenever you think fit, when the price has risen etc. Even at the time of the last rise I meant to tell you so. But I thought it wasn't necessary and that you were the best judge of the right opportunity. Since it's an excellent investment, there's in any case no harm in keeping them. What frightens me is their 'megalomania'. But still it's also a reason for the development from which the shareholders will profit. At all events if ever you have a mind to sell the Pennsylvanias or the Chicago Burlingtons or anything else, do as you think fit.

If you ever have reason to write to me and can glance at the share-lists you have in front of you (without bothering to search because it's not important), tell me what has happened to the 'Paras'.[1] Did the shares go up? Or did they fall? Did they produce any dividends? It's not that I still want to buy some, just that I don't take the newspapers that relate the story of their life.

You know that I still have some Packetfahrt to reinvest. But in the stupor of my worsening health and my forced labour I haven't been able to consult anyone.

Affectionately yours
Marcel Proust

1. In an earlier exchange Lionel Hauser had strongly advised against these highly speculative shares in a Brazilian port ('not exactly what you would call a gilt-edged investment').

352 *To Albert Nahmias*[1]

[November 1909]

Dear friend,

Since I believe that to live in accordance with sincere feelings is the only wisdom here on earth, I shall certainly not tax you with folly. I have too much respect for any true love for that.

Now that doesn't mean that you ought not to think, and I believe you have done so, about not distressing your girlfriend's family and your own. For, given the fact that your previous situation was in no way unpleasant and that you already had a fairly congenial intimacy with the person you loved, it would be wrong, for the sake of still greater happiness for you two, to cause unhappiness to seven people – I say seven and not eight, because I cannot believe that the eyes of Monsieur Magnus[2] would be changed into a torrent of tears. Apart from that, is it sensible from the practical point of view? And finally there is a third factor: society. You will tell me, my dear Monsieur Nahmias, that you don't give a hoot for society. And I frankly admit that such an attitude doesn't seem to me very unreasonable. Only you must tell yourself that in the event of the future Madame Albert Nahmias (for I assume that is your plan) wishing some day to go about a little, you have the duty to avoid putting her in advance, by acting rashly now, into an intolerable situation. Now society, which likes logic and virtue above all, finds it quite natural for a married woman to sleep with an old and ugly lover for money or from vice, but unpardonable for a girl who is genuinely in love to live openly with a man who is not yet her husband. So use a little discretion, Monsieur Nahmias. As long as you tell only me, no one will know, though I must point out to you that your envelope reached me open. But Nicolas is not inquisitive and moreover is discreet.

Dear Monsieur Nahmias, above all don't make your parents un-happy. As for the folly aspect, I cannot call that folly. I have known only one person wiser than you: yourself when you were giving advice to Aranyi.[3] Happy are those who are conquered by themselves alone. Love has been stronger than you, it is a powerful god and you are not the first to succumb to it. But still I must confess that I would have envisaged the future more tranquilly for you if it had all been in a few years' time. But I know that you are wise, brave and kind and that you have acted only after consulting your conscience and your heart.

Amen.

Your devoted but sick friend
Marcel Proust

1. Albert Nahmias (1886–1979), financial journalist, one of Proust's new young friends made at Cabourg where his family had a villa. Cf. letter 303, n. 2.

2. Louis Magnus, whose wife, *née* Esty Nahmias, was Albert's aunt, the seventh person referred to above. The others were the parents of the girl, Albert's parents and his two sisters, Anita and Estie. The Magnuses also lived in Cabourg.

3. Apparently the son of Dr Maximilien Aranyi who lived in the avenue Kleber.

353   *To Robert de Montesquiou*

[November 1909]

Dear Sir,

I cannot tell you how beautiful I find it,[1] vivid, profound (in places Bossuet-like, and perpetually like a very allegory of your life, your friendship and your sorrow), new, disturbing in its strangeness, its sadness, its lucid penetration into the mysteries of love and art and the phases of Destiny.

How I should like to see that portrait of Mme Béchevêt, having been so in love with her daughter.[2]

The card which you were kind enough to enclose with it fills me with profound gratitude. I shall do my best to deserve that Tsarist 'invariably' which you bestow on me.

I am struggling without making any progress, on the days when I'm not too unwell, with a novel which may perhaps give you a little more esteem for me if you have the patience to read it.

I shall make an effort to see you shortly, and thank you again with all my heart, with affectionate and admiring respect.

Marcel Proust

1. Montesquiou's preface to a catalogue of paintings and drawings by his friend Gustave Jacquet (1846–1909) which were to be sold after his recent death.

2. Mlle Anne de Béchevêt, whose portrait was painted by Helleu.

354   *To Georges de Lauris*

[Shortly before 27 November 1909]

My dearest Georges,

I was just going to write to you to explain a complicated matter from the fatigue of which I've been recoiling for three days. This is it. As a result of scruples which would take too long to explain, chiefly that some Cabourg people[1] who see me up and about there every day think it's out of snobbery that I don't see them in Paris, and that a person who is dear to me is connected with them (Georges, you will perhaps soon be hearing news of me, or rather I shall ask your advice: to make a very young and charming girl share my life, even if she's not afraid of doing so – would that not be a crime?[2]), I have promised, short of a too violent attack at the last minute, to take all the sons of these people (among others the young Plantevignes) to *Le Circuit* on Saturday.[3] And since it's the only time I shall be up, for I feel it's no longer possible for me to make fixed appointments, I wanted to take the opportunity to see some old friends, and I've invited Loche Radziwill and Christiane[4] and Emmanuel Bibesco;[5] also because these very sweet but too young people would have seemed a little boring . . . (One of them is *very* intelligent.[6]) Will you come? It will be a chance to see you and to make a rendezvous. Will you invite Bertrand and F. de Pâris on my behalf?[7] But I shall need yeses or noes because these boxes aren't infinitely extensible. Not Lucien Henraux, because of Christiane.[8]

Georges, I'm going to get down to work again because I've read my beginning to Reynaldo (200 pages!)[9] and his reaction has greatly encouraged me. Any evening you like I'll send it to you. I feel it's now my duty to subordinate everything else to trying to finish it. And afterwards I shall have only one aim, to try to see a lot of you.

Affectionately yours
Marcel

Could you lend me *Mâle* for twenty-four hours? What other books of yours have I, apart from *Port-Royal* which I shall return to you because I shan't be using it for several months?[10]

---

1. The families of his young friends from Cabourg (cf. letters 303, n. 2; 352).

2. She has never been identified! George Painter relates that she asked Antoine Bibesco, who published his correspondence with Proust, not to print her name.

3. Play by Georges Feydeau and Francis de Croisset which had just opened with a strong cast (including Geneviève Lantelme) at the Théâtre des Variétés; reviewers considered its seduction scene unusually frank.

4. His mistress (see letter 228, n. 6), also admired by Lucien Henraux.

5. In his invitation to Bibesco, Proust wrote of '. . . young people rather too young (not in the least Salaïstes!!) . . .', in other words not homosexual.

6. Probably Pierre Parent. See letter 349, n. 4.

7. Bertrand de Fénelon and François de Pâris.

8. See note 4 above.

9. The first version of *Combray*. Cf. letter 343, n. 1.

10. Emile Mâle's book *L'Art religieux de la fin du moyen âge*, which Proust probably needed for the description of churches, had been his New Year present to Lauris (see letter 316, n. 4); he had borrowed Sainte-Beuve's *Port-Royal* in January (see letter 320, n. 2).

355   *To Reynaldo Hahn*

[Friday 26 November 1909]

Vincht,

Yesterday the amiable brothers[1] (the emulators of what a handbook of literature which I possess calls the brothers who made an impact on the art of writing: the *brothers Pliny*! the brothers Corneille, the *brothers Racine*! the brothers Chénier) sent me a note saying they couldn't read me, and would prefer to carry on by taking it down in shorthand. So in spite of the state I was in I sent for them. They (he) left me at half past two. Afterwards I worked alone, demolishing what was done. And now the fog has begun to take hold of my heart, exhausted by this lack of rest; it's almost three o'clock in the afternoon and an attack seems to be imminent. All this to warn you that it might be wise to telephone before coming round.

Moreover since the brilliant Dioscuri of stenography can't deliver back the end (from the beginning of the first chapter,[2] already twice as long as Mme Daudet's new book) before Thursday at the earliest although it was promised for last Thursday, I couldn't show you the end before then, so don't put yourself out for me. Especially as if I have a severe attack and my heart is absolutely wrecked, I know how dangerous it would be to rise above it and see you earlier. But perhaps the crisis will subside between now and this evening. But I fear and fear. Schad because it's the firscht time I've felt so bad for a month.

Bonsoirs binibuls[3]

I've written to the Veaux[4] telling them that you would attend the Performance (like King Manuel[5] or the Moldavian students) on Saturday and they must at this moment be puffed up with pride! Don't say anything to Coco.[6]

1. 'Les aimables frères' was Sainte-Beuve's phrase for the Goncourts. The brothers referred to here were stenographers employed by Proust. He underlines the

'brothers' in his parenthesis because Pliny the Younger (one of his heroes since he was fifteen) was the nephew, not the brother, of Pliny the Elder, and Louis Racine was the son, not the brother, of the great dramatist.

2. See letter 354, n. 9.

3. Proust adds one of his long-nosed profiles (cf. letter 207, n. 8).

4. The bull-calves: Proust's nickname for the Plantevignes.

5. The eighteen-year-old King Manuel of Portugal, on a State visit to Paris, was not permitted to attend the improper play *Le Circuit* for which Proust had taken three boxes. See letter 354.

6. The painter Frédéric de Madrazo for whom Proust no longer had room.

356   *To Georges de Lauris*

[Early December 1909]

My dearest Georges,

I know it's to your blind friendship that I owe this divine letter, but that makes it all the more precious to me.[1] Since I have more affection than self-esteem, it gives me more pleasure to see that you're fond of me than to believe I have some talent. From the point of view of discretion you are quite free to say that I got you to read the beginning of my book and if anyone should consider – as I don't in the least flatter myself they will – that that is an exclusive privilege, I'm only too happy to proclaim and to underline my predilection for you. All I ask is that you shouldn't mention the subject or the title or indeed anything that might be indicative of what I'm up to (not that anyone would be interested). But more than that I don't want to be hurried, or pestered, or ferreted out, or anticipated, or copied, or discussed, or criticized, or knocked. There will be time enough when my thoughts have run their course to allow free rein to the stupidity of others.

According to the instructions I'm leaving you will be handed, together with the two exercise books (second and third), two pages of the first to replace those with the same pagination for a passage which will appear snobbish to a snobbish reader but in which you will recognize a praiseworthy effort to express things that are not readily expressed.[2] As for the exercise books I'm sending you, in spite of some crossings-out in ink done at random as I came across a few enormities, they remain far more the copyist's work than mine. So I'm counting on your intuitive and affectionate cooperation.

Your grateful
Marcel

I see that on page (115 I think) the copyist has repeated everything twice. It's absolute madness!

Don't infer that I like George Sand.[3] It isn't a piece of criticism. That's how it was at *that date*. The rest of the book will correct it.

1. Lauris's letter of congratulation on what he had read of *Combray*.

2. Perhaps the class comedy between the narrator's great-aunt, 'the only member of our family who could be described as a trifle "common"', and Swann, the loftiness of whose social position she has entirely failed to grasp (*RTP*, I, 16–21).

3. An allusion to the scene early in *Combray* where the young narrator's aching heart is soothed as his mother reads, leaving all the love-scenes out, from *François le Champi*, one of the pastoral novels of George Sand which his grandmother was to give him for his birthday (*RTP*, I, 42–6).

357 *To Georges de Lauris*

[Around 13 December 1909]

My dearest Georges,

I don't know how to thank you for your letters. I wanted to write to you to ask you something more, namely whether you think that if I were to die now without any further completion of the book, this part is publishable as a volume, and whether in that case you would look after it. But this fog has made me so ill that I couldn't write to you. However we can talk that over some other time; at the moment my work is interrupted and I'm incapable of even thinking about it.

Affectionately yours,

your extremely grateful

Marcel

I'm very pleased that you read it to your father; please give him my respectful thanks.

358 *To Robert de Montesquiou*

102 boulevard Haussmann
[Shortly before mid-December 1909]

Dear Sir,

I almost regretted that Helleu had told you he expected to see me, for, not being informed about my life, you may perhaps have deduced therefrom an improvement in my health which might make my silence seem like remissness in the maintenance of my admiration and attachment for you, in the practice of the cult. Not that I have 'left a letter unanswered' as you would have put it. If a letter from you has remained unanswered it's because I never received it. I don't know your new address and haven't dared ask for it since you didn't give it to me.[1] Otherwise, if I ever had a day's respite, I might make a sudden, affectionate and respectful irruption there.

I have undertaken a long work, a sort of novel the beginning of which will perhaps appear shortly.[2] Until its completion, if I can work, for my strength is perceptibly declining and I've been given the precise measure of it, I shall remain very preoccupied. But since my admiration and friendship for you languish from being so little able to manifest themselves, although I think of you so much, re-read you and memorize your words and deeds, I wanted to send you these few lines to forestall the expiration of your benevolence with regard to my affection.[3]

Marcel Proust

1. After giving a farewell party on 18 June at the Pavillon des Muses, Montesquiou had moved to the equally sumptuous Palais Rose, in the Paris suburb of Le Vésinet.

2. Proust assumed that he had agreed with Calmette on the serialization of his novel in *Le Figaro*. However, owing in part to the intervention of Beaunier, no extracts were to appear in *Le Figaro* until 1912. Cf. letter 345.

3. For part of Montesquiou's amiable reply (the letter is missing) see end of following letter.

359 *To Armand de Guiche*

[Wednesday, 15 December 1909]

Dear friend,

The chance of seeing you is always a great temptation for me. But really at the moment I fear it isn't feasible.

As regards hunting, briefly what I wrote to ask you was whether it would be a simple matter to get one of my friends who is unknown to you invited for a day's hunting at Vallière.[1] Then almost at once it happened that my request had ceased to have any object. So I wrote you a second letter asking you to ignore the first one. But after the second letter had gone I found the first which I had forgotten to have posted. You see it's a case of extreme complication in supreme triviality and I don't know whether this account, at once insipid and sibylline, demands more attention than it arouses interest.

However, in order not to add a further tedious entanglement to this vapid imbroglio, I omitted to explain its subtle inanity to the aspiring huntsman, who might have thought me guilty of insincerity and of not having approached you at all. So I would ask you simply to take good note of this otiose and daedalian myth, and, although it is improbable that you will ever meet the anonymous invitee, to remember that I asked you for an invitation to hunt, and that immediately afterwards I told you not to bother. Your memory is thus enriched with a precious burden.

One of my friends (in writing to tell me that your study[2] of the motion of a disc in the air, and the deduction you drew therefrom as to the formulae for the reaction of air over the surface of a moving plane, is of great interest from the point of view of the construction of aeroplanes) unwittingly depicts for me a Guiche reminiscent of Vinci, that aviator who also dabbled in physics, in painting, and even, so the Germans say, in sculpture.

Good-bye dear friend. I should like to think of you what R. de Montesquiou wrote to me this morning: 'There is permanence between us.'

Amen.

Your

Marcel Proust

I'm sad to hear that the Duchess is unwell,[3] and I beg you to lay at her feet the regards of her respectful servant M.P.

1. His family home. See letter 48, n. 2.

2. 'On the motion of a disc in a fluid' by A. de Gramont de Guiche, presented by M. Émile Picard. Published in the Weekly Review of the Meetings of the Academy of Sciences for 3 May 1909. Cf. letter 90, n. 2.

3. Proust was unaware that the Duchess had just given birth to twin sons. See following letter.

360   To Armand de Guiche

[Wednesday, 15 December 1909]

Dear friend,

I'm distressed that the criss-crossing of our letters (for which I indubitably possess an uncommon talent) caused me to write you those complicated frivolities at such a sweet and solemn moment.[1] The splendid, stirring news, that I am so touched that you should have announced to me personally, finds me sufficiently sensitive, sufficiently in tune with you, to make me wince – as at a painful and vulgar dissonance – at the thought of my confused babbling, which must have seemed to you even more jejune when set against the joyous solemnity of this event. What a Father! and what a Mother! as they say in *Athalie*.[2] My thoughts will turn unceasingly from now on to the frail and charming duchess who has shown herself so strong.

Dear friend, I don't want to abuse your patience by asking you for a few more lines. But by allowing me to attract the attention and perhaps the smile of the young mother by sending something pretty to her and the twins, you would give me great pleasure. It would be absurd of you to be bashful about it since it isn't for you. It's nothing to do with you; your role goes back many months and is now finished. So you can advise me disinterestedly, and make me genuinely happy.

Apart from *things*, would some fruit or flowers be welcome, and permitted by the doctor?

If you don't answer me I shall act on my own accord, and perhaps less appropriately.

I congratulate you, I envy you, I thank you, I admire you . . . I bore you!

Marcel Proust

1. See preceding letter, note 3.
2. Act II of Racine's play: scene 7, between Athalie and Joas.

361 *To Armand de Guiche*

[Shortly after 17 December 1909]

Dear friend,

Would you be kind enough to tell me (since I cannot come round to ask you in person) whether everything continues to go well with all your family.

As soon as I received your permission, I sent Madame la Duchesse last Friday an ingenious composition of fruit (I say ingenious because I devised it myself: it is thus, I believe, that authors are wont to speak of their works). Did she receive it safely? It was identifiable by the fact that I had arranged for a bunch of asparagus to be placed in the middle of it – a vegetable which I admire, as you will see from my book if you read it.[1] I don't know whether it was precisely as I've described, since I couldn't go myself.

I've just received (I don't know whether it was sent by you) a yearbook or brochure of the Polo Club of Cannes. But up to now my ambition and my instinct for debauch have been amply gratified by the one in Paris[2] and it isn't yet an article of export for me. I was also asked to send a photograph of myself on horseback for a book, and my appreciations of the various players, etc. You bring me some curious correspondence, not to mention the constant offers of ponies. One day I shall feel unable to refuse . . .

Your devoted
Marcel Proust

1. An allusion to the description of asparagus, and various other passages on the subject in *Combray*, culminating in the revelation of the connection of this vegetable with Françoise's jealous cruelty: 'We discovered that if we had been fed on asparagus day after day throughout that summer, it was because their smell gave the poor kitchen-maid who had to prepare them such violent attacks of asthma that she was finally obliged to leave my aunt's service' (*RTP*, I, 135; cf. 59, 63, 86 and 131).

In *The Guermantes Way*, the Duke and Duchess argue about Elstir's painting *A Bundle of Asparagus* (*RTP*, II, 520).

2. See letter 281, n. 2.

362   *To Max Daireaux*

[Around 31 December 1909]

My dear friend,

I am ending up the year so badly in every respect and moreover the novel I have settled down to at last tires my wrist so much that I no longer write letters. So I can offer myself the luxury of a 'New Year letter' since I feel I shall write only one. Allow it to be for you.

Without tiring myself out, I should like to say several things to you. First of all this: when you were asked by me to come one evening and to telephone, why did you never come? Although I told you that if I was too ill I wouldn't receive you, I couldn't ever bring myself to switch off my lights without leaving instructions that I was to be called if you were there. And no one came. If it was because you had other things to do you did right; if it was from indifference you did right; if it was from discretion you did wrong.

I went, as you probably know, to Cabourg. And I stayed there thinking every day that I would leave again the next day, so I didn't dare write to you. I had arranged to have some alterations which are essential for my peace and quiet done to my room in Paris, but as I kept writing to my architect 'I may come back tomorrow', they were unable to get down to work.[1] I went there very late, as it happened, thinking right up to the very morning of my departure that I wouldn't go. So that when M. Parent[2] came round the evening before to inquire after me, having come from Cabourg and being on the way back there, and wondering whether I was going, he was told that I was too unwell to see him and for the same reason wouldn't be going to Cabourg this year. I did in fact leave next morning, but I didn't expect to. He thought I wanted to keep it secret from him. Nothing could have been further from my mind. (I suddenly realize that all this sounds like *excuses*, as though to justify myself to you. What for? Nothing: there too nothing could be further from my thoughts; I'm simply telling you things about my life which have a connection with what you know. And my life is so bizarre and lamentable that whenever I talk about it I give the impression of apologizing.)

In Cabourg I didn't once go out on the front, my health having greatly deteriorated in the past year. I got up at half past nine or ten in the evenings on the days when I could get up at all, and I went down through the hotel, without going out, into the new Casino where I was immediately grabbed by a group of people who together with me represented every age of life. For on the one side there were some really young creatures[3] and on the other M. and Mme d'Alton and the Pontcharras,[4] Mme de Maupeou and Bertrand.[5] I don't know why Cabourg for me was inseparable from two people, Mme Montet[6] and you. You never came. Madame Montet came for a few days, but didn't ask to see me. I had sent her some flowers before her arrival. She sent me a note after her departure. I shall write a short story about these events. I should like to know about yours, not your events but your news, in every sense of the word. News of those you love, of your enchanting parents, your sister, your brothers, yourself.[7] And literary news.

I've sent the first part of what I've done to Calmette but I need him so badly at the moment for my own sake (if I weren't so tired I'd explain to you why but I feel my hand failing) that although I urgently need a note from him for his brother, a military doctor in Paris, in order to avoid medical examinations which are more exhausting than the 13 days,[8] I'd even prefer to go through with them rather than have to ask him so as to concentrate on the literary favour I want from him as I can't find a publisher.[9]

All this must sound as though I'm saying: 'Don't ask me for anything that involves him.' On the contrary; it's simply to tell you that I'd prefer to put you in touch with anyone else but him at the *Figaro*, but I'll certainly refer you to him if you don't find another outlet for your essays.[10] Tell me about all that, tell me if there's anything at all of any kind I can do for you, tell me about your life, tell me if you ever want anything from me and forgive me if my fingers refuse to write another line just as I was about to speak to you of my friendship.

Marcel Proust

Did you ever get to the bottom of the Mme d'Eichtal affair?[11]

1. The project was to have the walls of his bedroom lined with cork.

2. Louis Parent (1854–1909), an architect who had a villa at Cabourg (he had died aged fifty-six in November). His son Pierre was among Proust's young friends at Cabourg: see letter 349, n. 4.

3. See letter 354, n. 1.

4. Vicomte and Vicomtesse Charles d'Alton, whose two daughters, Hélène and Colette, were also among Proust's young Cabourg friends, and the Marquis and Marquise de Pontcharra; the latter had a villa near the Plantevignes in Cabourg.

5. Apparently Charles Bertrand, proprietor of the Grand Hôtel.

6. Mme E. Montet de la Chambatterie, a widow whose name crops up only in connection with Daireaux.

7. His sister was the Comtesse Lucien de Céligny; he had three older brothers.

8. The period of annual military training (Proust was in the territorial reserve) from which, the following year, Dr Emile Calmette managed to get him permanently exempted.

9. See letter 358, n. 2.

10. See letter 333, first letter to Daireaux.

11. See letter 336, second letter to Daireaux.

INDEX

Page references in bold refer to the first page of a letter from or to Proust. Identifying footnotes, both for Proust's correspondents and for other persons mentioned in the letters, are indicated with an italic *n* wherever more than one footnote reference occurs.